Deformation-enhanced Fluid Transport in the Earth's Crust and Mantle

Mineralogical Society Series

Series Editor

Dr A. P. Jones

The aim of the series is to provide up-to-date reviews through the selected but specialized contributions of leading experts. A particularly attractive feature of the series is the tutorial element, making it suitable for third year undergraduates and researchers at all levels not solely within the Earth Sciences. Each volume is purpose-designed, highly illustrated and serves as an excellent reference tool.

TITLES AVAILABLE

1 **Deformation Processes in Minerals, Ceramics and Rocks**
Edited by D. J. Barber and P. G. Meredith

2 **High-temperature Metamorphism and Crustal Anatexis**
Edited by J. R. Ashworth and M. Brown

3 **The Stability of Minerals**
Edited by Geoffrey D. Price and Nancy L. Ross

4 **Geochemistry of Clay-Pore Fluid Interactions**
Edited by D. A. C. Manning, P. L. Hall and C. R. Hughes

5 **Mineral Surfaces**
Edited by D. J. Vaughan and R. A. D. Pattrick

6 **Microprobe Techniques in Earth Sciences**
Edited by Philip J. Potts, John W. F. Bowles, Stephen J. B. Reed and Mark R. Cave

7 **Rare Earth Minerals**
Edited by Adrian P. Jones, Frances Wall and C. Terry Williams

8 **Deformation-enhanced Fluid Transport in the Earth's Crust and Mantle**
Edited by M. B. Holness

Deformation-enhanced Fluid Transport in the Earth's Crust and Mantle

Edited by Marian B. Holness
University Assistant Lecturer, Department of Earth Sciences,
University of Cambridge, UK

Published in association with
The Mineralogical Society of Great Britain and Ireland

CHAPMAN & HALL

London · Weinheim · New York · Tokyo · Melbourne · Madras

Published by Chapman & Hall, 2–6 Boundary Row, London SE1 8HN, UK

Chapman & Hall, 2–6 Boundary Row, London SE1 8HN, UK

Chapman & Hall GmbH, Pappelallee 3, 69469 Weinheim, Germany

Chapman & Hall USA, 115 Fifth Avenue, New York, NY 10003, USA

Chapman & Hall Japan, ITP-Japan, Kyowa Building, 3F, 2-2-1 Hirakawacho, Chiyoda-ku, Tokyo 102, Japan

Chapman & Hall Australia, 102 Dodds Street, South Melbourne, Victoria 3205, Australia

Chapman & Hall India, R. Seshadri, 32 Second Main Road, CIT East, Madras 600 035, India

First edition 1997

© 1997 The Mineralogical Society

Typeset in 10/12pt Times by AFS Image Setters Ltd, Glasgow
Printed in Great Britain by The University Press, Cambridge

ISBN 0 412 75290 5

A catalogue record for this book is available from the British Library

♾ Printed on permanent acid-free text paper, manufactured in accordance with ANSI/NISO Z39.48-1992 and ANSI/NISO Z39.48-1984 (Permanence of Paper)

Contents

CONTENTS

x

Contributors

Q. Bai   Institute of Geophysics and Planetary Physics, University of California, Riverside, CA 92521, USA.

M.J. Bickle   Department of Earth Sciences, University of Cambridge, Downing Street, Cambridge CB2 3EQ, UK.

S.L. Brantley   Department of Geosciences, Pennsylvania State University, University Park, PA 16802, USA.

M. Brown   Laboratory for Crustal Petrology, Department of Geology, University of Maryland, College Park, MD 20742-4211, USA.

M.B. Clark   Exxon Production Research Company, PO Box 2189, Houston, TX 77252-2189, USA.

J.D. Clemens   School of Geological Sciences, Kingston University, Penrhyn Road, Kingston upon Thames, Surrey KT1 2EE, UK.

C. Cole   Department of Geology and Geophysics, University of Edinburgh, King's Buildings, West Mains Road, Edinburgh EH9 3JW, UK.

M.J. Daines   Department of Geology and Geophysics, University of Minnesota, Minneapolis, Minnesota, USA.

P. Deines   Department of Geosciences, Pennsylvania State University, University Park, PA 16802, USA.

D.M. Fisher   Department of Geosciences, Pennsylvania State University, University Park, PA 16802, USA.

C.M. Graham   Department of Geology and Geophysics, University of Edinburgh, King's Buildings, West Mains Road, Edinburgh EH9 3JW, UK.

H.W. Green, II   Institute of Geophysics and Planetary Physics, University of California, Riverside, CA 92521, USA.

J. Grocott   School of Geological Sciences, Kingston University, Penrhyn Road, Kingston upon Thames, Surrey KT1 2EE, UK.

Z.-M. Jin   Division of Geomechanics, China University of Geosciences, Wuhan, China.

C.K. Mawer   Geoid Pty Ltd, PO Box 1113, Belconnen, ACT 2617, Australia.

G. Myers   310 8th Avenue, SW, Rochester, MN 55902, USA.

A.M. McCaig   Department of Earth Sciences, The University of Leeds, Leeds LS2 9JT, UK.

N. Odling   Nansen Environmental and Remote Sensing Centre, Edvard Griegsvei 3A, N-5037, Solheimsviken/Bergen, Norway.

N. Petford   School of Geological Sciences, Kingston University, Penrhyn Road, Kingston upon Thames, Surrey KT1 2EE, UK.

T. Rushmer   Department of Geology, University of Vermont, Burlington, VT 05405-0122, USA.

E.H. Rutter   Rock Deformation Laboratory, Department of Earth Sciences, University of Manchester, Manchester M13 9PL, UK.

A.D.L. Skelton   Department of Geology and Geophysics, University of Edinburgh, King's Buildings, West Mains Road, Edinburgh EH9 3JW, UK.

J. Wilson   School of Geological Sciences, Kingston University, Penrhyn Road, Kingston upon Thames, Surrey KT1 2EE, UK.

Preface

This volume is the result of a three-day meeting held in September 1995 in Edinburgh, under the aegis of the Mineralogical Society and the Geological Society of London. The meeting was sponsored by the Mineralogical Society and attended by an international gathering of 70 geologists. Presentations were given on the general topic of deformation-enhanced fluid flow over the entire range of interest of the earth sciences. Most of what you will read in this volume is based on the presentations given at the meeting, with the addition of two chapters to present what I must admit to be a personal, and perhaps non-comprehensive, view of the rapidly evolving field of deformation-enhanced fluid flow in rocks.

The work on this book was done while I was a Royal Society University Research Fellow at the University of Edinburgh.

Marian Holness
Edinburgh
September 1996

Acknowledgements

Putting together a volume of this type takes more than one person. Perhaps the first person to thank is Colin Graham, who was my co-convenor of the meeting, providing invaluable advice and support, for which I am very grateful. There are many other people I wish to thank for giving generously of their time and expertise in reviewing the submissions for the book. I would like to mention these people by name, as a small reparation for the unsung efforts they put into getting this volume on the shelves: Rainer Abart, Michael Atherton, Chris Barton, Stephen Brown, J.-P. Burg, Ian Cartwright, Mike Cheadle, James Connolly, Patience Cowie, Richard D'Lemos, Brian Evans, James Evans, Ulrich Faul, Gayle Gleason, Yves Guéguen, Greg Hirth, Simon Inger, Bertrand Maillot, Ian Main, Pat Meere, Nick Oliver, P.M. Piccoli, G.S. Solar, Ron Vernon, and Gordon Watt. Many of the contributing authors also were generous with their time, reading and commenting on other chapters, and I am grateful for their help: Mike Brown, John Clemens, Mike Bickle, Martha Daines, Colin Graham, Harry Green, Andrew McCaig, Noelle Odling, Nick Petford and Ernie Rutter. Financial support from the Royal Society of London during the preparation time is gratefully acknowledged.

We are grateful to the following individuals and organizations who have kindly given permission for the reproduction of copyright materials. (See figure captions and references at the end of each chapter for full details of sources.) Chapter 1: Figure 1.1(a) after Cooper and Hunter (1995), by permission of the Mineralogical Society; Figure 1.1(c) after Yardley and Lloyd (1989), by permission of Cambridge University Press; Figure 1.1(f) after Harte *et al.* (1991), by permission of Springer-Verlag. Chapter 2: Figures 2.1(a) and 2.3 (c, d) after Jin *et al.* (1994), by permission of *Nature*. Chapter 9: Figures 9.2, 9.7 and 9.8 after McCaig *et al.* (1990), by permission from Springer-Verlag; Figures 9.5, 9.6 and 9.11 after Bowman *et al.* (1994), by permission from *American Journal of Science*. Figure 9.10 after McCaig *et al.* (1995a), by permission from *American Journal of Science*. Chapter 10: Figure 10.2(e,f) after Fisher *et al.* (1995), by permission of American Geophysical Union.

Introduction

M.B. Holness

A significant fraction of the Earth's volume contains a fluid phase, either a melt or a H–O–C–N–S volatile phase. Movement of melts within the mantle results in the formation of oceanic crust and the differentiation of chemical and isotope compositions within the Earth. Fluids also move through the continental crust. During metamorphism, devolatilization reactions result in the release of volatiles which move upwards in response to buoyancy during compaction of the surrounding rocks. Large thermal gradients, such as those around igneous intrusions in the shallow crust, can generate significant fluid flow via the development of hydrothermal systems. Anatexis, the culmination of crustal metamorphism, also results in fluid flow, as granitic melts segregate from their source region and rise within the crust to their eventual resting place. The mechanisms by which these various fluids move within the Earth are poorly understood.

A significant proportion of the published work on rock permeability concentrates on rocks under hydrostatic stress (Holness, this volume), but the stress state of much of the Earth is not hydrostatic. Deviatoric stresses, resulting in deformation on all scales from large-scale faulting to intra-grain plastic deformation, are known to have a profound effect on rock permeability, controlling both its magnitude and the fluid flow mechanism (e.g. Etheridge *et al.*, 1983; Cooper, 1990; Knipe and McCaig, 1994; Oliver, 1996). An inverse relationship between deformation and fluid flow also exists, since deformation mechanisms and rates can be affected by the presence of a fluid phase (e.g. Kronenberg and Tullis, 1984; Paterson, 1989; Davidson *et al.*, 1994; Yardley, 1981) The ubiquity of deviatoric stresses in the Earth means that these complex interrelationships are important in all fluid-bearing environments, and is reflected in the breadth of the studies aimed at the problem.

A useful analogy for the current status of the problem of deformation-enhanced fluid flow is the story of three blind men encountering an elephant for the first time. They had only their hands to examine the long-suffering animal and were pawing at different parts of its body. The first one happened to be close to its head and touched the trunk: 'Ooh!' he said, 'This beast is like a snake!' 'I

Deformation-enhanced Fluid Transport in the Earth's Crust and Mantle. Edited by M.B. Holness.
Published in 1997 by Chapman & Hall, London. ISBN 0 412 75290 5.

don't think so,' said the second, who had encountered one of its legs: 'It's more like a tree.' The third one disagreed with them both, as he was in the process of stroking its flank, and he pronounced that an elephant resembles a wall. Thus all three were wrong in general, but right in detail. Given enough time, each would have assembled an accurate picture of the elephant (presuming, of course, that it was a patient elephant that would have stood still for long enough), but they could have arrived at the correct and complete picture in a fraction of the time if they had pooled their observations. Geologists trying to understand deformation-enhanced fluid flow are in an analogous situation, in that we are all trying to solve the same problem by approaching it from different directions. It will become apparent, while reading the various contributions to this book, that the subjects covered, be they mantle melting, the movement of granitic melts within the crust, or the passage of volatile fluids during metamorphism, each concentrate on different processes occurring at a whole variety of scales, from the grain-scale up to kilometre-scale.

A historical perspective

Flow of fluids within fractured rock is a familiar concept. Dykes are recognized as features formed by movement and subsequent freezing of molten rock in fractures, and mineral veins are understood to be the fossilized remains of crustal plumbing systems. It is only fairly recently that deformation-enhanced fluid flow has been recognized in environments with a more pervasive distribution of deformation, such as in the mantle and in high-temperature metamorphic environments up to, and including, the anatectic regime. Here I will give a brief overview of the development of our understanding of deformation-enhanced fluid flow, taking each sub-discipline in turn, to demonstrate the way in which we have separately approached the same problem.

A significant step forward in our understanding of fluid flow in the inaccessible Earth was the realization that interfacial energies can play a role in controlling rock permeability in high-temperature environments such as the mantle. Important experiments established that basaltic melt under hydrostatic stress occupies an interconnected network of pores in olivine aggregates (e.g. Waff and Bulau, 1979; Toramaru and Fujii, 1986). This permitted the erection of elegant models of permeability and melt flow within the mantle, enabling extraction rates and compositions of basaltic melt from a compacting mantle matrix to be well constrained (McKenzie, 1984, 1985; McKenzie & Bickle, 1988). The success of the basalt extraction model inspired a similar enthusiasm to examine the equilibrium (hydrostatic) topology of fluids within mineral aggregates of relevance to crustal anatexis and metamorphism (e.g. Jurewicz and Watson, 1984; Watson & Brenan, 1987; Holness, 1993).

This situation was short-lived as experiments designed to look at the influence of stress on melt topology in the mantle soon revealed that under non-hydrostatic stresses, the distribution of the melt phase differs significantly from the

2

hydrostatic case (e.g. Jin *et al.*, 1994), with thick films of fluid along grain boundaries, instead of channels along three-grain junctions. Similar fluid topologies have also been observed in deforming wet aggregates of evaporite minerals (Urai, 1983; Urai *et al.*, 1986), with profound implications for rock permeability. These observations are supported by a theoretical consideration of variation in interfacial energies with deviatoric stress (Heidug, 1991).

Geologists working on the movement of granitic melts in the crust also soon realized that understanding the hydrostatic case did not solve their problem. Although the majority of granitic melts form an interconnected network in crustal materials (e.g. Jurewicz and Watson, 1984; Laporte, 1994), the high viscosity of the melt means that granite segregation during gravity-driven compaction of the solid matrix cannot occur on geologically feasible timescales (McKenzie, 1985). This points to the primacy of deformation in facilitating granitic melt migration, which is confirmed by abundant field and experimental observations (Rutter, this volume; Brown and Rushmer, this volume). Furthermore, experimentally determined hydrostatic pore topologies in metamorphic rocks show that although in general the inferred relative permeability of various different lithologies can be predicted, the exceptions to the rule again demonstrate the important role played by deformation (Holness and Graham, 1995).

The role of deformation in determining rock permeability probably first became familiar to metamorphic petrologists, due to the temporal overlap between deformation and volatile fluid flow during regional metamorphism. The processes whereby deformation and fluid flow interact in metamorphic rocks are extremely complex because of several feedback mechanisms. Fluid flow is intimately linked with reaction, which can either release or consume fluid, resulting in changes in the effective pressure (the difference between lithostatic pressure and the pore fluid pressure). The effective pressure controls deformation mechanisms and rock strength (Secor, 1965; Phillips, 1972), which are also controlled by the grain-size, composition and texture of the rock. These three parameters can in turn be affected by fluid infiltration-driven reactions (e.g. White and Knipe, 1978; Rutter and Brodie, 1995). That fluid flow can be focused into localized deforming zones was first shown by observations of metasomatic alteration in shear zones (Beach, 1973; Fyfe *et al.*, 1978). Strain is focused into the reacting zone both by the positive feedback loop of reaction-controlled softening (Brodie & Rutter, 1985) and by the weakening caused by high pore fluid pressures (Etheridge *et al.*, 1984).

The final subsection of the broad subject of deformation-enhanced flow is that of flow in fractured rock. The concept of fluids taking advantage of high permeability channels is an old and well-accepted one, although recent field and experimental evidence has demonstrated that not all mineral-filled veins in metamorphic rocks represent the sites of major fluid flow channelways, but may have formed by local mass redistribution in the presence of small amounts of fluid (Fisher *et al.*, 1995; Brantley *et al.*, this volume). Our understanding of the

3

permeability of fractured and faulted terrains has been furthered recently with the application of powerful new concepts and a new generation of supercomputers. Modelling of flow within fractured rock now incorporates consideration of scaling laws describing the size distribution of faults and fractures, enabling prediction of the permeability of a terrain in which the limited scale of observation precludes the detection of all sizes of features (Heffer and Bevan, 1990; Gauthier and Lake, 1993). Percolation theory has also been applied with considerable success, resulting in the development of effective models of flow within fracture networks (Balberg *et al.*, 1991). Some of these ideas are introduced in the following chapter, but are reviewed in detail by Odling (this volume).

Differences in approach

The preceding section was intended to illustrate the different directions from which the problem of deformation-enhanced fluid flow has been approached. The appositeness of the analogy of the elephant story will become even more apparent if we appreciate that the differences in approach to the problem are a consequence of fundamental variations in the scale and nature of the field evidence.

Let us first consider the case of deformation-enhanced flow of melts in the mantle. Geologists interested in the formation of oceanic crust and the evolution of the mantle contend with the fact that most of the rocks they are interested in end up being subducted into the mantle. They must content themselves with scraps of mantle material thrust onto the continental crust in the form of ophiolites, or with material dredged up or drilled from the bottom of the ocean. They are fortunate, however, in that the mantle has a relatively simple composition, and that the basalt generated from partial melting of the mantle has a similarly restricted compositional range. The problem of extraction of basalt at mid-ocean ridges lends itself to elegant models and small-scale laboratory experiments. Geologists interested in deformation-enhanced flow of basalt thus obtain information about the underlying processes on the grain-scale, and can make predictions of extraction rates based on the scale of hundreds of kilometres.

The great efficiency with which basaltic melt is extracted from its source region (McKenzie, 1985) means that partial melts cannot be observed frozen *in situ* in mantle material. The grain-scale distribution of basalt within partially molten mantle can only be determined by laboratory experiments (Bai *et al.*, this volume; Daines, this volume), although this does create the not insignificant task of extrapolation of the experimental results to mantle conditions. The corresponding problem does not occur when examining the movement of granitic melts within the crust. Solidified pockets of granitic melt are common in outcrops of partially melted crustal rock, permitting inference of the effect of deformation on their distribution on all scales from millimetres to hundreds of kilometres (Brown and Rushmer, this volume; Grocott and Wilson, this volume). Experimental studies of the effect of deformation on the movement of granitic melts are not well advanced as they are beset with difficulties relating to the compositional and textural complexity of the rocks involved (e.g. van der Molen and Paterson, 1979; Rutter, this volume).

The approaches needed to understand the flow of volatiles through crustal materials are different again. Due to recrystallization, metamorphic rocks do not retain any significant porosity once the metamorphic event is over. In the absence of veins or metasomatism, the passage of volatile fluids may be marked only by variations in isotope or trace element composition (Knipe and McCaig, 1994), or by mineralogical alteration. Attempts have been made to constrain the total fluxes of fluid by the erection of models of fluid–solid exchange (e.g. Bickle and McKenzie, 1987; Bickle and Baker, 1990; Dipple and Ferry, 1992) to gain valuable insights into the geochemical consequences of fluid flow (McCaig, this volume; Graham et al., this volume), but they are reliant on assumptions of rock permeability. Experimental determinations of the permeability of deforming crustal materials are generally based on flow of volatile fluids in which the solid phase is not soluble (e.g. Fischer and Paterson, 1992; Zhang et al., 1994a, 1994b). Neither interfacial energies nor fluid–solid chemical interactions play an important role in these experiments, and so extrapolation of the results to metamorphic conditions is problematic.

In both convening the meeting from which this volume is derived and by compiling the volume itself, I aimed to bring together geologists looking at different parts of the same elephant to pool observations and increase understanding of the role that non-hydrostatic stresses play in controlling rock permeability. The insights gained from looking at the distribution of melts within experimentally deformed partially molten rock can surely be applied to metamorphic rocks in which the fluid phase is now absent but through which it undoubtedly passed. Similarly, the models and concepts developed to constrain fluid fluxes from metamorphic rocks could be applied to understanding the chemical changes inherent in movement of melts through the crust for example. I entertain great hopes that this important and fundamentally interdisciplinary problem can be solved by mutually rewarding collaboration between the different branches of geology.

References

Bai, Q., Jin, Z-M. and Green, H.W. II (this volume) Experimental investigation of the rheology of partially molten peridotite at upper mantle pressures and temperatures.

Balberg, I., Berkowitz, B. and Drachsler, G.E. (1991) Application of a percolation model to flow in fractured hard rocks. *Journal of Geophysical Research*, **96**, 10015–21.

Beach, A. (1973) The mineralogy of high temperature shear zones at Scourie, N.W. Scotland. *Journal of Petrology*, **14**, 231–48.

Bickle, M.J. and Baker, J. (1990) Migration of reaction and isotopic fronts in infiltration zones: assessments of fluid flux in metamorphic terrains. *Earth and Planetary Science Letters*, **98**, 1–13.

Bickle, M.J. and McKenzie, D.P. (1987) The transport of heat and matter by fluids during metamorphism. *Contributions to Mineralogy and Petrology*, **95**, 384–92.

INTRODUCTION

Brantley, S.L., Fisher, D.M., Deines, P. *et al.* (this volume) Segregation veins: evidence for the deformation and dewatering of a low grade metapelite.

Brodie, K.H. and Rutter, E.H. (1985) On the relationship between deformation and metamorphism, with special reference to the behaviour of basic rocks, in *Kinetics, Textures and Deformation* (eds Thompson, A.B. and Rubie, D.C.) *Advances in Geochemistry*, **4**, 138–79.

Brown, M. and Rushmer, T. (this volume) The role of deformation in the movement of granitic melt: views from the laboratory and the field.

Cooper, R.F. (1990) Differential stress-induced melt migration: an experimental approach. *Journal of Geophysical Research*, **95**, 6979–92.

Daines, M.J. (this volume) Melt distribution in partially molten peridotites: implications for permeability and melt migration in the upper mantle.

Davidson C., Schmid, S.M. and Hollister, L.S. (1994) Role of melt during deformation in the deep crust. *Terra Nova*, **6**, 133–42.

Dipple, G.M., and Ferry, J.M. (1992) Fluid flow and stable isotopic alteration in rocks at elevated temperatures with applications to metamorphism. *Geochimica et Cosmochimica Acta*, **56**, 3539–50.

Etheridge, M.A., Wall, V.J. and Vernon, R.H. (1983) The role of the fluid phase during regional metamorphism and deformation. *Journal of Metamorphic Geology*, **1**, 205–26.

Etheridge, M.A., Wall, V.J., Cox, S.J. and Vernon, R.H. (1984) High fluid pressures during regional metamorphism and deformation: implications for mass transport and deformation mechanisms. *Journal of Geophysical Research*, **89**, 4344–58.

Fischer, G.J. and Paterson, M.S. (1992) Measurement of permeability and storage capacity in rocks during deformation at high temperature and pressure, in *Fault Mechanics and Transport Properties of Rocks* (eds Wong, T-J. and Evans, B.), pp. 213–52.

Fisher, D.M., Brantley, S.L., Everett, M. and Dzvonik, J. (1995) Cyclic flow through a regionally extensive fracture network within the Kodiak accretionary prism. *Journal of Geophysical Research*, **100**, 12881–94.

Fyfe, W.S., Price, N.J. and Thompson, A.B. (1978) *Fluids in the Earth's Crust*, Elsevier, New York.

Gauthier, B.D.M. and Lake, S.D. (1993) Probabilistic modeling of faults below the limit of seismic resolution in Pelican Field, North Sea, Offshore UK. *AAPG Bulletin*, **77**, 761–77.

Graham, C.M., Skelton, A.D.L., Bickle, M. and Cole, C. (this volume) Lithological, structural and deformation controls on fluid flow during regional metamorphism.

Grocott and Wilson (this volume) Ascent and emplacement of granitoid plutonic complexes in subduction-related extensional environments.

Heffer, K.J. and Bevan, T.G. (1990) Scaling relationships in natural fractures: Data, theory and application. *Society of Petroleum Engineers*, Paper SPE 20981, 1–12.

Heidug, W.K. (1991) A thermodynamic analysis of the conditions of equilibrium at nonhydrostatically stressed and curved solid-fluid interfaces. *Journal of Geophysical Research*, **96**, 21909–21.

Holness, M.B. (1993) Temperature and pressure dependence of quartz-aqueous fluid dihedral angles: the control of adsorbed H_2O on the permeability of quartzites. *Earth and Planetary Science Letters*, **117**, 363–77.

Holness, M.B. (this volume) The permeability of non-deforming rock.

Holness, M.B. and Graham, C.M. (1995) P-T-X effects on equilibrium carbonate-H_2O-CO_2-NaCl dihedral angles: constraints on carbonate permeability and the role of deformation during fluid infiltration. *Contributions to Mineralogy and Petrology*, **119**, 301–13.

Jin, Z-M., Green, H.W. and Zhou, Y. (1994) Melt topology in partially molten mantle peridotite during ductile deformation. *Nature*, **372**, 164–7.

6

REFERENCES

Jurewicz, S.R. and Watson, E.B. (1984) Distribution of partial melts in a felsic system: the importance of surface energy. *Contributions to Mineralogy and Petrology*, **85**, 25–9.

Knipe, R.J. and McCaig, A. M. (1994) Microstructural and microchemical consequences of fluid flow in deforming rocks, in *Geofluids: Origin, Migration and Evolution of Fluids in Sedimentary Basins* (ed. Parnell, J.), *Geological Society Special Publication*, **78**, 99–111.
Kronenberg, A.K. and Tullis, J. (1984) Flow strengths of quartz aggregates: grain size and pressure effects related to hydrolytic weakening. *Journal of Geophysical Research*, **89**, 4281–97.

Laporte, D. (1994) Wetting behaviour of partial melts during crustal anatexis: the distribution of hydrous silicic melts in polycrystalline aggregates of quartz. *Contributions to Mineralogy and Petrology*, **116**, 486–99.

McCaig, A.M. (this volume) The geochemistry of volatile fluid flow in shear zones.
McKenzie, D.P. (1984) The generation and compaction of partially molten rock. *Journal of Petrology*, **25**, 713–65.
McKenzie, D.P. (1985) The extraction of magma from the crust and mantle. *Earth and Planetary Science Letters*, **74**, 81–91.
McKenzie, D.P., and Bickle, M.J. (1988) The volume and composition of melt generated by the extension of the lithosphere. *Journal of Petrology*, **29**, 625–79.

Odling, N. (this volume) Fluid flow in fractured rocks at shallow levels in the earth's crust: an overview.
Oliver, N.H.S. (1996) Review and classification of structural controls on fluid flow during regional metamorphism. *Journal of Metamorphic Geology*, **14**, 477–92.

Paterson, M.S. (1989) The interaction of water with quartz and its influence on dislocation flow – an overview, in *Rheology of Solids and of the Earth* (eds Karato, S.I. and Toriumi, M.), Oxford University Press, Oxford, pp. 107–42.
Phillips, W.J. (1972) Hydraulic fracturing and mineralization. *Journal of the Geological Society of London*, **128**, 337–59.

Rutter, E.H. (this volume) The influence of deformation on the extraction of crustal melts: a consideration of the role of melt-assisted granular flow.
Rutter, E.H. and Brodie, K.H. (1995) Mechanistic interactions between deformation and metamorphism. *Geological Magazine*, **30**, 227–39.

Secor, D.T. (1965) Role of fluid pressure in jointing. *American Journal of Science*, **263**, 633–46.

Toramaru, A. and Fujii, N. (1986) Connectivity of the melt phase in a partially molten peridotite. *Journal of Geophysical Research*, **91**, 9239–52.

Urai, J.L. (1983) Water-assisted dynamic recrystallisation and weakening in polycrystalline bischofite. *Tectonophysics*, **96**, 125–57.
Urai, J.L. Spiers, C.J., Zwart, H.J. and Lister, G.S. (1986) Weakening of rock salt by water during long-term creep. *Nature*, **324**, 554–7.

Waff, H.S. and Bulau, J.R. (1979) Equilibrium fluid distribution in an ultramafic partial melt under hydrostatic stress conditions. *Journal of Geophysical Research*, **84**, 6109–14.
Watson, E.B. and Brenan, J.M. (1987) Fluids in the lithosphere, I. Experimentally-determined wetting characteristics of CO_2-H_2O fluids and their implications for fluid transport, host-rock physical properties, and fluid inclusion formation. *Earth and Planetary Science Letters*, **85**, 497–515.

7

White, S.H. and Knipe, R.J. (1978) Transformation and reaction enhanced ductility in rocks. *Journal of the Geological Society of London*, **135**, 513–16.

Van der Molen, I. and Paterson, M.S. (1979) Experimental deformation of partially-melted granite. *Contributions to Mineralogy and Petrology*, **70**, 299–318.

Yardley, B.W.D. (1981) Effect of cooling on the water content and mechanical behaviour of metamorphosed rocks. *Geology*, **9**, 405–8.

Zhang, S., Paterson, M.S., and Cox, S.F. (1994a) Porosity and permeability evolution during hot isostatic pressing of calcite aggregates. *Journal of Geophysical Research*, **99**, 15741–60.

Zhang, S., Cox, S.F., and Paterson, M.S. (1994b) The influence of room temperature deformation on porosity and permeability in calcite aggregates. *Journal of Geophysical Research*, **99**, 15761–75.

CHAPTER ONE

The permeability of non-deforming rock

M.B. Holness

1.1 Introduction

Movement of a fluid phase through rock is of major importance in the Earth. At one end of the spectrum is the process of differentiation of the continental crust and the formation of oceanic crust via upwards flow of melt. At the other is the movement of aqueous fluids from the hydrosphere through the uppermost parts of the crust resulting in, for example, the formation of massive sulphide deposits at mid-ocean ridges, and transport of anthropogenic pollutants in groundwater. In order to understand the mechanics of these various processes we must constrain the basic fluid flow paths and mechanisms.

Fluid transport through rock necessitates the development and maintenance of a flow path. The critical parameter controlling the flow of fluid is thus the connectivity of the pore space, which is a function of rock type and history. Pore geometry will also be strongly affected by deviatoric stresses. In this chapter, I will review what is known about rock permeability, with an emphasis on rocks under hydrostatic stress. This will provide a basis with which the reader can assess the importance of deviatoric stress in enhancing rock permeability when reading the remainder of this volume.

1.2 Porosity

Rocks of all types contain pore space (Figure 1.1) which may contain oil, gas or subcritical aqueous fluid (in the uppermost crust), supercritical H–O–C–S fluids (in the case of metamorphic rocks undergoing reaction), or melt. Types of rock porosity fall into two broad categories: porosity in rocks at or near the surface; and that in deeply buried rocks. Direct observation of pore space is only possible for the first type, and although it is possible in some cases to infer the nature of melt-filled porosity from surface exposures (see later), in general we must rely on

Deformation-enhanced Fluid Transport in the Earth's Crust and Mantle. Edited by M.B. Holness. Published in 1997 by Chapman & Hall, London. ISBN 0 412 75290 5.

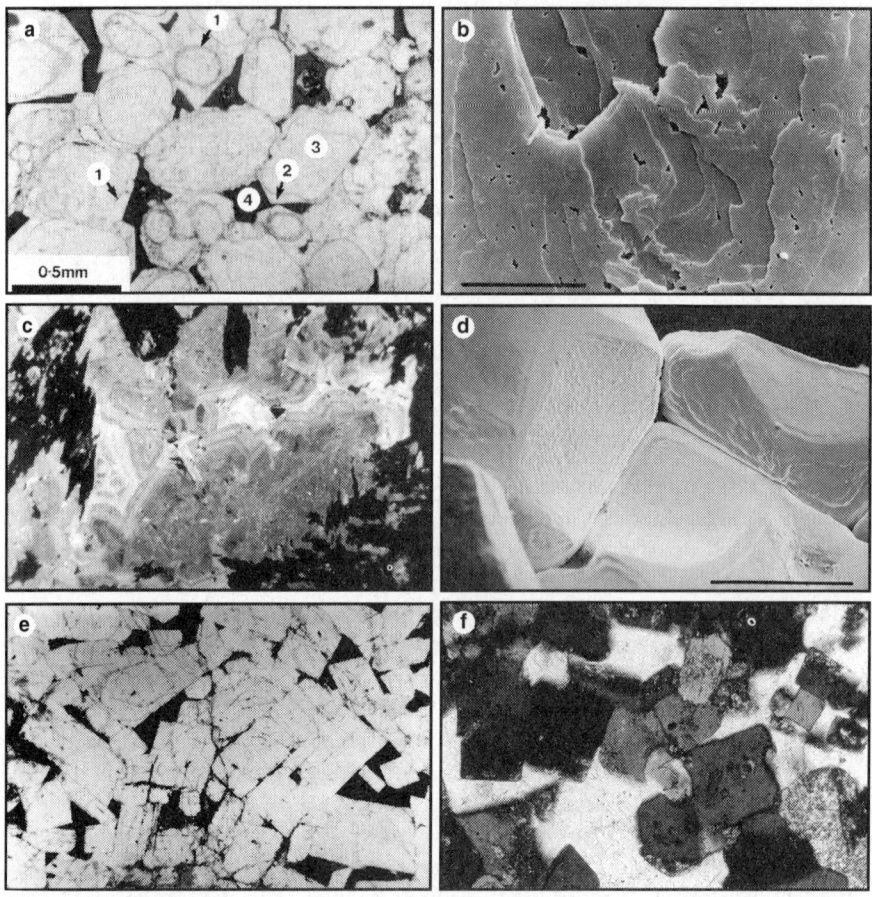

Figure 1.1 (a) Plane polarized light photomicrograph of a quartz-cemented sandstone. Outlines of detrital grains (3) are defined by a mixed layer of iron oxide and grain coating clays (1). The quartz overgrowths (2) have variable widths. The intergranular porosity (4) has been impregnated with epoxy resin. Reproduced from Cooper and Hunter (1995) with permission of the Mineralogical Society. (b) Secondary electron SEM image of a freshly created cleavage surface of alkali feldspar from the Lower Devonian Shap granite showing abundant micropores. The porosity of feldspars from the Shap granite reaches 1.03% (Walker *et al.*, 1995). Scale bar is 10 μm long. Photomicrograph kindly provided by Martin Lee. (c) Cathodoluminescence micrograph illustrating growth zoning of calcite (pale grey) which is apparently growing into void space near tremolite (black) in a tremolite-dolomite marble from Connemara, Ireland. The region of zoned calcite is interpreted to be an infilling of secondary metamorphic porosity. The width of the field of view is 2 mm. Reproduced from Yardley and Lloyd (1989), with permission from Cambridge University Press. (d) Scanning electron micrograph of porosity developed in a texturally equilibrated H_2O-bearing halite aggregate at 1.5 kbar and 400 °C. A fourth grain has been plucked away during sample mounting, revealing channels along each of the three-grain junctions and a large pore at the four-grain junction. Scale bar is 200 μm long. (e) Plagioclase–clinopyroxene orthocumulate from the Skaergard intrusion. The geometry of the final melt phase has been pseudomorphed by the dark pyroxene which infills the gaps between the euhedral–subhedral plagioclase crystals.

10

experimental evidence for the geometry of pore space in deeply buried rocks.

Original porosity in clastic sedimentary rocks comprises the voids between the grains (Figure 1.1(a)). The geometry and dimensions of the pores depends on grain size, grain sorting and the presence or absence of a pore-filling cement. Impregnation of sandstones with an acid-resistant material, followed by dissolution of the mineral matrix (Pittman and Duschatko, 1970) demonstrates that two kinds of pores occur: large, almost equant pores at the junction of four grains, and connective, sheet-like pores along two-grain faces. A third type of pore may also occur as tube-like channels along three-grain junctions (Bernabé, 1991). Porosity may be highly variable between beds in a layered sediment, reaching values as high as 80%, but generally ranging from 5% to 30% (Guéguen and Palciauskas, 1994). In contrast to clastic sediments, evaporites have some similarities to igneous rocks and commonly have a very low porosity (0.1%), although individual layers may have high porosities due to incomplete diagenesis (Yechieli et al., 1995).

Direct observation of void space in crystalline rocks presently exposed at the surface demonstrates that porosity occurs as small voids along grain boundaries and tubules along grain edges (e.g. Brace et al., 1972; White and White, 1981), although abundant scattered micropores occur within some grains, especially feldspar (Figure 1.1(b); Brace et al., 1972; Worden et al., 1990; Walker et al., 1995). Porosities in crystalline rocks are generally low (<1%, e.g. Norton and Knapp, 1977). Cracks and fractures at all scales are abundant in low porosity crystalline rocks, many of which are the result of tectonic or thermal stresses suffered subsequent to the final crystallisation/recrystallisation episode (see Odling, this volume).

Porosity present in rocks buried at depth will not in general be the same in either quantity or quality as that observed in surface exposures or in samples recovered from boreholes. During metamorphic events porosity can be generated by reactions which result in a reduction in total solid volume (Figure 1.1(c)). The amount of porosity present will be controlled by the relative rates of reaction and compaction (Graham et al., this volume). Pore geometry will be strongly influenced by the effective pressure, with low effective pressures facilitating fracture formation, as demonstrated by observation using cathodoluminescence (Sprunt and Nur, 1979), by trails of volatile fluid-filled inclusions formed by incomplete healing of microcracks (e.g. Roedder, 1984), and by the presence of abundant mineral-filled veins. Another mechanism for porosity formation during metamorphism is the non-uniform thermal expansion of neighbouring crystals, which has been termed 'thermal decompaction' by

Photomicrograph kindly supplied by Ben Harte. (f) Photomicrograph under crossed polars of a partially molten pelite from the Chaotic Zone of the Ballachulish contact aureole, Scotland. Euhedral–subhedral alkali feldspar grains (dark) lie in a matrix of optically continuous interstitial quartz (light) which is believed to have crystallized in and pseudomorphed a melt-filled porosity. The short axis of the figure is 1 mm. Reproduced from Harte et al. (1991) with permission from Springer-Verlag.

Zaraisky and Balashov (1995). Given sufficient time, pore space in a thermally decompacted rock will close by creep-controlled compaction, so significant porosity generation is most likely in environments undergoing rapid heating (such as thermal aureoles) in polymineralic rocks in which the solids have a low solubility in the fluid phase (V.N. Balashov, pers. comm., 1996).

The geometry of melt-filled porosity is controlled by a variety of factors such as crystal growth anisotropy and minimization of interfacial energies. Experimental studies demonstrate that basaltic melt occupies a pore structure in mantle materials that is controlled primarily by the minimization of interfacial energies as the rock approaches or attains textural equilibrium (e.g. Figure 1.1(d)). In crystallizing plutons the geometry of melt-filled pores during the last stages of crystallization can be inferred from the disposition of the last crystallizing phases (Figure 1.1(e); Hunter, 1987; Harte *et al.*, 1993; Bryon *et al.*, 1995, 1996). In pelitic rocks that were once partially molten, textural studies have shown that the melt-filled porosity can be pseudomorphed by minerals such as quartz or alkali feldspar (Figure 1.1(f); Harte *et al.*, 1991). The geometry of melt-filled pore space in anatectic rocks on a larger scale can be deduced from the distribution of leucosomes in migmatites (see Brown and Rushmer, this volume).

1.3 Permeability and Darcy's law

The permeability of a rock is the measure of how easily fluids flow through it. Permeability, k, being a transport property, relates the fluid flux to the fluid pressure gradient driving the flow. For most cases relevant to the flow of geological fluids through non-deforming materials, the fluid flux is linearly proportional to the pressure gradient, dP/dx, and is related via Darcy's law:

$$v = \frac{k}{\eta} \frac{dP}{dx} \qquad (1.1)$$

where v is the fluid flux (i.e. the volume of fluid per unit time passing a unit cross-sectional area perpendicular to the flow direction), and η is the viscosity of the fluid. The permeability coefficient is the scaling factor, k, and has dimensions of length^2. It can thus be thought of as an effective cross-section for flow. Alternatively, the square root of k can be seen as a characteristic length scale of the fluid flow. A commonly used unit for permeability is the darcy, where 1 darcy $= 0.987 \times 10^{-12}$ m^2.

Darcy's law is only applicable for situations in which the transport of momentum by the motion of the fluid is much less than that by diffusion (i.e. the Reynolds number is low). This translates to slow, steady, laminar flow. Darcy's law is not applicable to high permeability systems or to those with a low viscosity fluid in which flow is turbulent. The value of the critical Reynolds number over which Darcy's law no longer holds ranges from 0.1 to 75

(Scheidegger, 1960), depending on differences in pore structure such as surface roughness (Dullien, 1992). Darcy's law is also only applicable if the amount of fluid in the system remains constant. If there is a difference in fluid volume entering and leaving the system (i.e. during melting or devolatilization reactions) then the solid matrix must deform to accommodate the change in rock volume (Sleep, 1974).

Rock permeability in the very shallow crust can be measured directly from boreholes (Brace, 1984) or by measurement of laboratory samples (e.g. Brace, 1980; Bourbié and Zinszner, 1985; Shmonov *et al.*, 1995). Measurements from borehole data are about three orders of magnitude higher than the laboratory values (Clauser, 1992), and this discrepancy is generally attributed to the presence of joints and fractures at a scale larger than that of the laboratory samples (Brace, 1984; Guéguen *et al.*, 1996; Odling, this volume). Laboratory measurements are generally based on the flow of chemically inert fluids (such as argon), and most data are derived at low pressures and temperatures (although see Shmonov *et al.*, 1995, for data up to 600 °C and 200 MPa). Thus they apply only to rocks in which the transport properties are not controlled by the minimization of interfacial energies, and do not give any insight into the flow rates of melt or highly reactive fluids.

Published determinations of rock permeability from laboratory measurements span more than 10 orders of magnitude (Figure 1.2). Sedimentary rocks in general cover a wide range of permeabilities, from the 10^{-22} m^2 of shales and well-cemented limestones, to 10^{-12} m^2 of porous sandstones. Volcanic rocks (10^{-13}–10^{-17} m^2) are generally more permeable than plutonic rocks (10^{-16}–10^{-21} m^2). Metamorphic rocks have a permeability similar to that of plutonic rocks.

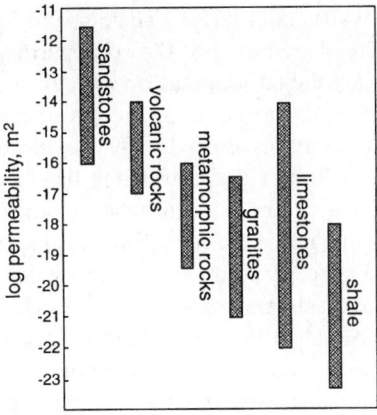

Figure 1.2 Laboratory measured permeabilities. Hydrostatic pressure <10 MPa and $T = 25$ °C. After Brace (1980).

13

When considering the permeability of rock, the question of scale is of paramount importance. The three orders of magnitude difference between borehole measurements and laboratory measurements of rock cores is a good example of this problem. Rock is highly heterogeneous on all scales, and the larger the volume under consideration, the more likely it is to contain major features like fractures and faults, that can act as high permeability channels for fluid flow. The common layered structure of rocks also means that permeability may be highly anisotropic. These considerations are currently of great importance in developing theoretical models of rock permeability (Odling, this volume; Guéguen et al., 1996).

1.4 Controls on fluid transport in rock

Direct measurement of rock transport properties is only possible at the surface or at very shallow depths. We do not have any data for deeply buried rocks or for rocks undergoing metamorphism or anatexis. At present we can only constrain the transport properties of these texturally and chemically evolving rocks using indirect arguments based on geochemical or textural observations, or by laboratory experiments. Either the porosity or the permeability of metamorphic rocks can be constrained if assumptions are made about the dependence of the permeability on porosity, using time-integrated fluid fluxes obtained either from consideration of alterations of stable isotope ratios in low permeability rocks (e.g. Bickle and Baker, 1990; Graham et al., this volume) or from mineral assemblages (e.g. Ferry, 1994). The observed distribution of metasomatic assemblages in hydrothermal zones can also be used to constrain rock transport properties at depth (e.g. Norton, 1988).

Attempts to infer the geometry of melt-filled pores from textural studies of crystallized melt-bearing rocks (e.g. Hunter, 1987; Harte et al., 1991; Bryon et al., 1996) are hampered by recrystallization and growth of the rims of grains comprising the pore walls. Similarly, the distribution of leucosomes in migmatites may not be representative of the rock permeability throughout the partial melting event if significant segregation of melt occurred. However, laboratory studies of melt migration have been successful in determining migration rates (Watson, 1982; Daines and Kohlstedt, 1993) in basalt-bearing systems.

Fluid-bearing rocks can be divided into three broad categories, albeit with inevitable overlap. The first category contains rocks in which pore geometry is determined by the prior history of the rock. Such rocks are generally at shallow levels in the crust, with a porosity controlled by processes such as sedimentation and diagenesis, or by fracturing associated with exhumation, tectonism or local thermal effects. The essential feature of rocks in this 'non-interactive' category is that the fluid occupies a pre-existing pore structure, although passage of the fluid may modify the pore structure to some extent (e.g. by mineralization or dissolution).

The second category comprises rocks from deeper and/or hotter parts of the crust, in which pore connectivity is determined by the minimization of interfacial free energies. In rocks of this category, the rate at which fluid–solid textural equilibration is attained is rapid relative to that of deformation or reaction/crystal growth. Rocks which fall into this texturally equilibrated category include some partially molten rocks (e.g. Jurewicz and Jurewicz, 1986; Hunter, 1987; Laporte, 1994), halite (rock salt) at shallow crustal levels (Lewis and Holness, 1996), and rocks undergoing high- to medium-grade metamorphism (Watson and Brenan, 1987; Holness and Graham, 1991, 1995; Holness, 1992, 1993). In such rocks the pore structure is a function of composition, pressure and temperature. The essential feature is that the fluid phase itself determines the structure of the porosity. Indeed, spontaneous fluid infiltration into previously dry rock can occur if it results in a decrease of the total internal energy of the system.

A third category, that of 'dynamic' permeability, includes rapidly reacting metamorphic rocks and some partially molten rocks, in which the pore geometry is controlled by the kinetics of reaction and crystal growth rather than minimization of interfacial energies. Textural disequilibrium during crystallization of granitic plutons has been inferred from detailed analysis of mineral textures (Bryon et al., 1996) which show that the pore geometry was controlled predominantly by variable crystal nucleation rates and anisotropy of grain growth rate. Melt geometry during crustal anatexis may similarly be controlled by kinetic processes (e.g. Wolf and Wyllie, 1991; Connolly et al., 1997). Thermal decompaction of metamorphic rocks coupled with rapid pore fluid generation would also fall into this category.

1.5 Theoretical permeability models

In order to derive a theoretical treatment of fluid flow in rocks we must first characterize or approximate the pore geometry. Fluid flow through the pore space is generally assumed to be steady, laminar and unidirectional, with a parabolic velocity distribution across the conduit (i.e. Poiseuille flow). Under certain restrictive conditions, simple relationships between permeability and porosity can be derived. For example, flow in parallel-sided cracks leads to a power-law scaling of permeability with porosity, i.e., $k \propto \phi^3$. Poiseuille flow in cylindrical tubes leads to a similar relationship: $k \propto \phi^2$ (e.g. Guéguen and Palciauskas, 1994). Although this kind of simple power-law relationship between permeability and porosity can be useful in some circumstances and is much used in theoretical models of fluid flow (e.g. Graham et al., this volume) it masks the fundamental fact that permeability is actually determined by pore microstructure, rather than porosity. It also ignores much of the complexity of rock structure. Rock heterogeneity often results in the formation of high conductance flow pathways, in which case the critical parameter controlling permeability will not be the total porosity but some network of flow paths. These considerations will be covered in the later part of this section, but I will first review the permeability of texturally equilibrated rocks as they contain potentially the simplest type of pore microstructure.

1.5.1 Models of texturally equilibrated aggregates

The pore topology and permeability of the simplest fluid-bearing texturally equilibrated system, comprising identical grains with no variation of surface energies with crystal orientation (i.e. the solid phase is isotropic), are well known (e.g. Beere, 1975; Cheadle, 1989; Wray, 1976). However, rocks do not closely resemble the 'ideal' geometry of such a simple system, primarily because they are polymineralic, but importantly also because rock-forming minerals are not isotropic. These departures from the ideal system and their implications for rock permeability will be discussed later.

A texturally equilibrated material is one in which the surface topology of its constituents is in mechanical and thermodynamic equilibrium, with the total interfacial free energy at a minimum. In the ideal aggregate, the interfacial free energies must balance at all grain intersections. These energies act in the plane of the surface and perpendicular to any point on the line of intersection. At the intersection, the following expression must hold:

$$\frac{\gamma_{23}}{\sin \Theta_1} = \frac{\gamma_{13}}{\sin \Theta_2} = \frac{\gamma_{12}}{\sin \Theta_3} \tag{1.2}$$

where γ_{xy} is the surface free energy of the interface between the two phases x and y, and Θ_z is the opposite angle (Figure 1.3(a)). For the case of a monomineralic aggregate containing a fluid phase, there are only two interfacial energies in the system: the grain boundary energy, γ_{gb}; and the energy of the fluid–solid interface, γ_{fs}. At each solid–solid–fluid junction, i.e. at each pore corner, the following relationship must hold:

$$\frac{\gamma_{gb}}{\gamma_{fs}} = 2 \cos \frac{\Theta}{2} \tag{1.3}$$

where Θ is the dihedral angle (Figure 1.3(b); Smith, 1948). The topology of the fluid phase can be determined from the magnitude of Θ (e.g. Smith, 1948; Beere, 1975; Park and Yoon, 1985; von Bargen and Waff, 1986; Cheadle, 1989).

If $\Theta \geq 60°$, the pore geometry in the fluid-bearing ideal rock will be that of isolated pores at four-grain junctions, unless the porosity exceeds a threshold value for coalescence and connectivity (Figure 1.3(c)). This threshold is a function of dihedral angle, ranging from 0.6% for $\Theta = 60°$ to 6–10% for dihedral angles of around 100° (Cheadle, 1989). These connected pores are unstable relative to the isolated geometry, and if the porosity exceeds the threshold value, fluid will be expelled from the system or accumulate in large pools in order to attain the minimum energy configuration.

If $\Theta < 60°$, the fluid phase forms a stable interconnected network of channels

16

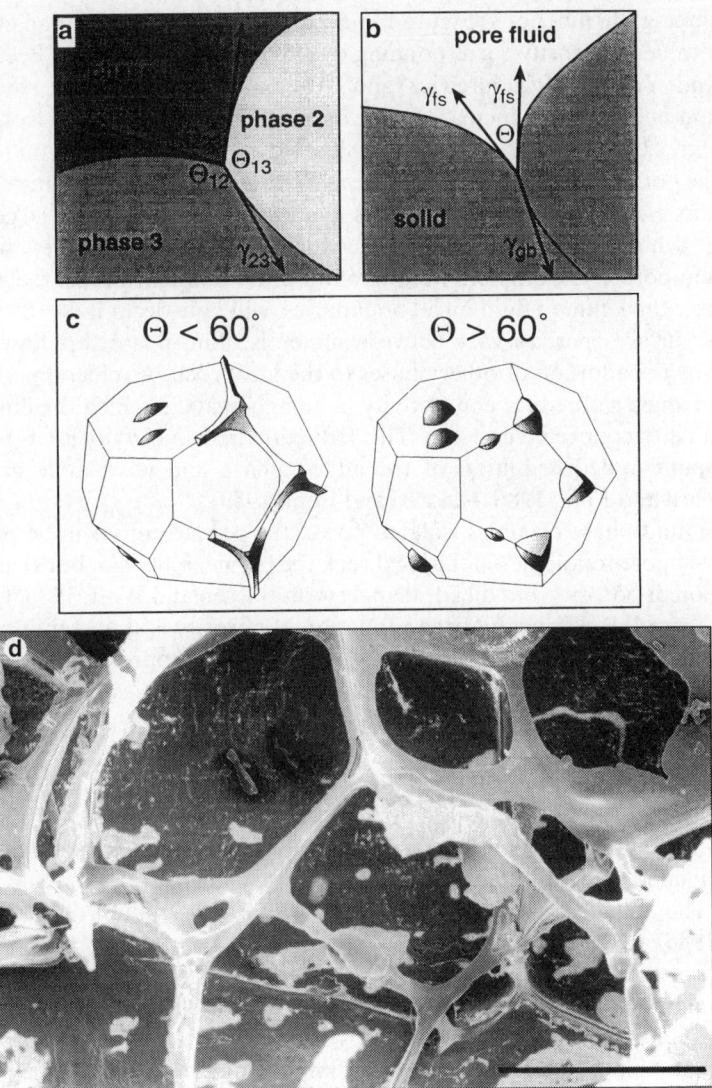

Figure 1.3 (a) The balancing of interfacial energies at the junction of three phases. The interfacial energies act in the plane of the interface and away from the three-phase junction. Characteristic angles develop at the junction. For the case of two phases, for example solid + fluid (b), the dihedral angle, Θ, develops at the corner of the pore and is a function of the chemical composition of the two phases involved. (c) The variation in pore topology as a function of dihedral angle. For all angles less than 60°, the fluid forms an interconnected network on grain edges. For angles greater than this critical value, the pores are isolated below a certain threshold porosity (after Watson and Brenan, 1987). (d) Scanning electron photomicrograph of interconnected porosity in a texturally equilibrated H_2O–halite experimental charge. The porosity was vacuum impregnated with epoxy which was allowed to set before the solid halite matrix was dissolved in water, leaving behind a three-dimensional mould of the pore network (compare with Figure 1.1(d)). Scale bar is 50 μm long.

along three-grain junctions (Figure 1.3(c), (d)). For all angles less than 60°, there exists a value of porosity corresponding to a minimum energy state (Beere, 1975; Park and Yoon, 1985; Jurewicz and Watson, 1985; Cheadle, 1989). This minimum energy value increases with decreasing dihedral angle, from 0% at $\Theta = 60°$ to 22% for $\Theta = 0°$ (Cheadle, 1989), and is the true equilibrium topology when the porosity is allowed to vary freely. For a rock containing more than its minimum energy porosity, the excess fluid will accumulate in large pools, leaving behind an interconnected network of grain-edge pores with the minimum porosity. Complete wetting of all grain boundaries, i.e. the presence of a stable thick film of fluid on all boundaries, will only occur if $\Theta = 0°$.

Since surface energies vary between minerals, fluid phase topology will be altered by the addition of other phases to the ideal rock. A mineral with a high dihedral angle will reduce connectivity in an aggregate in which the fluid phase forms an interconnected network. This reduction in connectivity is a function of the amount and distribution of the added phase and its relative grain size (Toramaru and Fujii, 1986; Nakano and Fujii, 1989).

If the fluid phase occupies isolated pores, the permeability will be zero. For connected pore topologies in the ideal rock the permeability can be calculated as a function of porosity and dihedral angle (von Bargen and Waff, 1986; Cheadle, 1989; Figure 1.4). Permeability as a function of porosity is almost indistinguishable for dihedral angles in the range 1–40°, and has the approximate form:

$$k = c\phi^2 \tag{1.4}$$

where c is a constant (von Bargen and Waff, 1986; Cheadle, 1989). This is the form expected for Poiseuille flow in a series of identical connected tubes (e.g. Frank, 1968; Guéguen and Dienes,1989), although the porosity is only closely approximated by connected tubes of constant diameter for $\phi < 1\%$ (Cheadle, 1989). The actual equations derived by Cheadle (1989) and von Bargen and Waff (1986) are given in Figure 1.4.

In theory it would be possible for a rock with $\Theta < 60°$ to be permeable at vanishingly small porosities, but in practice most rock-forming minerals have a finite degree of surface energy anisotropy and this will act to close off pores (see later). Also thin-film effects will become important in channels as they become smaller than the thickness of adsorbed fluid films on the fluid–solid interface.

1.5.2 Models of non-texturally equilibrated aggregates

The case for non-texturally equilibrated rocks is more complex, as the pore geometry cannot be derived analytically. Short of detailed characterization of every substance of interest, the way forward for such materials is to develop models relating permeability to porosity or to an assumed pore geometry, and to compare these with inferences of pore geometry in real rocks or with experimental determinations of transport properties. There are two basic approaches

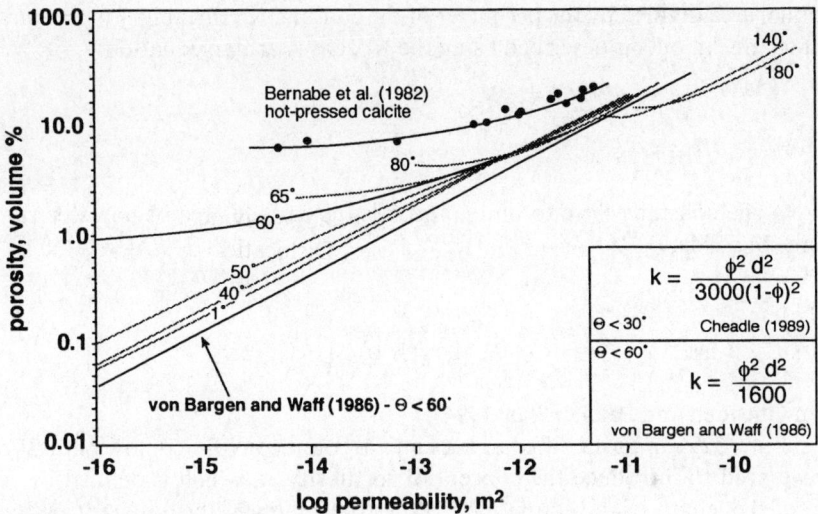

Figure 1.4 Analytic and numeric solutions for the permeability as a function of porosity and dihedral angle in a monomineralic, isotropic, texturally equilibrated aggregate with a uniform grain size of 1 mm (after Cheadle, 1989). The faint lines give the solutions of Cheadle (1989), and the heavy lines those of von Bargen and Waff (1986). The two equations give the analytic solutions for $\Theta<30°$ from Cheadle (1989), and for $\Theta<60°$ for von Bargen and Waff (1986). Symbols in the equations are as those in the text, with the addition of d, the grain diameter. The dots give the experimentally derived permeability of hot-pressed calcite aggregates from Bernabé *et al.* (1982), modified to take into account the smaller grain size (assumed constant at 2 μm) of the experimental charges, with the line of best fit from Cheadle (1989).

to the problem. The first assumes that the rock can be approximated by an equivalent porous medium in order to derive analytic solutions for permeability as a function of porosity, and the second is to erect numeric models to simulate rock permeability using networks of hydraulic bonds. The problem of rock heterogeneity can perhaps be most effectively tackled using numeric permeability models in which heterogeneities of either bond conductance or orientation can be included.

(a) Equivalent porous medium models
Perhaps the simplest variant of the equivalent porous medium approach is the **equivalent channel** model (Wyllie and Rose, 1950; Paterson, 1983; Walsh and Brace, 1984), which is based on the assumptions of no preferential flow paths and a completely connected pore space, i.e. the rock is completely homogeneous. The porous medium is presumed equivalent to a conduit with a highly complex cross-section of constant average area. The length scale governing the flow rate through the conduit is the hydraulic radius, r_H, which is defined as the pore volume divided by the area of the fluid–solid interface (or the flow cross-

19

sectional area divided by the perimeter of the conduit). Permeability in an equivalent channel model is described using the Kozeny-Carman equation:

$$k = \frac{a\phi^3}{S^2(1-\phi)^2} \tag{1.5}$$

where a is a constant close to unity, and S is the surface area of pores per unit volume of solids:

$$S = \frac{\phi}{r_H(1-\phi)} \tag{1.6}$$

(from Guéguen and Palciauskas, 1994).

Equation 1.5 can be modified to take into account convoluted flow paths. For this we need to introduce the concept of tortuosity, τ, which is defined by the ratio of the length of the actual path, L', to the length of the apparent path, L (Figure 1.5). The Kozeny-Carman equation then becomes:

$$k = \frac{1}{b\tau^2} \frac{\phi^3}{S^2(1-\phi)^2} \tag{1.7}$$

where b is a constant. The Kozeny-Carman model works well if the walls of the pores are smooth. This is particularly important for materials in which the parameter S is determined using gas adsorption, since the gas will 'see' a rougher surface compared to that 'seen' by the fluid flow. The equivalent channel model can be made more realistic by implementing an empirical weighting procedure

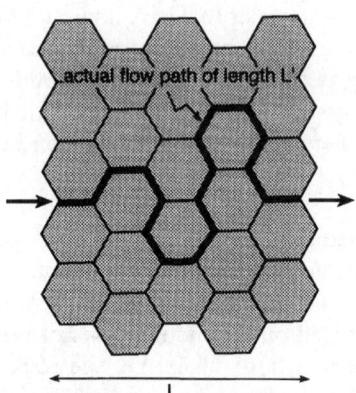

Figure 1.5 Flow in rocks generally involves a complex flow path, rather than flow in a straight line. The tortuosity, τ, is the ratio of the length of the flow path actually used, L', compared to the shortest line between the two end-points of flow, L. For a highly tortuous path, τ is high.

which reduces the effects of the dead-ends and other low-conductance pore space (e.g. Johnson and Sen, 1988; Schwartz *et al.*, 1989).

The main drawback of models of the equivalent channel type is that the pore space in real rocks is not homogeneous. Rock heterogeneity is accounted for in theories of the **effective medium** type (e.g. Kirkpatrick, 1973) by using a representative element volume (REV) which is big enough for the rock to appear homogeneous and the properties to be averaged out. Models of this type are useful only insofar as the assumption of statistical homogeneity is valid, and cannot be used if the scale of interest is insufficiently large to accommodate the heterogeneities of the pore space. Conduits in natural materials often have widely varying conductances (manifest, for example, by varying sizes or shapes), and effective medium-type models cannot accurately predict transport properties when the pore space involves a broad distribution of bond conductances (David *et al.*, 1990).

(b) Network models
Permeability models based on attempts to form simple physical analogues of rock pore structure approximate pore geometry by collections of tubes, penny-shaped cracks and spherical pores. The most complex of the models incorporate all three kinds of pore space (Figure 1.6; e.g. Bernabé, 1991). These models can be used to form either continuous three-dimensional networks of porosity to examine how variations in pore geometry can affect transport properties, or to examine the effect of variations in the connectivity of pore space (David *et al.*, 1990).

Network models are commonly formed on a regular grid, with individual porosity elements comprising equant pores connected by tubes and/or sheet-like bonds. The transport properties of the network are controlled by the connecting bonds, while the equant pores contribute to the storage capacity of the network

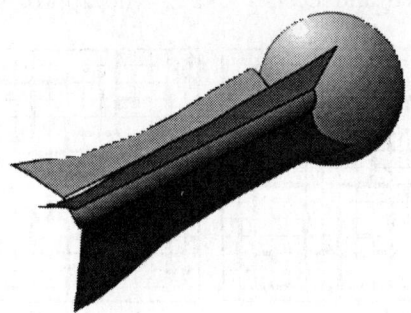

Figure 1.6 The three types of bonds used for the most realistic analogue models of rock porosity (e.g. Bernabé, 1991). The large spherical pores are assumed to occur at four-grain junctions, the tubes occur along three-grain junctions, and the planar features represent either porosity on grain boundaries or cracks.

21

(Bernabé, 1991). Bernabé (1995) generated a large set of networks with variations in pore size and connectivity and found that $k \propto \phi^{2.44}$, falling between the exponent of 3 for cracks and that of 2 for tubes.

Ascertaining the connectivity of pore space, whether cracks or grain-edge channels, is essential for determining permeability. Network models are used to examine the effect of varying connectivity using percolation theory (see Stauffer and Aharony (1994) for an introduction). The concepts behind percolation theory are easy to grasp. Figure 1.7 shows a square lattice such as those used in network models. The heavy black lines denote bonds between the lattice points, and are randomly distributed, with a probability, p, that any two lattice points are joined by a bond. Groups of contiguous bonds are termed 'clusters'. The number of bonds in Figure 1.7(b) is greater than in Figure 1.7(a), and in Figure 1.7(b), one cluster extends from one side of the lattice to the other. The percolation threshold is that value of p at which a percolating cluster first forms and connectivity is established (see Odling, this volume, for applications of the concepts of connectivity and the percolation threshold to flow in fracture networks).

The strength of network models lies in the ease with which highly heterogeneous porous media can be simulated. Models incorporating a broad range of heterogeneity of bond conductance show the development of preferential flow along sub-networks of high conductance pores which carry most of the flow (e.g. David, 1993). A broad range of conductances is likely to be found in fractured rock. In consequence, the permeability is controlled by a small fraction of the total interconnected pore space, emphasizing that variations in the bulk porosity may not be the only microstructural parameter controlling rock transport properties.

(c) Statistical models

A different approach to modelling rock permeability is the statistical model developed by Guéguen and Dienes (1989), who approximated pore space in

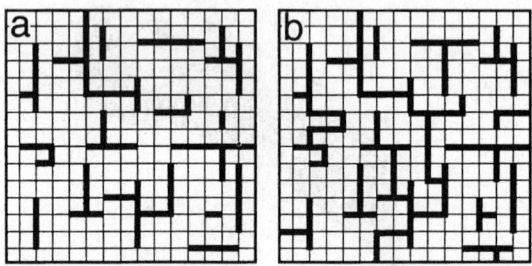

Figure 1.7 A square lattice on which lattice points are joined by bonds, randomly distributed. The bonds form clusters. (a) The clusters are isolated. (b) The probability, p, of two points being joined by bonds, is higher than in (a), and one of the clusters is now big enough to span the dimension of the lattice. Percolation from one side of the lattice to the other is possible in this case. The percolation threshold is that value of p for which a percolating cluster is first formed.

terms of the distribution of randomly oriented individual tubes or cracks of varying sizes (Figure 1.8). If all the pore elements form a connected network in three dimensions, it is possible to write expressions relating the permeability to three independent microvariables: the average bond length; the shape of the elements; and their spacing. Using the usual assumption of Poiseuille flow, the permeability is a simple cubic function of the average crack width for the crack model, and a quadratic function of the average pipe radius for the tube model. This model works best for a narrow range of microvariables.

1.6 Permeability–porosity relationships in rocks

The preceding sections covered the theoretical models used to describe and quantify rock permeability. What is known about the permeability of real rocks? In this section, I cover the three types of fluid-bearing rock identified earlier, detailing results and conclusions from field observations and experimental studies.

1.6.1 Experimental studies of P–T controls on rock permeability in the shallow crust

Rocks in which the fluid plays only a minor role in determining pore topology are generally in the shallow crust, where temperatures are low and the predominant processes affecting rock permeability are pore collapse due to increases in confining pressure related to burial, and porosity reduction by cementation during diagenesis. The bulk of experimental investigations into permeability of rocks in the shallow crust have concentrated on determining the effects of

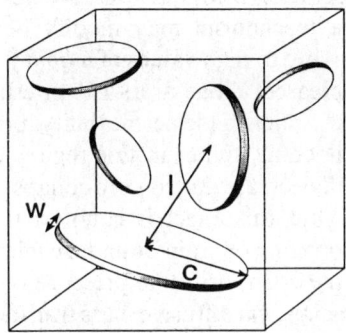

Figure 1.8 A collection of randomly oriented and distributed variously sized penny-shaped cracks used to model rock permeability. This statistical model (Guéguen and Dienes, 1989) can be used to generate expressions for permeability in terms of the microvariables w (the average crack width), c (the average crack radius), and l (the average spacing between crack centres). It has the advantage of being able to incorporate both heterogeneity in pore size and variations in pore connectivity. A similar model using tubes has also been developed (Guéguen and Dienes, 1989).

increasing pressure and temperature, simulating changes in rock permeability and porosity during increasing burial.

David *et al.* (1994) recognize three separate mechanisms of permeability decrease with increasing pressure. The first is related to the elastic and reversible closure of cracks and occurs in low porosity crystalline rocks. Experimental data for the permeability–porosity relationship as a function of increasing pressure in crystalline rocks demonstrate a power-law relationship with $n \approx 3$ (Brace *et al.*, 1968) in agreement with the simple models of flow in cracks (e.g. Guéguen and Dienes, 1989). This power-law k–ϕ relationship is only applicable in the shallow crust because asperities on the crack walls start to come into contact with increasing pressure, increasing both the tortuosity of the flow path and the porosity exponent. For rocks containing pores of higher aspect ratio the effect of pressure will not be so marked, due to the combined effect of a smaller pressure sensitivity of both spherical pores compared with flat cracks and the major contribution flat cracks make to permeability. Comparison of the pressure sensitivity of permeability of natural sandstones with predictions based on network models shows consistency with observed proportions of spherical, tabular and tubular pores (Bernabé, 1991).

The second mechanism is that of grain rearrangement, and occurs in poorly consolidated, high porosity materials. Permeability reduction with increasing pressure is generally minor compared to low porosity materials. A power-law k–ϕ relationship is again observed, but the value of n in this field can vary widely from 1 to 25 (David *et al.*, 1994). Poorly consolidated materials generally have a low exponent. The high values may correspond to a porosity structure where much of the porosity does not contribute to the permeability (*ibid.*). The sandstones examined by David *et al.* (1994) contain a highly heterogeneous porosity structure with large equant pores and micropores coexisting with large numbers of grain boundary microcracks. If compaction involves grain movements then it is plausible that such a mechanism may modify permeability more than it changes the porosity, leading to high values of n (*ibid.*).

As the pressure is increased above a critical threshold, grain crushing and pore collapse start to occur. No simple permeability–porosity relationships have been observed in experimental studies in this regime (*ibid.*). The permeability during grain crushing will decrease due to pore collapse and infilling of available pore space with debris, but this effect is counterbalanced by an increase in permeability due to the opening of grain boundary microcracks.

At greater depths in the crust, or in the presence of a fluid, compaction can take place by chemical means, via diffusive mass transfer, crystal plasticity, pressure solution, and solution–reprecipitation processes. These processes occur during textural equilibration, but the discussion of experimental studies of chemical compaction belongs here because textural equilibrium is only achieved during the final stages of these experiments (Olgaard and FitzGerald, 1993). Chemical compaction is similar to hot isostatic pressing (HIPing) – a manufacturing technique used for materials that cannot be cast or machined. HIPing is generally divided into three stages. Grains in the loose powder comprising the starting material are angular and non-cohesive during the first stage, and undergo relative movement and comminution. The second stage occurs when the rounded grains undergo densification by plastic flow and/or diffusive mass transport. The final stage involves disconnection of the isolated voids, and further densification occurs via plastic collapse of the pore space.

Several recent laboratory studies have examined the evolution of permeability and porosity during HIPing of calcite and quartz (Bernabé et al., 1982; Zhang et al., 1994; Lockner and Evans, 1995). All three studies demonstrate that there are three regions in the development of the porosity structure (Figure 1.9), which are thought to correspond to the intermediate and final stages of HIPing (Zhu et al., 1995). The first, at high porosities, shows a power-law dependence of permeability with an exponent of 3. The second, which starts at some critical porosity in the range of 7–15%, has a greater sensitivity of permeability to decreasing porosity, but has no simple power-law dependence, although Zhang et al. (1994) demonstrate that the cubic k–ϕ relationship observed in their experimental data extends to porosities of 7% (i.e. throughout regions 1 and 2) if only **connected** porosity is considered.

The third stage is when the permeability falls below the limit of measurability, and is probably related to the reduction of pore connectivity near a percolation threshold which occurs at porosities between 3% and 7% (Zhang et al., 1994; Zhu et al., 1995; Knackstedt and Cox, 1995). The treatment of the

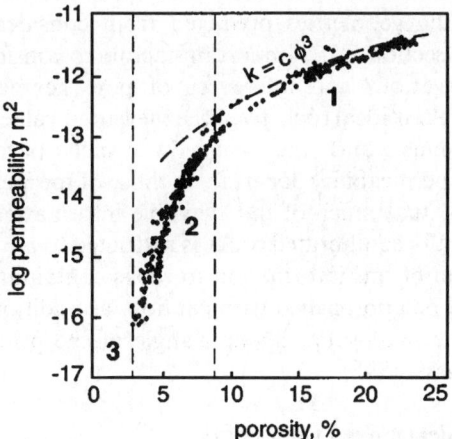

Figure 1.9 Evolution of permeability as a function of porosity in naturally lithified Fontainebleau sandstone (after Zhu et al. (1995), based on the experimental data of Bourbié and Zinszner (1985)). The dashed line represents a power-law permeability–porosity relationship with $n = 3$ where c is a constant. Three different regimes are marked: (1) the region in which the power law applies; (2) for porosities lower than about 9%, an accelerated reduction in permeability is observed; (3) as the porosity is reduced below about 4%, the permeability becomes too small to measure, implying an approach to the percolation threshold with loss of connectivity in the pore space.

compacting aggregates as a percolation problem necessitates a two-stage model involving bond shrinkage followed by cutting of bonds with progressive compaction, resulting in a steep decrease of permeability to the percolation threshold (e.g. Zhu *et al.*, 1995). Direct observation of pore closure during HIPing (Bernabé *et al.*, 1982; Olgaard and FitzGerald, 1993) gives support for a two-stage model. In contrast, Lockner and Evans (1995) succeeded in reproducing their experimentally observed changes in electrical conductivity with porosity with a one-stage network model in which porosity reduction was simulated by the deposition of material evenly on all pore walls. The entire porosity network remains interconnected in this model, until the cross-sectional area of the connecting tubular bonds becomes vanishingly small and the network ceases to be permeable.

A similar relationship between permeability and porosity is observed in (undeformed) Fontainebleau sandstone (Figure 1.9; Bourbié and Zinszner, 1985). In this case the changes are due to varying degrees of silicification during diagenesis, and is a demonstration of the potential application of percolation theory to developing our understanding of the permeability of sediments in the shallow crust, in which the controls on permeability are chemical cementation processes rather than the compaction observed in the above experimental studies.

1.6.2 Texturally equilibrated rocks

In this section, I will discuss how closely pore space in hydrostatically stressed rocks approaches the geometries predicted from consideration of the ideal system described in section 1.5.1. Important factors to consider include whether experimental observations and inferences of pore geometry correspond to predictions based on our ideal rock given the measured values of dihedral angles in geological systems, and the ways in which pore shape and the predicted/observed permeability depart from those of the ideal system. It should also be remembered that much of the available information on pore shape in fluid-bearing texturally equilibrated rocks is restricted to mineralogically simple systems. Application of this information to rocks containing large numbers of solid phases should be approached with caution, as addition of further phases can greatly alter the fluid–solid dihedral angle of the primary phase at high temperatures (Holness, 1995).

(a) Experimentally determined values of Θ

Experimentally determined values for the equilibrium fluid–solid dihedral angle in geological systems fall in the range 40–100° for H_2O–CO_2–NaCl fluids in monomineralic quartz, carbonate, feldspar and halite aggregates, and in the range 20–60° for basic and silicic melts in, or close to, chemical equilibrium with their silicate host (Figure 1.10). Dihedral angles are a function of fluid composition, pressure and temperature, via a control of adsorption of surface-active

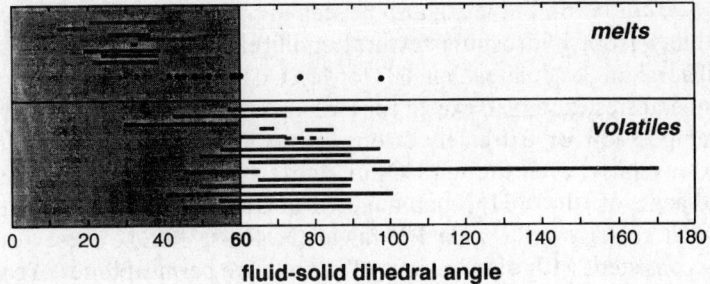

fluid-solid dihedral angle

Figure 1.10 Experimentally determined fluid–solid dihedral angles in geological systems (after Holness, 1996). Dots represent single reported values, and lines show the ranges observed in systems as a function of P, T and fluid compositional variation. All melt data are ≤60° apart from two isolated pyroxene values (Toramaru and Fujii, 1986) which were later shown to be too high (Fujii *et al.*, 1986; von Bargen and Waff, 1988). All volatile angles occupy a tight group between about 40° and 100°, apart from one study (Hay and Evans, 1988) with angles ranging from 97° to 171°. The measuring technique is thought to have been inappropriate and the samples poorly equilibrated (Holness and Graham, 1995). The only study with volatile angles almost exclusively <60° was that of Lee *et al.* (1991), and their experiments were probably not fully texturally equilibrated (Brantley, 1992).

species on the fluid–solid interface and the grain boundaries. At any given pressure and temperature, the lowest angles are observed in systems in which either the fluid phase contains highly surface-active species (e.g. brines in quartz aggregates) and/or the fluid phase has a structure and composition similar to that of the solid (e.g. basaltic melt in olivine aggregates). In the former case (i.e. systems with a highly surface-active fluid), the lower the angle, the greater the sensitivity of the angle to changes in pressure and temperature (e.g. Holness, 1993; Agee and Shannon, 1996).

In experimental charges in which the dihedral angle is less than 60° the fluid phase occupies a network of channels on three-grain edges (e.g. Vaughan *et al.*, 1982; Daines and Richter, 1988; Holness, 1992; Laporte, 1994; Figure 1.3(d)), in contrast to the isolated pores observed in charges in which the dihedral angle is greater than 60°, in agreement with the predicted geometries. Predictions concerning the minimum energy porosity have also been shown to be valid. In charges containing a large volume fraction of silicic melt, the majority of the melt was observed to accumulate in scattered large pores, leaving behind a connected network of channels on three-grain junctions corresponding to the minimum energy porosity (Jurewicz and Watson, 1985).

(b) Is the field evidence consistent with the predictions?
Field evidence for rock permeability during metamorphism shows that carbonate rocks are impermeable compared to pelitic or calc-silicate rocks (e.g. Nabelek *et al.*, 1984; Bickle and Baker, 1990). This observation has been ascribed to differences in the amount of reaction-generated porosity (e.g.

Richards *et al.*, 1996), but it can also be seen in the context of the predictions of permeability from hydrostatic textural equilibrium experiments. Carbonates have dihedral angles greater than 60° for H_2O–CO_2–NaCl fluids at all pressures and temperatures examined except for P<2 kbar if the fluids are of intermediate X_{CO_2} composition or extremely strong brines (Holness and Graham, 1991, 1995). Conversely, even though CO_2 in quartz aggregates has a PT-insensitive dihedral angle of ≈100°, H_2O-rich fluids form a connected porosity network in quartz-rich rocks over a wide PT range (Holness, 1993). These results are broadly consistent with observations of the relative permeabilities of carbonate and pelitic rocks.

There appears to be no firm evidence for pervasive flow of fluids along a stable grain-edge pore network. Arita and Wada (1990) demonstrate sub-grainscale variations in stable isotope composition of calcite, which they attribute to grain-edge flow of aqueous fluids, but the chemical composition of the fluid is not known. More circumstantial evidence that experimental determinations of dihedral angle can be used to predict pore fluid behaviour comes from a consideration of the compositions of fluid inclusions in quartz-rich rocks. Aqueous fluids apparently preferentially move through the grain boundary regions while CO_2-rich fluids are trapped as inclusions (Craw and Norris, 1993; Johnson and Hollister, 1995), consistent with predictions of isolated porosity for CO_2-rich fluids and a connected grain-edge porosity for aqueous fluids (Holness, 1993).

(c) Direct experimental determinations of permeability in texturally equilibrated rocks

Attempts to determine experimentally permeability–porosity relationships for texturally equilibrated fluid-bearing rock have led to ambiguous results. Maaløe and Scheie (1982) attempted a direct test of equation 1.4 using compressed glass spheres as an analogue for texturally equilibrated, two-phase aggregates, and found that the equation holds in the range $0.5\% < \phi < 25\%$. In contrast, Riley and Kohlstedt (1991) inferred a linear dependence of permeability on porosity (i.e. $n = 1$) in experiments designed to look at migration rates of basalt from a high porosity source region into an initially dry, non-porous olivine aggregate. Daines and Kohlstedt (1993) developed this further, using a melt sink initially containing about 20% of the total melt fraction, and demonstrated that it is not possible to distinguish between k–ϕ power law relationships with exponents ranging from 1 to 3, since the larger the value of ϕ, the smaller the change in permeability as the exponent increases. The exponent only has a major change on the modelled porosity profiles for very small porosities.

(d) How close to ideal are texturally equilibrated rocks?

On a gross level, experiments and field observations demonstrate that real rocks will behave in a similar fashion to the ideal system. In detail, however, there are some very important departures from ideality. For example, the low fluid–solid dihedral angle in both the olivine-carbonate melt and the quartz–H_2O systems would imply that connectivity of the fluid phase will occur at vanishingly low porosities. However, diffusion rate experiments in these systems demonstrate that complete connectivity only occurs for porosities greater than about 0.3% (Brenan, 1993; Minarik and Watson, 1995). This cut-off in connectivity at low porosities is attributed to the stabilization of dry three-grain junctions due to crystalline anisotropy.

All rock-forming minerals show some degree of anisotropy of surface energies. The degree of anisotropy increases from the relatively isotropic mineral group scapolite–quartz–calcite–garnet–feldspar, through pyroxenes and amphiboles, to micas and sillimanite (Kretz, 1966; Vernon, 1968). Anisotropy produces torque forces acting to rotate grain boundaries and fluid-solid interfaces into lower energy configurations (Herring, 1951), leading to the development of planar, rational, low-energy crystal faces in contact with fluid. These coexist with smoothly curved solid–fluid interfaces in a variety of geological systems (olivine–basalt: Waff and Faul, 1992, and Daines, this volume; quartz–silicic melt: Laporte, 1994; amphibole–silicic melt: Wolf and Wyllie 1991; biotite–amphibole–silicic melt: Laporte and Watson, 1995; pyroxene–H_2O–CO_2: Watson and Lupulescu, 1993; quartz–H_2O: Laporte and Watson, 1991). Laporte and Watson (1995) point out that the most anisotropic minerals observed in crustal materials are generally those with high melting points. Melt geometries during crustal anatexis will thus be controlled by anisotropic biotite rather than the more isotropic quartz or feldspar.

One of the effects of stabilization of planar fluid–solid interfaces is to generate porosity as fluid films on two-grain junctions. These have been observed in quartz-silicic melt (Laporte, 1994) and olivine-basalt systems (Waff and Faul, 1992; Faul et al., 1994). The majority of melt in olivine aggregates is contained within large, elongate melt pockets commonly bounded by more than three grains and extending deep within grain boundaries (Figure 1.11), with only 20% of the melt contained within the network of grain-edge channels predicted from consideration of the isotropic case (Faul et al., 1994). This will have an important effect on permeability since although an interconnected melt path is provided by the grain-edge tubes, if the majority of the melt occupies non-contiguous large pockets the permeability will be lower than the value predicted assuming all melt were in grain-edge channels (Faul et al., 1994).

Rocks are polymineralic, and show heterogeneity on a variety of scales. Crystallographic preferred orientations are common, especially in metamorphic rocks and mantle materials. Preferred crystal orientations lead to preferred orientations of fluid-filled pores if fluid–solid interfacial energies are anisotropic. These issues are specifically addressed by Laporte and Watson (1995), who looked at the distribution of silicic melt in partially molten biotite-bearing rocks (Figure 1.12), and the implications of their results for the development of melt distribution in migmatites are discussed by Brown and Rushmer (this volume). Daines (this volume) extends the discussion of the effect of preferred crystal orientations in melt distribution to mantle materials.

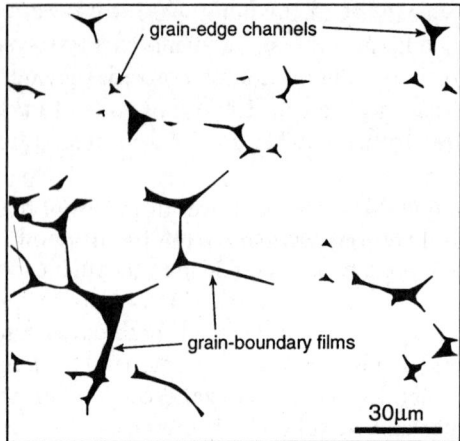

Figure 1.11 Two-dimensional section through an olivine-basalt experimental charge. The black areas represent quenched melt. Triple-junction melt tubes can be seen at the top of the figure, in accord with predictions of pore topology from the ideal, isotropic system (see Figure 1.3), but at the bottom of the figure are large pores along grain boundaries and in low aspect ratio inclusions in which the majority of the glass resides (after Faul *et al.*, 1994).

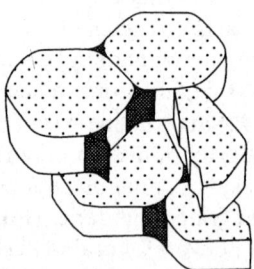

Figure 1.12 Schematic grain-scale distribution of silicic melt in a biotite aggregate in which the biotite grains have a preferred orientation. The melt-filled pores are shown by the dark shading, and the (001) faces of the biotite grains are shown by the pale shading. The melt-filled pores are truncated by the (001) biotite faces (after Laporte and Watson, 1995). The pore geometry is very different from that predicted for an isotropic mineral, and would result in a highly anisotropic permeability.

A question which should be addressed (and which leads straight into the following section) is the extent to which textural equilibrium is attained in rocks and in experimental charges. High-grade metamorphic rocks and many plutonic rocks generally display low energy grain configurations (e.g. Kretz, 1966; Vernon, 1968; Hunter, 1987), but these are solid aggregates. We need to know the timescales during which textural equilibrium can be established in fluid-bearing systems. The experimental work done so far demonstrates that the pore geometry in charges run for hours or days at high homologous temperatures is in, or close to, textural equilibrium (Watson and Brenan, 1987; Holness and Graham, 1991). Perhaps more importantly, it has been shown that reliable and reproducible values of the equilibrium dihedral angle can be obtained from incompletely equilibrated experimental charges (e.g. Holness, 1992; Laporte and Watson, 1995) because textural equilibration is first established at pore corners and propagates outwards with time (Figure 1.13). It is possible that equilibrium needs only to be established over distances of the order of the pore channel diameter (i.e. 1–10 μm) in order for minimization of interfacial energies to control connectivity in low porosity rocks.

1.6.3 Dynamic permeability

In rocks with a rapidly changing porosity structure the changes may occur at rates faster than those for textural equilibration. This can occur in rapidly melting or crystallizing systems, in rocks undergoing metamorphic devolatilization reactions (in which volatiles are released so rapidly that the build-up in pore pressure results in fracture), or in systems in which chemical changes occur at rates too rapid for textural equilibrium to be attained. Cementation and solution reprecipitation processes occurring during diagenesis (e.g. Bourbié and Zinszner, 1985) fall into this category as the low temperatures mean that textural equilibration is prohibitively slow. Pore geometry and the permeability of rocks in this category are poorly understood.

Chemical equilibrium is not achieved during melting experiments on rock cores (e.g. Brearley and Rubie, 1990; Wolf and Wyllie, 1991; Rutter and Neumann, 1995), in agreement with the commonly observed chemical disequilibrium of trace element and isotopic composition during anatexis (e.g. Sawyer, 1991; Harris *et al.*, 1992; Watt and Harley, 1993). Wolf and Wyllie (1995) argue that chemical equilibration of major elements is achieved faster than textural equilibration, and suggest that textural equilibrium may be very rarely attained during anatexis. During experimental melting of natural rock cores, melt first appears at the junctions of grains taking part in the reaction (e.g. Brearley and Rubie, 1990; Wolf and Wyllie, 1991). The distribution of the melt phase, at least on the time-scale of these experiments, is primarily controlled by the kinetics of the melting reaction and the distribution of reactants, rather than minimization of surface energies.

Chemically reactive fluids play an important role in permeability formation. An example of flow of chemically reactive fluid in a metamorphic environment is provided by retrograde fluids in the Dalradian of the Scottish Highlands (Fein *et al.*, 1994). Aqueous brines moved along grain edges in calcite-rich rocks, resulting in replacement of the calcite by dolomite. This retrogression took place at a pressure and temperature for which brines have a dihedral angle >60° in calcite aggregates. The pervasive flow resulted from dissolution of calcite on three-grain junctions, and the total distance moved by the retrogressive fluids within each carbonate-rich layer shows a direct correlation with calcite content.

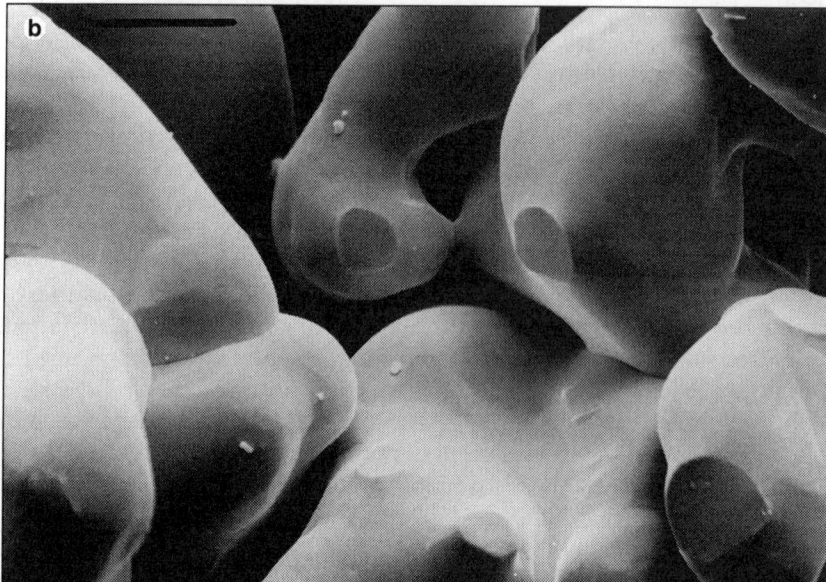

Figure 1.13 (a) Scanning electron photomicrograph of quartz grains equilibrated at 8 kbar and 800 °C in the presence of H_2O. The scale bar is 10 μm long. Note that the grains are facetted to within 3–5 μm of the junctions of two grains, and within this distance the grain surfaces are smoothly curved. The facets were formed during growth of the crystals from gel, and textural equilibration occurred subsequently, with the smooth rounded surfaces propagating outwards from grain junctions. With time, the grains become smoothly curved all over (b) and lose their growth-controlled facets. This sample of well-rounded grains is quartz equilibrated with an equimolar mixture of H_2O and CO_2 at 7 kbar and 800 °C. The scale bar is 20 μm long.

Another example of flow of a fluid which is not in chemical equilibrium with its matrix is the recent suggestion that the movement of basalt through the mantle may occur by focused porous flow along high-permeability channels (Daines and Kohlstedt, 1994; Kelemen *et al.*, 1995a; Aharonov *et al.*, 1995) resulting in the formation of elongate dunite bodies (Kelemen *et al.*, 1995b). Textural and chemical equilibrium are probably attained within the channels once they are established however.

1.7 Discussion and conclusions

It is relatively easy to develop and test permeability models on surface rocks, either by direct measurement in the field or in the laboratory. Rock transport properties at depth are a different matter, and it is important to constrain the various controls on pore geometry in these inaccessible regions. A major problem with our current understanding of rock permeability is that we do not know much about the transitions between the three categories of porous rock. The underlying reason for this is that we cannot yet confidently extrapolate the experimental results to geological time-scales.

The results of rock compaction experiments are of limited applicability to progressive burial in the shallow crust unless it can be shown that much longer time-scales and the presence of a chemically active fluid (such as H_2O) have no effect on the mechanisms involved. The flow of fluids through a rock compacting by grain crushing and pore collapse in the crust is likely to have a major role in redistribution of material and alteration of the pore structure on the time-scales at which compaction occurs in the crust. The applicability of the HIPing studies to the development of metamorphic porosity also is critically determined by the choice of pore fluid and the assumption of a constant mechanism at all time-scales. Compaction rates in the laboratory experiments are much faster than those likely to be encountered in the crust, and it is common to use a chemically inert fluid phase such as air (Bernabé *et al.*, 1982; Olgaard and FitzGerald, 1993) or argon (Zhang *et al.*, 1994).

The most important gap in our knowledge is thus one of kinetics. The third category of porous rock, that in which the pore geometry is modified by the fluid phase but is controlled by reaction kinetics rather than minimization of interfacial energies, is currently poorly defined because we know little about the relative rates of the two processes. How quickly does a melt topology move from one governed by the kinetics of reaction to one governed by minimization of interfacial energies, for example? Until this is known, we cannot assess the significance of this category.

The remainder of this volume is dedicated to the question of how deviatoric stresses can modify rock permeability, and the problem of kinetics is fundamental here too. How quickly, for example, can a fluid-bearing rock at high temperatures regain the pore topology representing the lowest energy configuration for a hydrostatically stressed rock after or during deformation? The mechanism and rates controlling the formation of the grain boundary fluid films discussed in the following two chapters are still unknown, although they must surely involve processes occurring at the fluid–solid interface. Similarly the rate of fracture healing is controlled by surface chemical processes. Perhaps the key to understanding the importance of deformation in enhancing rock permeability in the Earth lies in determining the rates of processes occurring on the fundamental scale of the grain boundary and the fluid–solid interface.

Acknowledgements

My thanks are due to Mike Bickle, James Connolly, Bertrand Maillot, Yves Guéguen, Noelle Odling, Mike Brown, and Bruce Yardley for critically reading the manuscript. Their comments significantly improved an earlier draft, but any remaining mistakes and misconceptions are my own. Thanks are also due to Martin Lee, Bruce Yardley and Ben Harte for providing photomicrographs.

References

Agee, C.B. and Shannon, M.C. (1996) Wetting properties of molten Fe-alloy at high pressure. *Terra Abstracts*, **8**, 1.

Aharonov, E., Whitehead, J.A., Kelemen, P.B. and Spiegelman, M. (1995) Channeling instability of upwelling melt in the mantle. *Journal of Geophysical Research*, **100**, 20433–50.

Arita, Y. and Wada, H. (1990) Stable isotopic evidence for migration of metamorphic fluids along grain boundaries of marbles. *Geochemical Journal*, **24**, 173–86.

Beere, W.A. (1975) A unifying theory of the stability of penetrating liquid phases and sintering pores. *Acta Metallurgica*, **23**, 131–8.

Bernabé, Y. (1991) Pore geometry and pressure dependence of the transport properties in sandstones. *Geophysics*, **56**, 436–46.

Bernabé, Y. (1995) The transport properties of networks of cracks and pores. *Journal of Geophysical Research*, **100**, 4231–41.

Bernabé, Y., Brace, W.F. and Evans, B. (1982) Permeability, porosity and pore geometry of hot-pressed calcite. *Mechanics of Materials*, **1**, 173–83.

Bickle, M.J. and Baker, J. (1990) Advective diffusive transport of isotopic fronts: an example from Naxos, Greece. *Earth and Planetary Science Letters*, **97**, 78–93.

Bourbié, T. and Zinszner, B. (1985) Hydraulic and acoustic properties as a function of porosity in Fontainebleau Sandstone. *Journal of Geophysical Research*, **90**, 11524–32.

Brace, W.F. (1980) Permeability of crystalline and argillaceous rocks. *Int. J. Rock Mech. Min. Sci. Geomech. Abstr.*, **17**, 241–51.

Brace, W.F. (1984) Permeability of crystalline rocks: new in-situ measurements. *Journal of Geophysical Research*, **89**, 4327–30.

Brace, W.F., Walsh, J.B. and Frangos, W.T. (1968) Permeability of granite under high pressure. *Journal of Geophysical Research*, **73**, 2225–36.

Brace, W.F., Silver, E., Hadley, K. and Goetze, C. (1972) Cracks and pores: a closer look. *Science*, **178**, 162–4.

Brantley, S.L. (1992) The effect of fluid chemistry on quartz microcrack lifetimes. *Earth and Planetary Science Letters*, **113**, 145–56.

Brearley, A.J. and Rubie, D.C. (1990) Effects of H_2O on the disequilibrium breakdown of muscovite and quartz. *Journal of Petrology*, **31**, 925–56.

REFERENCES

Brenan, J.M. (1993) Diffusion of chlorine in fluid-bearing quartzite: effects of fluid composition and total porosity. *Contributions to Mineralogy and Petrology*, **115**, 215–24.

Brown, M. and Rushmer, T. (this volume) The role of deformation in the movement of granitic melt: views from the laboratory and the field.

Bryon, D.N., Atherton, M.P. and Hunter, R.H. (1995) The interpretation of granitic textures from serial thin-sectioning, image analysis and three-dimensional reconstruction. *Mineralogical Magazine*, **59**, 203–11.

Bryon, D.N., Atherton, M.P., Cheadle, M.J. and Hunter, R.H. (1996) Melt movement and the occlusion of porosity in crystallizing granitic systems. *Mineralogical Magazine*, **60**, 163–71.

Cheadle, M.J. (1989) Properties of texturally equilibrated two-phase aggregates. Unpublished PhD thesis, University of Cambridge.

Clauser, C. (1992) Permeability of crystalline rocks. *EOS Transactions AGU*, **73** (21), 237–8.

Connolly, J.A.D., Holness, M.B., Rubie, D.M. and Rushmer, T. (1997) Reaction-induced micro-cracking: an experimental investigation of a mechanism for anatectic melt extraction. *Geology* (in press).

Cooper, M.R. and Hunter, R.H. (1995) Precision serial lapping, imaging and three-dimensional reconstruction of minus-cement and post-cementation intergranular pore-systems in the Penrith Sandstone of north-western England. *Mineralogical Magazine*, **59**, 213–20.

Craw, D. and Norris, R.J. (1993) Grain boundary migration of water and carbon dioxide during uplift of garnet-zone Alpine Schist, New Zealand. *Journal of Metamorphic Geology*, **11**, 371–8.

Daines, M.J. (this volume) Melt distribution in partially molten peridotites: implications for perme-ability and melt migration in the upper mantle.

Daines, M.J. and Kohlstedt, D.L. (1993) A laboratory study of melt migration. *Philosophical Transactions of the Royal Society of London*, **A342**, 43-52.

Daines, M.J. and Kohlstedt, D.L. (1994) The transition from porous to channelized flow due to melt-rock reaction during melt migration. *Geophysical Research Letters*, **21**, 145–8.

Daines, M.J. and Richter, F.M. (1988) An experimental method for directly determing the intercon-nectivity of melt in a partially molten system. *Geophysical Research Letters*, **15**, 1459–62.

David, C. (1993) Geometry of flow paths for fluid transport in rocks. *Journal of Geophysical Research*, **98**, 12267–78.

David, C., Guéguen, Y. and Pampoukis, G. (1990) Effective medium theory and network theory applied to the transport properties of rock. *Journal of Geophysical Research*, **95**, 6993–7005.

David, C., Wong, T.-F., Zhu, W. and Zhang, J. (1994) Laboratory measurement of compaction induced permeability change in porous rocks: implications for the generation and maintenance of pore pressure excess in the crust. *Pure and Applied Geophysics*, **143**, 425–56.

Dullien, F.A.L. (1992) *Porous Media: Fluid Transport and Pore Structure*. Academic Press, San Diego, California.

Faul, U.H., Toomey, D.R. and Waff, H.S. (1994) Intergranular basaltic melt is distributed in thin elongated inclusions. *Geophysical Research Letters*, **21**, 29–32.

Fein, J.B., Graham, C.M., Holness, M.B. *et al.* (1994) Controls on the mechanisms of fluid infiltra-tion and front advection during regional metamorphism: a stable isotope and textural study of retrograde Dalradian rocks of the SW Scottish Highlands. *Journal of Metamorphic Geology*, **12**, 249–60.

Ferry, J.M. (1994) Overview of the petrologic evidence for fluid flow during regional metamorphism in northern New England. *American Journal of Science*, **294**, 905–88.

Frank, F.C. (1968) Two component flow model for convection in the Earth's upper mantle. *Nature*, **220**, 350–2.

Fujii, N, Osamura, K. and Takahashi, E. (1986) Effect of water saturation on the distribution of partial melt in the olivine-pyroxene-plagioclase system. *Journal of Geophysical Research*, **91**, 9253–9.

Graham, C.M., Skelton, A.D.L., Bickle, M. and Cole, C. (this volume) Lithological, structural and deformation controls on fluid flow during regional metamorphism.

Guéguen, Y. and Dienes, J. (1989) Transport properties of rocks from statistics and percolation. *Mathematical Geology*, **21**, 1–13.

Guéguen, Y. and Palciauskas, V. (1994) *Introduction to the Physics of Rocks*. Princeton University Press, Princeton, New Jersey.

Guéguen, Y., Gavrilenko, P. and Le Ravalec, M. (1996) Scales of rock permeability. *Surveys in Geophysics*, **17**, 245–63.

Harris, N.B.W., Gravestock, P. and Inger, S. (1992) Ion-microprobe determinations of trace-element concentrations in garnets from anatectic assemblages. *Chemical Geology*, **100**, 41–9.

Harte, B., Hunter, R.H. and Kinny, P.D. (1993) Melt geometry, movement and crystallization, in relation to mantle dykes, veins and metasomatism. *Philosophical Transactions of the Royal Society of London*, **A342**, 1–21.

Harte, B., Pattison, D.R.M. and Linklater, C. M. (1991) Field relations and petrography of partially melted pelitic and semi-pelitic rocks, in *Equilibrium and Kinetics in Contact Metamorphism: The Ballachulish Igneous Complex and its Aureole* (eds Voll, G., Töpel, J., Pattison, D.R.M. and Seifert, F.), Springer-Verlag, Berlin and Heidelberg, pp. 181–210.

Hay, R.S. and Evans, B. (1988) Intergranular distribution of pore fluid and the nature of high-angle grain boundaries in limestone and marble. *Journal of Geophysical Research*, **93**, 8959–74.

Herring, C. (1951) Surface tension as a motivation for sintering, in *The Physics of Powder Metallurgy* (ed. Kingston, W.E.), McGraw Hill Inc, New York, pp 157–79.

Holness, M.B. (1992) Equilibrium dihedral angles in the system quartz-CO_2-H_2O-NaCl at 800 °C and 1–15 kbar: the effects of pressure and fluid composition on the permeability of quartzites. *Earth and Planetary Science Letters*, **114**, 171–84.

Holness, M.B. (1993) Temperature and pressure dependence of quartz-aqueous fluid dihedral angles: the control of adsorbed H_2O on the permeability of quartzites. *Earth and Planetary Science Letters*, **117**, 363–77.

Holness, M.B. (1995) The effect of feldspar on quartz-H_2O-CO_2 dihedral angles at 4 kbar, with consequences for the behaviour of aqueous fluids in migmatites. *Contributions to Mineralogy and Petrology*, **118**, 356–64.

Holness, M.B. (1996) Surface chemical controls on pore fluid connectivity in texturally equilibrated materials, in *Fluid Flow and Transport in Rocks: Mechanisms and Effects* (eds Jamtveit, B.D. and Yardley, B.W.D.), Chapman & Hall, London, pp. 149–70.

Holness, M.B. and Graham, C.M. (1991) Equilibrium dihedral angles in the system H_2O-CO_2-NaCl-calcite, and implications for fluid flow during metamorphism. *Contributions to Mineralogy and Petrology*, **108**, 368–83.

Holness, M.B. and Graham, C.M. (1995) P-T-X effects on equilibrium carbonate-H_2O-CO_2-NaCl dihedral angles: constraints on carbonate permeability and the role of deformation during fluid infiltration. *Contributions to Mineralogy and Petrology*, **119**, 301–13.

Hunter, R.H. (1987) Textural equilibrium in layered igneous rocks, in *Origins of Igneous Layering* (ed. Parsons, I.), Reidel, Dortrecht, pp. 473–503.

Johnson, D.L. and Sen, P.N. (1988) Dependence of the conductivity of a porous medium on electrolyte conductivity. *Physics Reviews*, **B37**, 3502–10.

Johnson, E.L. and Hollister, L.S. (1995) Syndeformational fluid trapping in quartz: determining the pressure-temperature conditions of deformation from fluid inclusions and the formation of pure CO_2 fluid inclusions during grain boundary migration. *Journal of Metamorphic Geology*, **13**, 239–49.

Jurewicz, S.R. and Jurewicz, A.J.G. (1986) Distribution of apparent angles on random sections with emphasis on dihedral angle measurements. *Journal of Geophysical Research*, **91**, 9277–82.

Jurewicz, S.R. and Watson, E.B. (1985) The distribution of partial melt in a granitic system: the application of liquid phase sintering theory. *Geochimica et Cosmochimica Acta*, **49**, 1109–21.

REFERENCES

Kelemen, P., Whitehead, J., Aharonov, E. and Jordahl, K. (1995a) Experiments on flow focusing in soluble porous media with applications to melt extraction from the mantle. *Journal of Geophysical Research*, **100**, 475–96.

Kelemen, P.B., Shimizu, N. and Salters, V.J.M. (1995b) Extraction of mid-ocean-ridge basalt from the upwelling mantle by focused flow of melt in dunite channels. *Nature*, **375**, 747–53.

Kirkpatrick, S. (1973) Classical transport in disordered media: scaling and effective-medium theories. *Physics Reviews Letters*, **27**, 1722–5.

Knackstedt, M.A. and Cox, S.F. (1995) Percolation and the pore geometry of crustal rocks. *Physical Review*, **51E**, R5181–4.

Kretz, R. (1966) Interpretation of the shape of mineral grains in metamorphic rocks. *Journal of Petrology*, **7**, 68–94.

Laporte, D. (1994) Wetting behaviour of partial melts during crustal anatexis: the distribution of hydrous silicic melts in polycrystalline aggregates of quartz. *Contributions to Mineralogy and Petrology*, **116**, 486–99.

Laporte, D. and Watson, E.B. (1991) Direct observation of near-equilibrium pore geometry in synthetic quartzites at 600–800 °C and 2–0.5 kbar. *Journal of Geology*, **99**, 873–8.

Laporte, D. and Watson, E.B. (1995) Experimental and theoretical constraints on melt distribution in crustal sources: the effect of crystalline anisotropy on melt interconnectivity. *Chemical Geology*, **124**, 161–84.

Lee, V.W., Mackwell, S.J. and Brantley, S.L. (1991) The effect of fluid chemistry on wetting textures in novaculite. *Journal of Geophysical Research*, **96**, 10023–37.

Lewis, S. and Holness, M.B. (1996) Equilibrium halite-H_2O dihedral angles: high rock-salt permeability in the shallow crust? *Geology*, **24**, 431–4.

Lockner, D. and Evans, B. (1995) Densification of quartz powder and reduction of conductivity at 700 °C. *Journal of Geophysical Research*, **100**, 13081–92.

Maaløe, S. and Scheie, Å (1982) The permeability controlled accumulation of primary magma. *Contributions to Mineralogy and Petrology*, **82**, 350–7.

Minarik, W.G. and Watson, E.B. (1995) Interconnectivity of carbonate melt at low melt fraction. *Earth and Planetary Science Letters*, **133**, 423–37.

Nabelek, P.I., Labotka, T.C., O'Neil, J.R. and Papike, J.J. (1984) Contrasting fluid-rock interaction between the Notch Peak granitic intrusion and argillites and limestones in western Utah: evidence from stable isotopes and phase assemblages. *Contributions to Mineralogy and Petrology*, **86**, 25–34.

Nakano, T. and Fujii, N. (1989) The multiphase grain control percolation: its implication for a partially molten rock. *Journal of Geophysical Research*, **94**, 15653–61.

Norton, D. (1988) Metasomatism and permeability. *American Journal of Science*, **288**, 604–18.

Norton, D. and Knapp, R. (1977) Transport phenomena in hydrothermal systems: the nature of porosity. *American Journal of Science*, **277**, 913–36.

Odling, N. (this volume) Fluid flow in fractured rocks at shallow levels in the Earth's crust: an overview.

Olgaard, D.L. and FitzGerald, J.D. (1993) Evolution of pore microstructures during healing of grain boundaries in synthetic calcite rocks. *Contributions to Mineralogy and Petrology*, **115**, 138–54.

Park, H.-H. and Yoon, D.N. (1985) Effect of dihedral angle on the morphology of grains in a matrix phase. *Metallurgical Transactions*, **16A**, 923–8.

Paterson, M.S. (1983) The equivalent channel model for permeability and resistivity in fluid-saturated rock – a reappraisal. *Mechanics of Materials*, **2**, 345–52.

Pittman, E.D. and Duschatko, R.W. (1970) Use of pore casts and scanning electron microscope to study pore geometry. *Journal of Sedimentary Petrology*, **40**, 1153–7.

Richards, I.J., Labotka, T.C. and Gregory, R.T. (1996) Contrasting stable isotope behaviour between calcite and dolomite marbles, Lone Mountain, Nevada. *Contributions to Mineralogy and Petrology*, **123**, 202–21.

Riley, G.N. and Kohlstedt, D.L. (1991) Kinetics of melt migration in upper mantle-type rocks. *Earth and Planetary Science Letters*, **105**, 500–21.

Roedder, E. (1984) Fluid inclusions. *Reviews in Mineralogy, Mineralogical Society of America*, **12**, 644.

Rutter, E.H. and Neumann, D.H.K. (1995) Experimental deformation of partially molten Westerly granite under fluid-absent conditions, with implications for the extraction of granitic magmas. *Journal of Geophysical Research*, **100**, 15697–715.

Sawyer, E.W. (1991) Disequilibrium melting and rate of melt-residuum separation during migmatization of mafic rocks from the Grenville front, Quebec. *Journal of Petrology*, **32**, 701–38.

Scheidegger, A.E. (1960) *The Physics of Flow through Porous Media*, University of Toronto Press, Toronto.

Schwartz, L.M., Sen, P.N. and Johnson, D.L. (1989) Influence of rough surfaces on electrolytic conduction in porous media. *Physics Reviews*, **B40**, 2450–8.

Shmonov, V.M., Vitovtova, V.M. and Zarubina, I.V. (1995) Permeability of rocks at elevated temperatures and pressures, in *Fluids in the Crust: Equilibrium and Transport Properties* (eds Shmulovich, K.I., Yardley, B.W.D. and Gonchar, G.G.), Chapman & Hall, London.

Sleep, N.H. (1974) Segregation of magma from a mostly crystalline mush. *Geological Society of America Bulletin*, **85**, 1225–32.

Smith, C.S. (1948) Grains, phases and interfaces: an interpretation of microstructure. *Transactions of the Metallurgical Society of the AIME*, **175**, 15–51.

Sprunt, E.S. and Nur, A. (1979) Microcracking and healing in granites: new evidence from cathodoluminescence. *Science*, **205**, 495–7.

Stauffer, D. and Aharony, A. (1994) *Introduction to Percolation Theory*, 2nd edn, Taylor & Francis, London, pp. 181.

Toramaru, A. and Fujii, N. (1986) Connectivity of the melt phase in a partially molten peridotite. *Journal of Geophysical Research*, **91**, 9239–52.

Vaughan, P.J., Kohlstedt, D.L. and Waff, H.S. (1982) Distribution of the glass phase in hot-pressed olivine-basalt aggregates: an electron microscopy study. *Contributions to Mineralogy and Petrology*, **81**, 253–61.

Vernon, R.H. (1968) Microstructures of high-grade metamorphic rocks at Broken Hill, Australia. *Journal of Petrology*, **9**, 1–22.

Von Bargen, N. and Waff, H.S. (1986) Permeabilities, interfacial areas and curvatures of partially molten systems: results of numerical computations of equilibrium microstructures. *Journal of Geophysical Research*, **91**, 9261–76.

Waff, H.S. and Faul, U.H. (1992) Effects of crystalline anisotropy on fluid distribution in ultramafic partial melts. *Journal of Geophysical Research*, **97**, 9003–14.

Walker, F.D.L., Lee, M.R. and Parsons, I. (1995) Micropores and micropermeable texture in alkali felspars – Geochemical and geophysical implications. *Mineralogical Magazine*, **59**, 505–34.

Walsh, J.B. and Brace, W.F. (1984) The effect of pressure on porosity and the transport properties of rock. *Journal of Geophysical Research*, **89**, 9425–31.

Watson, E.B. (1982) Melt infiltration and magma evolution. *Geology*, **10**, 236–40.

REFERENCES

Watson, E.B. and Brenan, J.M. (1987) Fluids in the lithosphere, 1. Experimentally-determined wetting characteristics of CO_2-H_2O fluids and their implications for fluid transport, host-rock physical properties, and fluid inclusion formation. *Earth and Planetary Science Letters*, **85**, 497–515.

Watson, E.B. and Lupulescu, A. (1993) Aqueous fluid connectivity and chemical transport in clinopyroxene-rich rocks. *Earth and Planetary Science Letters*, **117**, 279–94.

Watt, G.R. and Harley, S.L. (1993) Accessory phase controls on the geochemistry of crustal melts and restites produced during water-undersaturated partial melting. *Contributions to Mineralogy and Petrology*, **114**, 550–66.

White, J.C. and White, S.H. (1981) On the structure of grain boundaries in tectonites. *Tectonophysics*, **78**, 613–28.

Wolf, M.B. and Wyllie, P.J. (1991) Dehydration-melting of solid amphibolite at 10 kbar. Textural development, liquid interconnectivity and applications to the segregation of magma. *Mineralogy and Petrology*, **44**, 151–79.

Wolf, M.B. and Wyllie, P.J. (1995) Liquid segregation parameters from amphibolite dehydration melting experiments. *Journal of Geophysical Research*, **100**, 15611–21.

Worden, R.H., Walker, F.D.L., Parsons, I. and Brown, W.L. (1990) Development of microporosity, diffusion channels and deuteric coarsening in perthitic alkali felspars. *Contributions to Mineralogy and Petrology*, **104**, 507–15.

Wray, P.J. (1976) The geometry of two-phase aggregates in which the shape of the second phase is determined by its dihedral angle. *Acta Metallurgica*, **24**, 125–35.

Wyllie, M.R.J. and Rose, W.D. (1950) Some theoretical considerations related to the quantitative evaluation of the physical characteristics of reservoir rock from electrical log data. *Transactions of the American Institute of Mechanical Engineering*, **189**, 105–18.

Yardley, B.W.D. and Lloyd, G.E. (1989) An application of cathodoluminescence microscopy to the study of textures and reactions in high-grade marbles from Connemara, Ireland. *Geological Magazine*, **126**, 333–7.

Yechieli, Y., Ronen, D. and Berkowitz, B. (1995) Are sedimentary salt layers always impermeable? *Geophysical Research Letters*, **22**, 2761–4.

Zaraisky, G.P. and Balashov, V.N. (1995) Thermal decompaction of rock, in *Fluid in the Crust: Equilibrium and Transport Properties* (eds Shmulovich, K.I., Yardley. B.W.D. and Gonchar, G.G.), Chapman & Hall, London, pp. 253–84.

Zhang, S., Paterson, M.S. and Cox, S.F. (1994) Porosity and permeability evolution during hot isostatic pressing of calcite aggregates. *Journal of Geophysical Research*, **99**, 15741–60.

Zhu, W., David, C. and Wong, T.-F. (1995) Network modelling of permeability evolution during cementation and hot isostatic pressing. *Journal of Geophysical Research*, **100**, 15451–64.

CHAPTER TWO

Experimental investigation of the rheology of partially molten peridotite at upper mantle pressures and temperatures

Quan Bai, Zhen-Ming Jin and Harry W. Green, II

2.1 Introduction

The vast majority of igneous activity in the Earth is produced by decompression melting during upwelling of mantle peridotite beneath ocean ridges, accompanied and followed by separation and extraction of basalt to form the oceanic crust. Recent years have seen extensive geochemical analysis of mid-ocean ridge basalts (MORBs) and dredged samples of oceanic peridotites (Salters and Hart, 1989; Johnson *et al.*, 1990), field studies of ophiolites, especially in Oman (e.g. Ildefonse *et al.*, 1993), experimental investigations of the behaviour of partially molten mantle rocks and analogues (e.g. Cooper and Kohlstedt, 1984, 1986; Bussod and Christie, 1991; Beeman and Kohlstedt, 1993; Jin *et al.*, 1994; Hirth and Kohlstedt, 1995a, 1995b), and computer modelling of upwelling at ridges (e.g. Scott and Stevenson, 1986; Buck and Su, 1989; Sparks and Parmentier, 1991). In addition, mid-ocean ridges are the current focus of much geophysical investigation in connection with the RIDGE programme (e.g. Forsyth, 1992; Sleep, 1993; Forsyth and Chave, 1994). Despite this broad array of studies, several aspects of the basic process of flow and melt separation remain poorly understood. For example, the composition of MORBs, including their trace element geochemistry, has been interpreted to require that melting begins in the garnet peridotite facies and that melt is extracted virtually as soon as it is generated (Salters and Hart, 1989; Johnson *et al.*, 1990). That is, the data suggest that at any one time only about 1% melt resides within the upwelling source rock as it is progressively depleted from a garnet pyrolite composition (with a few per cent garnet, 10–15% clinopyroxene (cpx) and perhaps 30% orthopyroxene (opx)) to that of

Deformation-enhanced Fluid Transport in the Earth's Crust and Mantle. Edited by M.B. Holness. Published in 1997 by Chapman & Hall, London. ISBN 0 412 75290 5.

a spinel harzburgite (with a few per cent spinel and cpx and 10–15% opx).

On the other hand, numerous early experimental studies of melt topology in mantle materials indicated that melt in equilibrated partially molten synthetic aggregates of olivine + basalt and peridotite is situated in narrow tubes along three-grain intersections (e.g. Waff and Bulau, 1979, 1982; Vaughan et al., 1982), although not necessarily all of them (Toramaru and Fujii, 1986). These results have been challenged recently, with important implications for melt extraction processes. Waff and Faul (1992) and Faul et al. (1994) have shown that some low-index olivine faces are fully wetted by the melt and, despite the small number of grain boundaries wetted, the majority of melt is contained in these few, widely dispersed, boundaries (Faul et al., 1994). If one assumes that this geometry obtains in the source rock for MORB, it is extremely difficult to envision how melt can be extracted at very small melt fractions. Nevertheless, experimental evidence that extraction is possible at low melt fractions has been reported by Riley and Kohlstedt (1991; see also Kohlstedt, 1992, for a review). These workers performed experiments on infiltration of melt into olivine and derived a relation in which the permeability was related to the first power of the melt fraction rather than to the second power, as is expected in Poiseuille flow in a system of interconnected tubules (Cheadle, 1989; see Scott, 1992, for a review). An extension of this study (Daines and Kohlstedt, 1994) confirmed the results of Riley and Kohlstedt (1991) for circumstances in which the sink of the melt migration couple had no initial melt, but could not do so unequivocally in specimens containing initial melt in the sink. Linear dependence of permeability on melt fraction produces relatively high permeabilities at low melt fractions, potentially resolving the problem of rapid melt extraction in the mantle. However, the underlying mechanism for linear dependency of permeability on melt fraction is presently unknown, pointing to the need for further experimentation and theoretical investigation.

A basic peculiarity about melting of peridotite is that the denser phases contribute preferentially to early melting. As a consequence, both the melt and the residuum are lighter than the original source rock. In the hot environment beneath ocean ridges, that means that both melt and residuum will rise diapirically, with melt rising faster if and when it separates from the source. The very poor thermal conductivity of peridotite results in upwelling material remaining hotter than its surroundings, thereby increasing the tendency for the residuum to rise. An inescapable consequence of this melting behaviour in a gravitational field is that in addition to the microscopic flow that must take place in the solid matrix to allow melt extraction, bulk flow will also take place as upwelling proceeds. It is not clear, therefore, whether the equilibrium microstructure that is now well characterized from static experiments (Faul et al., 1994) is the microstructure that develops in the partially molten, upwelling, mantle. If the microstructure is different during dynamic partial melting, that could be a clue to understanding the process by which melt can be extracted rapidly.

41

Two recent discoveries suggest that bulk deformation during melting may be crucial to this process. The first of these grew out of a series of preliminary experiments (Borch and Green, 1990; Borch et al., 1991) that exhibited microstructures suggesting that, despite the fact that basaltic melt does not wet olivine and pyroxene grain boundaries under static conditions, specimens subjected to low stress flow under mantle conditions developed melt films on curved and irregular boundaries of no particular crystallographic orientation. More recently, we have confirmed the phenomenon of 'dynamic wetting' of grain boundaries and suggested that in the mantle this process may be associated with facilitating melt extraction (Jin et al., 1994). The second discovery comes from the field. Kelemen and Dick (1995) have demonstrated a correlation between shear zones in the basal portions of the Josephine ophiolite (Oregon, USA) and dunites which represent the pathways of melt migration through the mantle (Nicolas, 1986; Kelemen et al., 1995 – see also Daines and Kohlstedt, 1994, for related evidence). Kelemen and Dick (1995) concluded that the shearing probably localized the melt pathways, with implications for the correlation between bulk deformation and enhancement of melt extraction.

The lower portions of exposed ophiolite complexes and mantle xenoliths recovered from volcanic vents on oceanic islands such as Hawaii, provide sample cross-sections of oceanic lithosphere. The former often contain evidence of a late deformation superposed on the microstructures developed in the upper mantle, but many of the latter (the protogranular texture of Mercier and Nicolas, 1975) appear to be samples of the residuum of the early stages of melting which have been little affected by thermal or mechanical events since their formation at a spreading centre. These xenoliths are typically very coarse-grained (5–15 mm), are generally much less depleted chemically than the bulk of the uppermost oceanic mantle (i.e. they are lherzolites rather than harzburgites), and are the chemically most equilibrated mantle xenoliths (i.e. they have the greatest homogeneity of phases). However, this class of xenoliths is also characterized by extraordinarily curved and convoluted grain boundaries; the grains often display re-entrant angles and develop 'orientation families' (amoeboid and disconnected portions of the same crystal – Urai, 1983) in which individual olivine and pyroxene crystals are complexly intertwined (Urai and Green, unpublished results) – a texture requiring extensive grain-boundary migration and interaction processes. Thus, the apparently most primitive microstructure from the mantle, which probably had its origin beneath ocean ridges, argues strongly for microstructural disequilibrium being maintained under conditions leading to an extremely close approach to total chemical equilibrium. As we will see below, these microstructures are typical of dynamic partial melting. The description of similar textures by Kelemen and Dick (1995) in the dunites through which they believe significant amounts of melt have passed lend support to this interpretation.

In addition to studies of melt topology in partially molten ultramafic rocks, a class of experiments that is potentially very instructive for our understanding of

42

these processes is the series of investigations on a variety of salts with small amounts of intergranular brine that were carried out by J. Urai and colleagues in the 1980s. These studies involved both two-dimensional deformation studies with real-time microscopic observation and time-lapse videotaping (Urai, 1983, 1985), and specimens deformed in bulk followed by sectioning and microscopic examination (Urai, 1985, 1987; Urai *et al.*, 1986, 1987). In real-time and time-lapse studies of bischoffite and carnalite, small quantities of brine resident in triple junctions and grain-boundary bubbles were observed to spread out onto grain boundaries during deformation, coupled with extensive and rapid grain-boundary migration; the grain-boundary brine returned to surface-energy-controlled features after cessation of deformation. It was in these experiments that Urai first observed the development of 'orientation families' of complexly intergrown crystals. The strength of halite (NaCl) is greater than that of bischoffite and carnalite, hence all halite experiments were performed in bulk. In specimens of halite that contained grain-boundary brine bubbles similar to those in the other salts, grain coarsening was similarly greatly enhanced during deformation compared to dry halite, and grain-boundaries were smooth and featureless after deformation. However, portions of the deformed halite specimens examined one year later, exhibited once more the grain-boundary bubbles characteristic of the starting material (Urai *et al.*, 1987). Thus, in all three of the salt-brine mixtures studied, an intergranular fluid that does not wet the grain boundaries under conditions of textural equilibrium was demonstrated to emerge from its equilibrium 'home', forming thin films covering a large fraction of grain boundaries. Formation of these films correlated with greatly enhanced grain-boundary migration and formation of amoeboid and disconnected portions of the same crystal (orientation families). Upon cessation of deformation, the fluid spontaneously formed rolls and bubbles as surface-energy forces reasserted themselves and initiated the return toward textural equilibrium. In the bischoffite and carnalite systems, the kinetics of the process produced rapid development of bubbles on grain boundaries; in halite, the process took months. We have previously shown (Jin *et al.*, 1994) and we will summarize below that microstructures indistinguishable from those in these salts develop during low stress dynamic partial melting and annealing of peridotite under the pressure and temperature conditions typical of melting in the mantle.

These previous studies suggest that macroscopic rock deformation, melt distribution and melt migration are interdependent processes. In this article, we report experimental results on microstructural observations of pyrolite deformed while partially molten, and mechanical data for this rock-melt aggregate. Although the effect of dynamic wetting of grain boundaries on the rheology of partially molten aggregates is still not fully characterized, we will also show below that it appears to facilitate flow and thus to enhance super-solidus weakening. An unresolved difference exists between the experimental results described here and those of D.L. Kohlstedt and colleagues over the last 10 years (e.g. Cooper and Kohlstedt, 1984, 1986; Hirth and Kohlstedt, 1995a,

1995b). The latter studies have not found the dynamic wetting phenomenon that characterizes our specimens and controls their mechanical behaviour. Potential explanations and geophysical consequences of this discrepancy are discussed after presentation of our results.

2.2 Materials and methods

2.2.1 Starting materials

We prepared a chemically related series of peridotites from a single large spinel lherzolite xenolith from Damaping, eastern China. The xenolith was gently ground to disaggregate it into individual crystals and carefully hand-picked to separate into individual mineral fractions. The crystals were then recombined in various mineral proportions to constitute a pyrolite, a lherzolite, and a harzburgite. Each powder was ground to ensure that all particles were less than 50 µm diameter and was isostatically hot pressed commercially in molybdenum-lined steel cans at a temperature (T) of 1475 K and a pressure (P) of 100 MPa for 1 h. Prior to hot pressing, each container filled with powder was cold-compacted, baked at 800 K, flushed with dry argon, and baked again before evacuation and sealing. In all three cases, the resulting synthetic rocks exhibited a grain size of 10–50 µm and a porosity of less than 5% throughout. The pyrolite sample discussed here showed a dominant grain size of approximately 20–30 µm. For experimentation, samples 3 mm diameter by 5–7 mm long were cored from the synthetic rocks, dried for >100 h at 383 K and encapsulated in platinum. After specimen encapsulation, entire assemblies were stored in a vacuum oven at 383 K for $\geq$48 h before deformation in molten salt in a modified Griggs-type apparatus (Green and Borch, 1989; Tingle et al., 1993) Both annealing (12–40 h) and constant-strain-rate $(7 \times 10^{-6} – 7 \times 10^{-4} \text{ s}^{-1})$ deformation experiments were conducted at P = 1.5 GPa and T = 1400–1500 K, yielding 0–7% melt. All deformed specimens were pre-annealed for 12–14 h at the run P and T to assure that they began with an approximately equilibrated microstructure. We placed Ni foil inserts within the Pt capsules in order to buffer the oxygen fugacity, fO_2, of the samples near Ni/NiO. The drying procedures followed during synthetic rock fabrication and specimen preparation were intended to ensure that our specimens were effectively dry. The fact that the experimental specimens were surrounded outside the Pt capsule by a molten alkali halide pressure medium that had been pre-dried along with the rest of the specimen assembly should provide further assurance that the specimens are dry because these molten salts are very efficient hydrogen 'getters'. Fourier transform infrared spectroscopic analysis performed on polished slices from samples GL600 and GL605 demonstrated that the hydroxyl concentration was below the resolution of the spectrometer (~10 H/10^6 Si). Thus, our specimens should be on the weak side of olivine rheology due to the positive effect of fO_2 on strain rate (Bai et al., 1991), but should be virtually unaffected by hydrolytic weakening. Chemical composi-

44

tion measured by electron probe microanalyser is reported in Table 2.1 for the hot-pressed starting material.

2.2.2 Apparatus improvements

These high-pressure, low-stress, experiments are made possible by the molten salt cell (Green and Borch, 1989, 1990; Tingle et al., 1994). When this project began, we had difficulty conducting long-term experiments above 1500 K with the molten salt cell because of problems with degradation of the carbon resistance furnace that usually restricted such experiments to less than 20 h. During the early stages of this work, a major effort was made to develop longer-lived assemblies (Tingle et al., 1994). As a consequence of those improvements and switching to a higher grade of graphite for furnace construction, we now can routinely conduct experiments of several days' duration at temperatures up to 1600 K. In addition, we initiated the use of tapered furnaces which reduce the temperature variation within individual specimens to less than ~15 K as measured by a pair of thermocouples located on opposite sides of the Pt capsule near its ends. These developments contributed considerably to our ability to demonstrate that the dynamic microstructures of these experiments (Jin et al., 1994) are not artefacts due to insufficient time at temperature to achieve textural equilibrium.

2.3 Results

2.3.1 Microstructures

Throughout this paper, the term 'melt' will be used to refer to both the existence of a melt phase at high temperature and to the glass in quenched samples that

Table 2.1 Composition of starting materials

SiO_2	46.92 (wt %)
TiO_2	0.09
Al_2O_3	3.92
Cr_2O_3	0.52
MgO	39.20
CaO	3.13
MnO	0.12
FeO	7.27
Na_2O	0.18
K_2O	0.01
Total	101.36
olivine	52.0 (vol. %)
opx	31.0
cpx	13.9
spinel	3.1

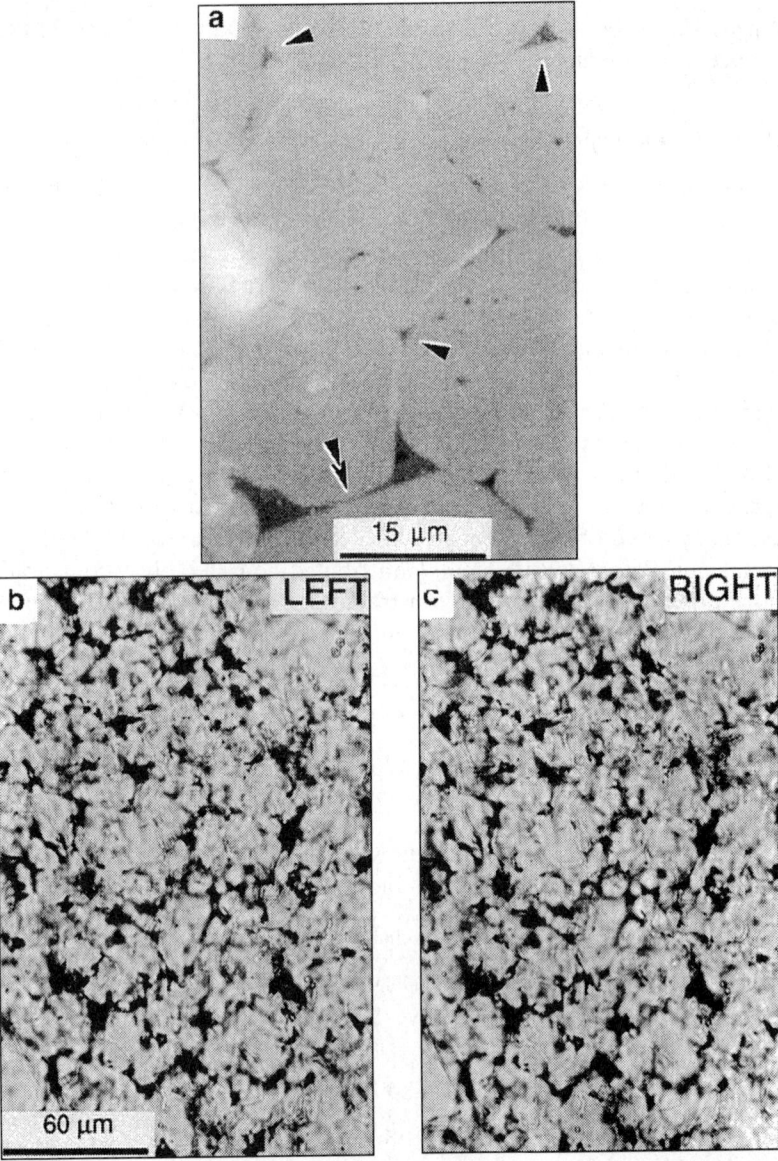

Figure 2.1 Annealing microstructures. (a) Reflection optical micrograph of pyrolite specimen GL525 (annealed 40 h at 1.5 GPa, 1475 K), showing equilibrated microstructure consisting of polygonal crystals and melt tubules of triangular cross-section along three-grain intersections (arrowheads). Note that near the bottom a flat wetted facet (double arrowheads) has developed, which is similar to the static annealing results of Faul *et al.* (1994). (b,c) Stereoscopic pair of optical micrographs showing the three-dimensional distribution of melt in pyrolite specimen GL501, annealed identically to the preanneal of GL506; even with this shorter annealing time, melt is primarily restricted to three-grain-intersections and small pools which may be the sites of melt generation. (Panel (a) after Figure 1 of Jin *et al.*, 1994.)

represents the former existence of such a melt phase. Static annealing above the solidus yields a distribution of melt tubules with an approximately triangular cross-section (Figure 2.1(a)) along three-grain intersections (Waff and Bulau, 1982; Toramaru and Fujii, 1986; von Bargen and Waff, 1986; Watson *et al.*, 1990; Waff and Faul, 1992), but with the dominant volume of melt in scattered pockets (Figure 2.1(b,c)). In fully equilibrated material, these pockets are bordered by low-index crystal facets (Faul *et al.*, 1994). By contrast, deformation at differential stresses of 5–50 MPa after establishment of such an annealed (i.e. equilibrium) microstructure results in a melt configuration in which triangular triple-junction tubules are still present, but the majority of grain boundaries exhibit melt along all or part of their area (Figure 2.2).

To test whether this unexpected result was an artefact due to lack of time to achieve microstructural equilibrium, we re-encapsulated half of one deformed specimen (GL506), and reannealed it for 40 h as experiment GL509. If the microstructural differences between static and dynamic experiments were due to factors other than deformation, the second anneal should have yielded microstructures similar to those of Figure 2.1(a) or Figure 2.2(a), or somewhere in between. That is, one should find that the melt returned to triple-junctions, remained on the grain boundaries, or perhaps partially withdrew toward triple-junctions. Such was not the case; the melt remained in place on grain boundaries but broke up into rolls and droplets (Figure 2.3(a,b)) characteristic of healing of thin films (Smith and Evans, 1984; Wanamaker and Evans, 1985; Hickman and Evans, 1987) due to surface-energy minimization (Nichols and Mullins, 1965a, 1965b) which would eventually lead back to containment of the films at triple-junctions. Specimen GL506 was deformed for only four hours, less than half the time it was pre-annealed, and only 10% of the time that part of it was re-annealed as experiment GL509. The observation that a large fraction of grain boundaries became covered with melt during the 4 h deformation while the melt films had only begun to re-equilibrate at the end of the 40 h re-anneal demonstrates that spreading of melt onto boundaries occurs much faster than its removal.

Evidence of grain-boundary migration was prominent in all specimens deformed in the partially molten regime: (i) the grain size approximately doubled during 4 h of deformation (compare Figures 2.1(b,c) with Figures 2.2(d,e)); (ii) grain boundaries were highly sinuous, and apparent indentations of one grain by another were abundant (arrowheads, Figure 2.3(d)); (iii) cusps were common on grain boundaries (double arrowheads, Figure 2.3(c,d)), indicating local pinning of advancing boundaries by impurities. In many cases the pinning objects were observed to be small, euhedral, chromite crystals produced in the melting process as chromium-rich clinopyroxene is consumed. By contrast, re-annealing after deformation produces no change in grain size after 40 h, but grain-shape irregularities appear to be reduced (compare Figures 2.2(d,e) with Figure 2.3(a,b)).

To investigate the mechanism of deformation in the solid matrix, we oxidized

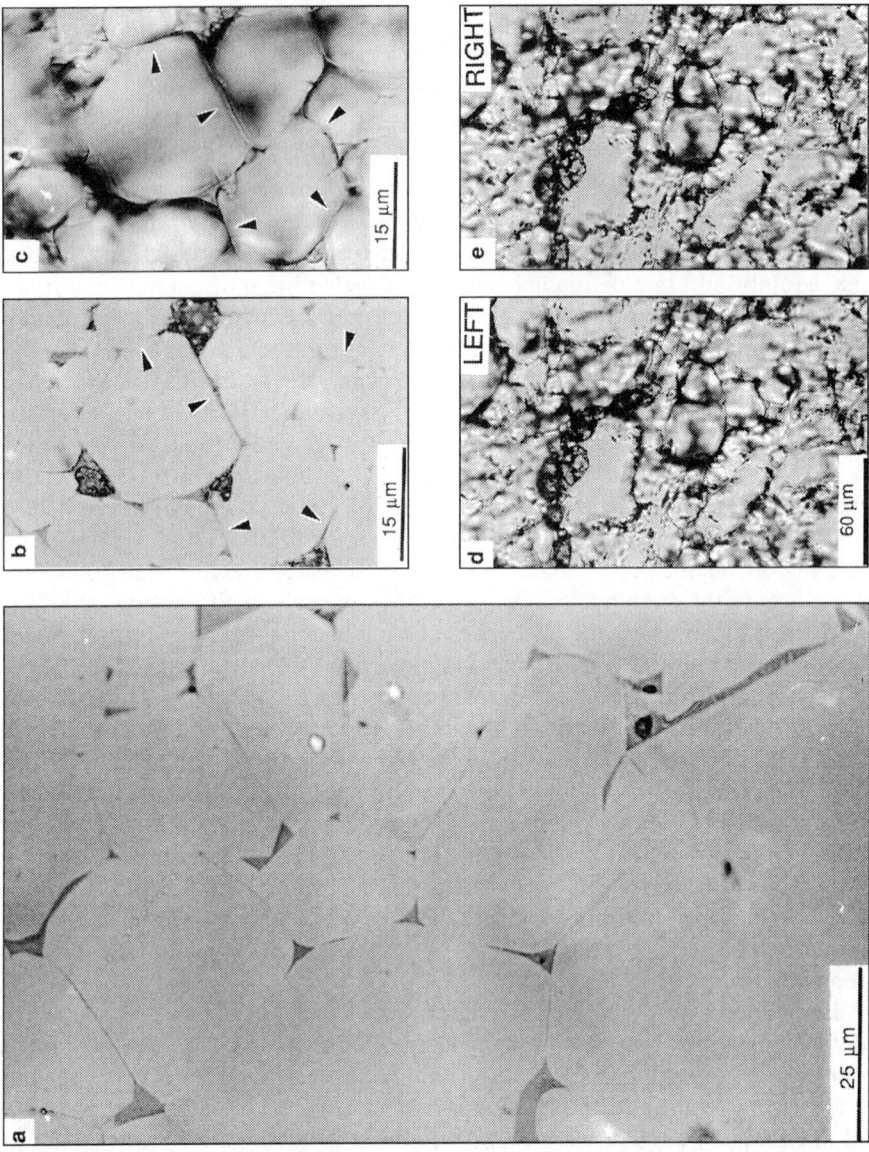

Figure 2.2 Dynamic microstructures. (a) Reflection optical micrograph of pyrolite specimen GL469 (annealed 12 h at 1.0 GPa, 1500 K and then deformed at 3×10^{-5} s^{-1}), showing euhedral crystal facets facing into melt pockets. Many grain boundaries also contain melt. (b,c) Reflection (b) and transmission (c) micrographs of the same area of pyrolite specimen GL506 (annealed 13 h at 1.5 GPa, 1475 K and then deformed at 3×10^{-5} s^{-1}) showing enhanced dynamic 'wetting' of grain boundaries during deformation (arrowheads). Irregular black areas in (b) are places where glass plucked out of the surface. (d,e) Stereoscopic pair of optical micrographs showing the three-dimensional distribution of melt in specimen GL506 in the vicinity of the detail shown in (b) and (c); in addition to three-grain-intersections, melt forms films on many grain boundaries.

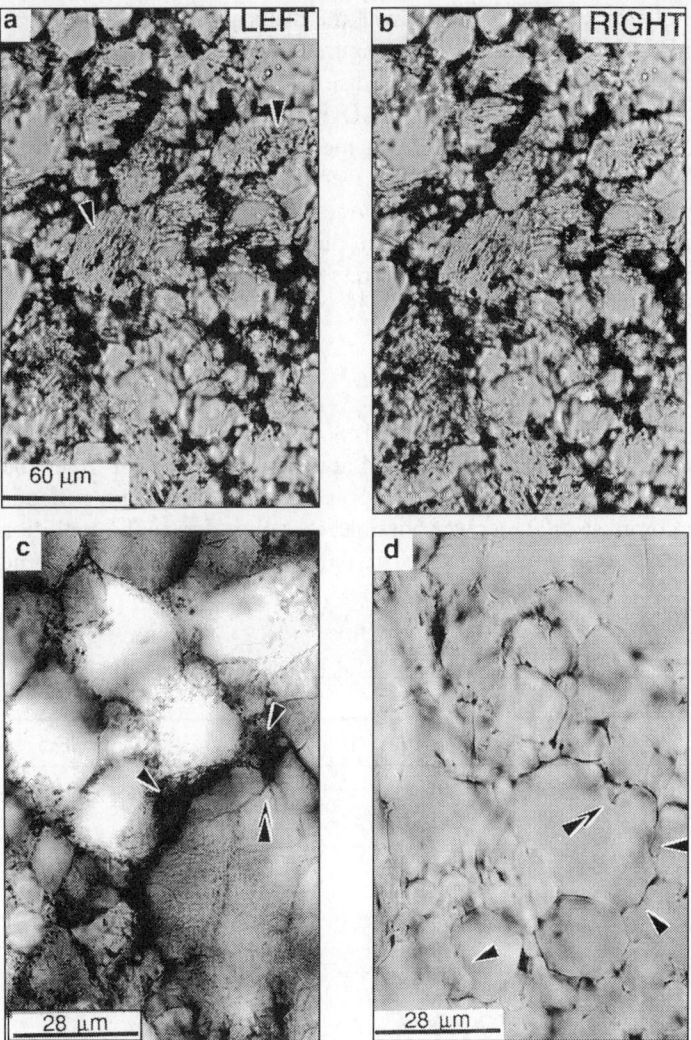

Figure 2.3 (a,b) Stereoscopic pair of optical micrographs showing the three-dimensional distribution of melt in specimen GL509 (half of specimen GL506 that was reannealed for 40 h at 1.5 GPa, 1475 K). Striations (arrowheads) and spots abundant on grain boundaries are tubes and droplets of melt that have developed as the melt films on grain boundaries have begun to decompose in response to surface energy forces. (c) Optical micrograph of a thin-section of an oxidized portion of specimen GL502 showing a very uniform density of dislocations in olivine and oxidation of melt on grain boundaries. A highly irregular, melt-filled boundary of a large olivine crystal displays a cusp (double arrowhead) where it is retarded in completing its consumption of another olivine grain (arrowheads). Pyroxenes do not oxidize and therefore appear clear, but their boundaries, like the olivine boundaries, are decorated by iron oxides in the devitrified melt. (d) Optical transmission image of a thin-section of specimen GL506 showing irregular grain boundaries with cusps (double arrowhead) and indentations (arrowheads) that are characteristic of dynamic partial melting. (Panels (c) and (d) from Jin *et al.*, 1994.)

a portion of our specimens to decorate the dislocations in olivine (Kohlstedt *et al.*, 1976; Zeuch and Green, 1977) (Figure 2.3(c)). Olivine crystals displayed a homogeneous distribution of free dislocations, with subgrain boundaries present in various stages of formation. Oxidation made it much easier to investigate what kinds of boundaries became melt-covered, because pyroxenes do not oxidize. Most grain boundaries were covered with oxidized, partially devitrified, glass, including olivine–olivine, olivine–pyroxene and pyroxene–pyroxene boundaries. Dislocation densities (ρ) in olivine varied with the stress supported by the specimens (e.g. GL506: flow stress = 10 MPa, $\rho = 5$–8×10^6 cm^{-2}; GL502: flow stress = 30 MPa, $\rho = 1$–2×10^7 cm^{-2}).

2.3.2 Mechanical results

The mechanical results of 15 successful experiments are summarized in Table 2.2 and Figure 2.4. Included in Figure 2.4 are the results of four other rheological studies on dry olivine +/– melt, extrapolated to the conditions of our experiments for comparison (wherever possible, extrapolation to the conditions of our experiments was performed using the parameters reported in the study being extrapolated).

Table 2.2 Summary of experimental conditions and results for the deformation tests of pyrolite at 1.5 GPa total pressure

Run #	T (K)	Strain rate (s^{-1})	Stress (MPa)	Melt fraction	Salt cell
GL502	1475	3×10^{-5}	55	0.065	E2*
GL506	1475	3×10^{-5}	30	0.07	E2
GL548	1475	1×10^{-5}	8	0.07	CsCl
GL546	1475	6.7×10^{-4}	190	0.07	E2
	1475	3.5×10^{-5}	30		
GL569	1475	3×10^{-4}	130		CsCl
	1475	1×10^{-4}	90		
	1475	3×10^{-5}	50	0.065	
GL581–1	1400	3×10^{-5}	760	–	CsCl
GL581–2	1450	3×10^{-5}	200	0.03	
GL590	1400	3×10^{-5}	808	–	CsCl
GL595	1500	3×10^{-5}	29	–	CsCl
GL599	1400	3×10^{-5}	720	–	CsCl
GL600	1400	3×10^{-5}	581	–	CsCl
GL605	1475	3×10^{-5}	68	0.07	CsCl
GB40	1500	1.4×10^{-5}	8	–	E2
GL607–1	1450	3×10^{-5}	241	–	E2
		8.3×10^{-5}	316	–	
GL607–2	1435	1×10^{-5}	264	–	E2
		6.6×10^{-6}	216	–	
		3.4×10^{-5}	434	–	
		9.3×10^{-5}	545	–	

*The term 'E2' refers to a 1:1 salt mixture of NaCl and KCl.

summary of mechanical results

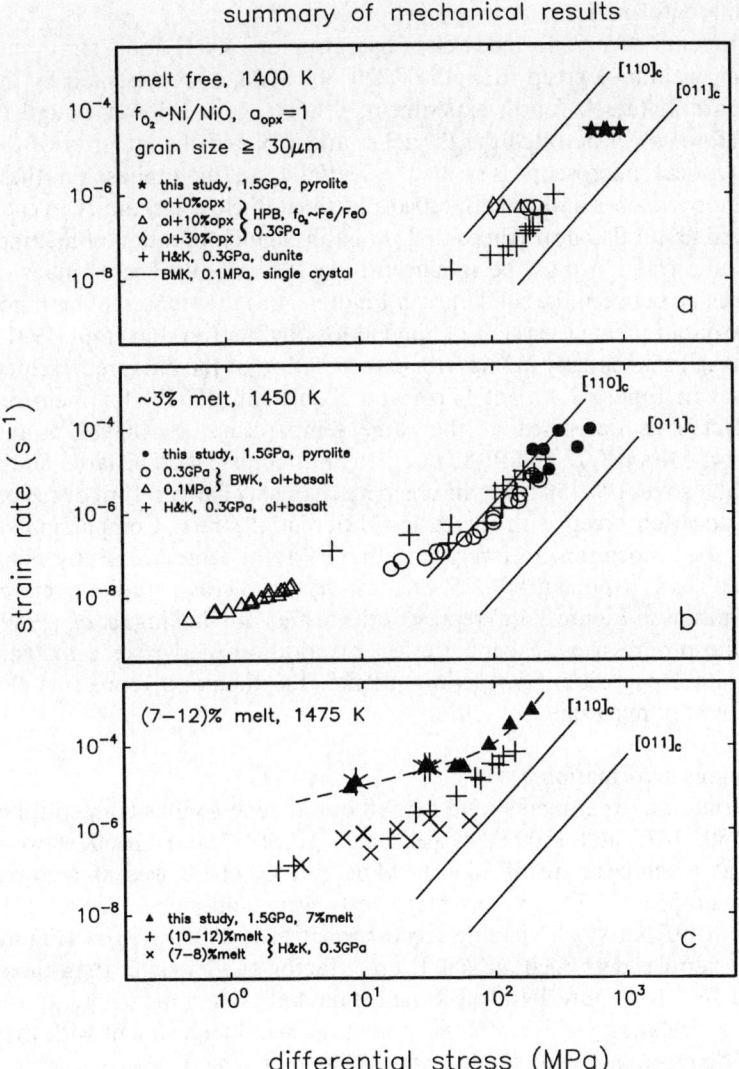

Figure 2.4 Summary of creep data of this and four other studies on (a) melt-free samples at 1400 K, (b) samples with ~3 vol.% melt at 1450 K and (c) samples with 7–12 vol.% melt at 1475 K. The data are from (1) HPB: Hitchings *et al.* (1989) on olivine+opx at P = 0.3 GPa, (2) H&K: Hirth and Kohlstedt (1995a, 1995b) on olivine+(3–12)% MORB at P = 0.3 GPa, (3) BWK: Bai *et al.* (1996) on olivine+3% MORB at P = 0.1 and 300 MPa, and (4) BMK: Bai *et al.* (1991) on olivine single crystals deformed along the weakest orientation [110]$_c$ and the strongest orientation [011]$_c$. The orthopyroxene activity, a_{opx}, is unity and the oxygen fugacity is set at the Ni/NiO buffer for all samples except those set at Fe/FeO of HPB. For deformation in the dislocation creep regime, the average grain size of samples is 30 μm or larger in all studies. In the diffusion creep experiments of Hirth and Kohlstedt (1995a), the sample grain size is 9–12 μm; these data are plotted normalized to a grain size of 30 μm, using a grain size exponent of –3.2 (Hirth and Kohlstedt, l995a).

(a) Subsolidus deformation

Three experiments were performed at a temperature of 1400 K and a strain rate of $3 \times 10^{-5}\,s^{-1}$, yielding a creep strength of 720–808 MPa, as represented by the stars in Figure 2.4(a). A fourth experiment, GL600, gave a lower bound of 580 MPa before salt penetrated the Pt jacket and violated the integrity of the specimen. Optical microscopy revealed essentially no melt phase in these samples; clinopyroxenes and spinels, especially when in close proximity to each other, showed textural abnormalities that probably signify initiation of melting, but the volume fraction must be significantly less than 0.01. These values of creep strength lie between the solid lines in Figure 2.4(a) that represent extrapolation of the data for single crystals of San Carlos olivine from Bai *et al.* (1991). Further, our data lie directly in line with extrapolation of the data, represented by the pluses in Figure 2.4(a) of Hirth and Kohlstedt (1995b) for melt-free olivine polycrystals deformed at the same temperature, $P = 300$ MPa and stress = 50–245 MPa (PI-72 and PI-82 of Hirth and Kohlstedt). The latter study found that at a stress of ~150 MPa the creep mechanism changes from diffusion creep to dislocation creep with increasing differential stress. Comparison of their data in the dislocation creep regime with ours at the same stresses by using the measured stress exponent of ~3.5, the data sets of the two studies overlap. The open symbols in Figure 2.4(a) represent the results of Hitchings *et al.* (1989) and show the progressive weakening effect of addition of pyroxene to their starting dunite composition (circles are dunite; the diamond represents the greatest amount of pyroxene, i.e. ~30%).

(b) Supersolidus deformation

Eleven deformation experiments were carried out at supersolidus temperatures of 1435, 1450, 1475 and 1500 K. Experiments GL581-2 and GL607-1 were performed at a temperature of 1450 K while run GL607-2 was at a lower temperature (1435 K). These deformation tests were done at strain rates of $6.6 \times 10^{-6}\,s^{-1}$ to $9.3 \times 10^{-5}\,s^{-1}$, yielding creep strengths of 200–545 MPa (Figure 2.4(b)). The samples deformed at 1450 K are a factor of ~2 weaker than those deformed at 1400 K (Figure 2.4(a)). Fitting the data at each of the temperatures to a power law relation yields a stress exponent of 3–4, in agreement with that for dislocation creep in olivine single crystals (Bai *et al.*, 1991). Microstructural analysis showed that the samples contain ~3 vol.% glass after quenching from 1450 K and correspondingly reduced percentages of the crystalline phases, principally clinopyroxene. Melting of spinel is microstructurally very obvious as well, but the volumetric amount is not reduced greatly – spinel is enriched in Cr and reduced in Al as low-melting components from spinel and clinopyroxene enter the melt and refractory components are concentrated into residual spinel.

Our data (dots) are compared in Figure 2.4(b) with those (pluses) of Hirth and Kohlstedt (1995b) and those (circles and open triangles) of Bai *et al.* (1996) for the deformation of partially molten ultramafic aggregates with ~3 vol.% basaltic melt. The experiments (PI-67 and PI-38) of Hirth and Kohlstedt were

52

performed on synthetic aggregates of olivine + MORB, mostly in the dislocation creep regime, at a temperature of 1523 K, pressure of 300 MPa and oxygen fugacity set at Ni/NiO. Those of Bai *et al.* (1996) were carried out on a similar material at temperatures of 1503–1623 K, pressures of 0.1 and 300 MPa and oxygen fugacities set at Ni/NiO. To compare with results of the present study, the data were normalized to a temperature of 1450 K using an activation energy for creep of 540 kJ/mol (Bai *et al.*, 1996).

Both Bai *et al.* (1996) and Hirth and Kohlstedt (1995a) determined an activation energy for diffusion creep of $Q = 530$–570 kJ/mol. This value probably overestimates the activation energy because melt increased from its initial amount with increasing temperature during measurement of Q (Hirth and Kohlstedt, 1995a). Thus, normalization using the Q value of 540 kJ/mol corrected both effects of temperature and melt volume change in the diffusion creep regime so that the comparison of data can be made at the same temperature and the same melt fraction. In the dislocation creep regime, the results of Bai *et al.* (1996) suggested that the rate-controlling step for deformation of partially molten olivine at low melt fraction is associated with dislocation movement in the hardest slip system, (101)[001]. Therefore, an activation energy of 540 kJ/mol for this slip system (Bai *et al.*, 1991), which is the same as that measured for coarse-grained olivine polycrystals deformed at subsolidus temperatures (Chopra and Paterson, 1984), was used in the data normalization.

In both of the previous studies, the flow mechanism changed from diffusion creep to dislocation creep at a stress near 100 MPa; our new data, with a stress exponent of 3–4, fall in the regime of dislocation creep. Both the values of strength and the strain rate sensitivity of strength for our samples are in good agreement with those of the other two studies. Finally, the results of all three studies are in agreement that a few per cent melt weakens the material slightly, bringing strength into agreement with the weakest olivine slip system in the power-law creep regime and resulting in significant weakening of very fine-grained aggregates in the diffusion creep regime.

In this study, six experiments at 1475 K and three experiments at 1500 K (Table 2.2) were performed at strain rates of $1 \times 10^{-5}\,\text{s}^{-1}$ to $6.7 \times 10^{-4}\,\text{s}^{-1}$. Normalized to 1475 K, the strength data of 1500 K plotted in the figure were reduced by a factor of 1.11, according to the measured effect of temperature on strength at supersolidus temperatures, which accounted for effects on stress induced by changes in both temperature and melt fraction. After deformation at 1475 K, the samples showed a melt fraction of ~7 vol.%. The mechanical data appear to fall into regimes of both diffusion creep and dislocation creep and the transition of the two mechanisms occurs at a stress near 100 MPa. As described above, the olivine grains, even in the very low stress experiments, display low and homogeneous dislocation densities, reflecting some contribution of dislocation creep, even when diffusion creep dominates bulk flow. Deconvolution calculations based on a combination of two power laws (i.e. Bai *et al.*, 1996) yielded stress exponents of 0.8 and 4 for the diffusion creep

53

mechanism and the dislocation creep mechanism, respectively. The dashed curve in Figure 2.4(c) represents the resultant curve fitting of the stress–strain rate data.

A comparison of the results of this study with those of other studies is also given in Figure 2.4(c) for deformation at 1475 K. Despite the reasonable agreement between our results and those of Hirth and Kohlstedt (1995a, 1995b) at smaller melt fractions (1450 K), results of experiments at higher degrees of partial melting show significant disagreement between the two studies. The data from Hirth and Kohlstedt (1995a, 1995b) are for olivine aggregates with melt fraction of 7–12 vol.% and grain size of 9–12 μm deformed at 1523 K. We normalized these data to a grain size of 30 μm and a temperature of 1475 K using their grain size exponent of –3.2 and activation energy of 575 kJ/mol for diffusion creep. Again, this Q value accounts for both effects of temperature and melt fraction changes and is significantly larger than that (~315 kJ/mol) for bulk and grain boundary diffusion of the major ions in olivine. Since our extrapolation of data in temperature is small (<50 K), using either value, 575 or 315 kJ/mol, would only introduce a difference in strain rate of a factor of <2. For dislocation creep, an activation energy for creep of 540 kJ/mol was used for the data normalization. All of the data for the partially molten polycrystalline samples fall at lower stresses than olivine single crystals of all three 45° orientations, but our partially molten pyrolite results are much weaker than the olivine + melt experiments of Hirth and Kohlstedt (1995a, 1995b), especially at the lowest stresses (in the diffusion creep regime).

2.4 Discussion

The rheology of subsolidus peridotite is widely believed to be controlled by the plasticity of olivine which is both the most abundant and the weakest phase. It would thus be expected that the results from the polycrystalline samples would fall between the rheology of the weakest and strongest 45° orientations of olivine single crystals. The excellent agreement between these studies, shown in Figures 2.4(a) and 2.4(b), is a strong argument that the differences shown in Figure 2.4(c) are real and must correspond to fundamental differences in the experimental materials or conditions. Indeed, Hirth and Kohlstedt (1995a, 1995b) reported that above 5% melt, the creep behaviour and the distribution of melt in their specimens is substantially different from expectations based on available theoretical models for the weakening effect of melt. Thus, the weakening effects of larger melt fraction are also seen in their data, but to a lesser degree than in ours. The similarity between our mechanical results and those of Hirth and Kohlstedt (1995a, 1995b) for melt fractions less than 5% suggests that wetting of grain boundaries is more enhanced at larger melt fractions; the greater weakening in our specimens could depend sensitively on effects of chemical or physical environment.

It is clear that the spreading of melt onto grain boundaries observed in our experiments is a dynamic effect. If that were not the case, the re-annealed specimen would not have shown the melt topology starting back toward microstructural equilibrium. It is important to recall that this behaviour (fluid in triple-junction tubes under static conditions, in grain-boundary films during deformation, and roll/bubble formation during re-annealing) is closely similar to that previously observed in salt-brine experiments (Urai, 1983, 1985, 1987; Urai *et al.*, 1986, 1987; Spiers and Schutjens, 1990). Stress-induced wetting of grain boundaries has also been recognized in feldspar (Tullis *et al.*, 1993). The observed large increase of grain boundary migration activity in our dynamic experiments and in those on wet salts suggests that the fluid reduces resistance to flow by facilitating grain-boundary sliding accommodated by dissolution in regions of stress concentration and/or local boundary migration.

Dynamic wetting of grain boundaries and induction of rapid grain-boundary migration is the principal difference between this study (including the results of Jin *et al.*, 1994) and that of Hirth and Kohlstedt (1995a, 1995b), who found that grain-boundary melt inhibits boundary migration. We tentatively attribute the difference in the two sets of mechanical results to the transition between power-law and diffusion creep occurring at higher strain rates in our experiments due to the solvent effect of the melt on the rapidly migrating boundaries. Further experiments are needed to identify the causes of the discrepancy. Candidates for the differences between our data set and theirs at the higher degrees of melting are differences in bulk chemistry, melt chemistry, volumetric proportions of the various solid phases, and the topology of melt distribution.

Although it is unlikely that the stresses in our experiments are as low as those present in the mantle during partial melting, at the higher temperatures they are comparable to or lower than those inferred from the dislocation substructure of olivine in normal mantle xenoliths from the oceanic upper mantle (e.g. Guéguen, 1977; Jin *et al.*, 1989 – not to be confused with high-stress mantle xenoliths deformed during the extraction process (Jin *et al.*, 1989; Green and Guéguen, 1974; Green, 1976)). Similarly, the microstructures of the experiments discussed here correspond closely to those found in studies of fossil spreading centres preserved in ophiolites (Nicolas, 1989). In addition, as stated in section 2.1 above, 'orientation families' observed to develop in the rapid grain-boundary migration regime of flowing brine-saturated salts (Urai, 1983) have also been found in mantle xenoliths from the oceanic lithosphere (J. Urai and H. W. Green, unpublished data), providing indirect evidence that dynamic wetting of peridotite grain boundaries by basaltic melt may occur in nature. Such orientation families have not been observed in our experiments, nor were they observed by Urai *et al.* (1986) in similar bulk experiments on halite. Development of that microstructure may require large strains such as were produced in Urai's 'see-through' experiments.

The driving force for dynamic wetting remains elusive. The correlation with rapid boundary migration suggests cause and effect. In the case of our experiments, we know that dislocations are generated even at the lowest stresses achieved. The stored strain energy associated with the dislocations is almost certainly the driving force for recrystallization. If dissolution of strained material from a crystal being consumed combined with reprecipitation of unstrained material on its growing neighbour is a more rapid process than is direct grain-boundary migration, then spreading of melt onto grain boundaries during deformation could be a spontaneous process. Upon cessation of deformation, as soon as the driving force for recrystallization is removed by this and other recovery processes, the melt would seek an equilibrium configuration.

If the scenario presented here is correct, one might expect that spontaneous grain-boundary wetting could occur for other boundary-migration driving forces as well. Martha Daines (pers. comm., 1996) has observed in her static experiments that when abnormal growth of a few grains is observed, in many cases the boundaries of the large grains are completely wetted and they contain numerous melt inclusions. The driving force for exaggerated grain growth (secondary recrystallization) is the energy contained in the grain-boundary network of the smaller grains. Again, if the presence of melt on boundaries facilitates the rate at which boundaries can migrate, wetting of the larger grains could be spontaneous. Should the larger grains collectively consume the finer-grained matrix, the new, coarser, aggregate should then attempt to equilibrate its boundaries through development of bubbles and rolls of melt and, ultimately, expulsion of melt to the new web of triple junctions.

We have observed in our experiments that development of larger grains is not uncommon, and that they tend to be covered with melt. One such larger crystal is visible in Figure 2.2(d,e). Oxidation of the dislocations within these larger grains generally shows them to contain very low dislocation densities in their rims, with the dislocations in these regions primarily oriented normal to the crystal boundaries – very possibly representing screw dislocations that facilitate growth from the melt phase. Perhaps the development of orientation families is a combination of the two modes of recrystallization.

The effective viscosity (stress divided by strain rate) of our partially molten specimens exhibiting dynamic wetting of grain boundaries is very low, suggesting that flow of the residuum during melt extraction beneath oceanic ridges could be significantly easier than that implied by olivine rheology at the same temperature, so long as sufficient melt is present to permit dynamic grain-boundary wetting. This effect could be a contributor to the separation of melt and residuum at very low melt fractions as inferred from the trace-element compositions of ridge basalts (e.g. Johnson *et al.*, 1990). That is, in the Earth, melt may be extracted as soon as it becomes sufficiently abundant to spread onto grain boundaries, thereby restricting melt fractions in nature to below those of this study. These observations suggest that in the Earth, spreading of melt onto grain boundaries could facilitate melt extraction and enhancement of upwelling very close to spreading axes (Sparks and Parmentier, 1991).

2.5 Summary and conclusions

Although this work shows that dynamic wetting of grain boundaries occurs in the peridotite-melt system just as it does in salt-brine systems, we still do not know the responsible fundamental driving force. It clearly is associated with rapid grain-boundary migration and is a disequilibrium process because when deformation stops, these systems start back toward textural equilibrium. Melting in our specimens begins at four-phase intersections; during the initial (static) anneal the melt moves out along three-grain intersection lines by capillarity. Under these conditions, melt never goes onto boundaries (except for specific low-index crystal faces; Waff and Faul, 1992) because it is not stable there. After being spread across boundaries during deformation, the development of rolls and bubbles during subsequent annealing results from approach to a local minimum in free energy (Nichols and Mullins, 1965b; Wanamaker and Evans, 1985; Hickman and Evans, 1987). Complete healing of the boundaries (in which the fluid would return to the triple junctions) is retarded because the energy saving over the bubble-and-roll structure is small and the activation barrier is large. At present, it appears most likely to us that the spreading of fluid onto boundaries and concomitant acceleration of grain-boundary migration must be a mechanism whereby recrystallization is facilitated (note that in static experiments this happens only during exaggerated grain growth, so it cannot be simply a mechanism for grain-coarsening or achievement of chemical equilibrium). We surmise that grain-boundary sliding, probably assisted by diffusion creep through the intermediary of the fluid, is greatly enhanced by this process.

We believe that dynamic wetting and consequent grain-boundary sliding is responsible for the enhanced weakening seen in our specimens at melt fractions above ~0.05 (especially in the diffusion creep regime), but additional data are necessary before a firm conclusion can be reached. If that is the case, it is conceivable that in the Earth, where differential migration of melt and residuum is possible (in contrast to the 'undrained' dynamic melting experiments conducted thus far), the same process that enhances grain-boundary migration and weakening in our specimens could contribute to separation of melt and residuum, perhaps at very low melt fractions.

Extrapolation of these or any laboratory data to conditions obtaining in the Earth necessarily involves extrapolation over many orders of magnitude in strain rate and two or more orders of magnitude in grain size. Such extrapolation can only be made with certainty if one understands the physics involved. For example, if dynamic wetting of grain boundaries is a recovery mechanism as we hypothesized above, it could operate under any conditions of stress above that needed to generate dislocations or under conditions where coarsening occurs by exaggerated grain growth. Thus, lack of understanding of the reason for dynamic wetting of grain boundaries remains a distinct problem for prediction of processes accompanying melting beneath oceanic ridges.

Acknowledgements

We thank Greg Hirth and Martha Daines for constructive reviews. We especially thank Martha Daines for relating her observation of melt-covered grains during exaggerated grain growth in static experiments. This work was supported by NSF Grants OCE90-12941 and DMF93-96262.

References

Bai, Q., Mackwell, S.J. and Kohlstedt, D.L. (1991) High-temperature creep of olivine single crystals 1. mechanical results for buffered sample. *J. Geophys. Res.*, **96**, 2441–63.

Bai, Q, Wang, Z.-C. and Kohlstedt, D.L. (1996) High-temperature creep of an olivine-basalt aggregate at total pressures of 0.1 and 300 MPa. *Phys. Earth Planet. Inter.* (in press).

Beeman, M.L. and Kohlstedt, D.L. (1993) Deformation of olivine-melt aggregates at high temperatures and confining pressure. *J. Geophys. Res.*, **98**, 6443–52.

Borch, R. S. and Green, II, H.W. (1990) Experimental investigation of the rheology and structure of partially molten lherzolite deformed under upper mantle pressures and temperatures. *EOS, Trans. Am. Geophys. Union*, **71**, 629.

Borch, R. S., Green, II, H.W. and Jin, J.-M. (1991) Rheology and microstructure of lherzolite experimentally deformed at upper mantle pressures, hypersolidus temperatures and low melt fractions. *Proc. IASPEI Symposium*, Quadrenniel Mtg. Int. Union Geod. and Geophys., Vienna, Austria.

Buck, W.R. and Su, W. (1989) Focused mantle upwelling below mid-ocean ridges due to feedback between viscosity and melting. *Geophys. Res. Lett.*, **16**, 641–4.

Bussod, G.Y. and Christie, J.M. (1991) Textural development and melt topology in spinel lherzolite experimentally deformed at hypersolidus conditions. *J. Petrol.*, Special Lherzolite issue, 17–39.

Cheadle, M.J. (1989) Properties of texturally equilibrated two-phase aggregates, PhD thesis, University of Cambridge, 156pp.

Chopra, P.N. and Paterson, M.S. (1984) The role of water in the deformation of dunite. *J. Geophys. Res.*, **89**, 7861–76.

Cooper, R.F. and Kohlstedt, D.L. (1984) Solution-precipitation enhanced diffusional creep of partially molten olivine-basalt aggregates during hot-pressing. *Tectonophysics*, **107**, 207–33.

Cooper, R.F. and Kohlstedt, D.L. (1986) Rheology and structure of olivine-basalt partial melts. *J. Geophys. Res.*, **91**, 9315–23.

Daines, M.J. and Kohlstedt, D.L. (1994) The transition from porous to channelized flow due to melt/rock reaction during melt migration. *Geophys. Res. Lett.*, **21**, 145–8.

Faul, U.H., Toomey, D.R. and Waff, H.S. (1994) Intergranular basaltic melt is distributed in thin, elongated inclusions. *Geophys. Res. Lett.*, **21**, 29-32.

Forsyth, D.W. (1992) Geophysical constraints on mantle flow and melt generation beneath mid-ocean ridges. *Geophysical Monograph* **71** (eds Morgan, Blackman and Sinton), 1–65.

Forsyth, D.W. and Chave, A.D. (1994) Experiment investigates magma in the mantle beneath mid-ocean ridges. *EOS, Trans. Am. Geophys. Union*, **75**, 537–40.

Green, II, H.W. (1976) Plasticity of olivine in peridotites, in *Electron Microscopy in Mineralogy* (eds H.-R. Wenk *et al.*), Springer-Verlag, Berlin and Heidelberg, pp. 443–62.

Green, II., H.W. and Borch, R.S. (1989) A new molten salt cell for precision stress measurement at high pressure. *Euro. J. Mineral.*, **1**, 213–19.

Green, II., H.W. and Borch, R.S. (1990) High pressure and temperature deformation experiments in a liquid confining medium, *The Brittle-Ductile Transition in Rocks. The Heard Volume, Geophys. Monogr.* (eds A.G. Duba *et al.*), **56**, 195–200, AGU, Washington, D.C.

Green, II, H.W. and Guéguen, Y. (1974) Origin of kimberlite pipes by diapiric upwelling in the upper mantle. *Nature*, **249**, 617–20.

Guéguen, Y. (1977) Dislocations in mantle peridotite nodules. *Tectonophysics*, **39**, 231–54.

Hickman, S. H. and Evans, B. (1987) Influence of geometry on crack healing rate in calcite. *Phys. Chem. Minerals*, **15**, 91–102.

Hirth, G. and Kohlstedt, D.L. (1995a) Experimental constraints on the dynamics of the partially molten upper mantle: Deformation in the diffusion creep regime. *J. Geophys. Res.*, **100**, 1981–2001.

Hirth, G. and Kohlstedt, D.L. (1995b) Experimental constraints on the dynamics of the partially molten upper mantle 2: Deformation in the dislocation creep regime. *J. Geophys. Res.*, **100**, 15441–9.

Hitchings, R.S., Paterson, M.S. and Bitmead, J. (1989) Effects of iron and magnetite addition in olivine-pyroxene rheology. *Phys. Earth Planet. Inter.*, **55**, 277–91.

Ildefonse, B., Nicolas, A. and Boudier, F. (1993) Evidence from the Oman ophiolite for sudden stress changes during melt injection at oceanic spreading centres. *Nature*, **366**, 673–5.

Jin, Z.-M. Green, II, H.W. and Borch, R.S. (1989) Microstructure of olivine and stress in the upper mantle beneath Eastern China. *Tectonophysics*, **169**, 23-50.

Jin, Z.-M, Green, II, H.W. and Zhou, Y. (1994) Melt topology in partially molten mantle peridotite during ductile deformation. *Nature*, **372**, 164–7.

Johnson, K.T.M., Dick, H.J.B. and Shimizu, N. (1990) Melting in the oceanic upper mantle: An ion microprobe study of diopsides in abyssal predotite. *J. Geophys. Res.*, **95**, 2661–78.

Kelemen, P.B. and Dick, H.J.B. (1995) Focused melt flow and localized deformation in the upper mantle: Juxtaposition of replacive dunite and ductile shear zones in the Josephine peridotite, SW Oregon. *J. Geophys. Res.*, **100**, 423–38.

Kelemen, P.B., Whitehead, J.A., Aharonov, E. and Jordahl, K.A. (1995) Experiments on flow focusing in soluble porous media, with applications to melt extraction from the mantle. *J. Geophys. Res.*, **100**, 475–96.

Kohlstedt, D.L. (1992) Structure, rheology and permeability of partially molten rocks at low melt fractions. *Geophysical Monograph*, **71** (eds Morgan, Blackman and Sinton),103–21.

Kohlstedt, D.L., Goetze, C., Durham, W.B. and Vauder Sande, J.B. (1976) New technique for decorating dislocations in olivine. *Science*, **191**, 1045–91.

Mercier, J.C.C. and Nicolas, A. (1975) Textures and fabrics of upper-mantle peridotites as illustrated by xenoliths from basalts. *Journal of Petrology*, **16**, 454–87.

Nicolas, A. (1986) A melt extraction model based on structural studies in mantle peridotites. *Journal of Petrology*, **27**, 999-1022.

Nicolas, A. (1989) *Structure of Ophiolites and Dynamics of Oceanic Lithosphere*, Kluwer, Dordrecht.

Nichols, F. A. and Mullins, W.W. (1965a) Morphological changes in a surface of revolution due to a capillarity-induced surface diffusion. *J. Appl. Phys.*, **36**, 1826–35.

Nichols, F. A. and Mullins, W.W. (1965b) Surface (interface) and volume diffusion contributions to morphological changes driven by capillarity. *Trans. AIME*, **233**, 1940–8.

Riley, G.N., Jr. and Kohlstedt, D.L. (1991) Kinetics of melt migration in upper mantle-type rocks, *Earth Planet. Sci. Lett.*, **105**, 500-21.

Salters, V.J.M. and Hart, S.R. (1989) The hafnium paradox and the role of garnet in the source of mid-ocean-ridge basalt. *Nature*, **342**, 420–2.

Scott, D.R. (1992) Small-scale convection and mantle melting beneath mid-ocean ridges. *Geophysical Monograph*, **71** (eds. Morgan, Blackman and Sinton), 327–52.

Scott, D.R. and Stevenson, D.J. (1986) Magma ascent by porous flow. *J. Geophys. Res.*, **91**, 9283–96.

Sleep, N.H. (1993) Upwelling beneath oceanic ridges. *Nature*, **366**, 635–6.

Smith, D. L. and Evans, B. (1984) Diffusional crack healing in quartz. *J. Geophys. Res.*, **89**, 4125–35.

Sparks, D.W. and Parmentier, E.M. (1991) Melt extraction from the mantle beneath spreading centers. *Earth Planet. Sci. Lett.*, **105**, 368–77.

Spiers, C. T. and Schutjens, P.M.T.M. (1990) *Deformation Processes in Minerals, Ceramics and Rocks* (eds D.J. Barber and P.G. Meredith), Unwin Hyman, London, pp. 334–53.

Tingle, T.N., Green, II, H.W., Scholz, C.H. and Koczynski, T.A. (1993) The rheology of faults triggered by the olivine-spinel transformation in Mg_2GeO_4 and its implications for the mechanism of deep-focus earthquakes. *J. Struct. Geol.*, **15**, 1249–56.

Tingle, T.N., Green, II, H.W., Young, T.E. and Koczynski, T.A. (1994) Improvements to Griggs-type apparatus for mechanical testing at high pressures and temperatures. *Pageoph.*, **141**, 523–43.

Toramaru, A. and Fujii, N. (1986) Connectivity of melt phase in a partially molten peridotite. *J. Geophys. Res.*, **91**, 9239–52.

Tullis, J., Yund, R. and Farver, J. (1993) Interaction of deformation, fluid distribution and bulk transport in feldspar aggregates. *EOS, Trans. Am. Geophys. Union*, **74**, 611.

Urai, J.L. (1983) Water assisted dynamic recrystallization and weakening in polycrystalline bischofite. *Tectonophysics*, **96**, 125–57.

Urai, J.L. (1985) Water-enhanced dynamic recrystallization and solution transfer in experimentally deformed carnallite. *Tectonophysics*, **120**, 285–317.

Urai, J.L. (1987) Development of microstructure during deformation of carnallite and bischofite in transmitted light. *Tectonophysics*, **135**, 251–63.

Urai, J.L., Spiers, C.T., Zwart, H.W. and Lister, G.S. (1986) Weakening of rock salt by water during long-term creep. *Nature*, **324**, 554–7.

Urai, J.L., Spiers, C.J., Peach, C.J. *et al.* (1987) Deformation mechanisms operating in naturally deformed halite rocks as deduced from microstructural investigations. *Geol. Mijnb.*, **66**, 165–76.

Vaughan, P.J., Kohlstedt, D.L. and Waff, H.S. (1982) Distribution of the glass phase in hot-pressed, olivine-basalt aggregates: an electron microscope study. *Contrib. Mineral. Petrol.*, **81**, 253–61.

Von Bargen, N. and Waff, H.S. (1986) Permeabilities, interfacial areas and curvatures of partially molten systems: Results of numerical computations of equilibrium microstructures. *J. Geophys. Res.*, **91**, 9261–76.

Waff, H.S. and Bulau, J.R. (1979) Equilibrium fluid distribution in an ultramafic partial melt under hydrostatic stress conditions. *J. Geophys. Res.*, **84**, 6109–14.

Waff, H.S. and Bulau, J.R. (1982) Experimental determination of near-equilibrium textures in partially molten silicates at high pressures, in *Advances in Earth and Planetary Sciences, High Pressure Research in Geophysics* (eds S. Akimoto and M.H. Manghnani), Center for Academic Publications, Tokyo, pp. 229–36.

Waff, H.S. and Faul, U.H. (1992) Effects of crystalline anisotropy on fluid distribution in ultra-mafic partial melts. *J. Geophys. Res.*, **97**, 9003–14.

Wanamaker, B.J. and Evans, B. (1985) Experimental diffusional crack healing in olivine, in *Point Defects in Minerals* (ed. R.N. Schock), AGU, Washington, D.C., pp. 194–210.

REFERENCES

Watson, E.B., Brenan. J.M. and Baker, D.R. (1990) Distribution of fluids in the continental mantle, in *Continental Mantle* (ed. M.A. Menzies), Clarendon Press, Oxford, pp. 111–24.

Zeuch, D. H. and Green, II, H.W. (1977) Naturally decorated dislocations in olivine from peridotite xenoliths. *Contrib. Mineral. Petrol.*, **62**, 141–51.

CHAPTER THREE

Melt distribution in partially molten peridotites: implications for permeability and melt migration in the upper mantle

M.J. Daines

3.1 Introduction

Melt migration rates depend on the permeability and rheology of the partially molten rock through which melt is percolating. Permeability and rheology in turn depend on the distribution of melt in the rock. Under hydrostatic stress conditions, the distribution of melt in a partially molten rock is controlled by the dihedral angle (Holness, this volume). However, crystalline anisotropy with respect to interfacial energies leads to the development of planar rather than smoothly curved mineral–melt interfaces which results in a combination of channel and pore shapes and various degrees of grain boundary wetting and melt connectivity. Non-hydrostatic stress conditions may also affect the melt network. First, stress may affect the chemical potential of interfaces depending on the orientation of the interface with respect to the applied stress. The resulting changes in interfacial curvature could lead to an anisotropic melt distribution wherein melt is preferentially located in melt channels parallel to the maximum compressive stress (von Bargen and Waff, 1986). Second, differential stress may cause deformation of the rock, with a concomitant redistribution of the melt. For example, rapid grain boundary migration during dynamic recrystallization can lead to complete wetting of grain boundaries, irrespective of orientation in the stress field (Urai *et al.*, 1986; Jin *et al.*, 1994; Bai, Jin and Green, this volume). In contrast, microcracking during deformation may occur as a result of a reduction in effective confining pressure due to the presence of melt or fluid (van der Molen and Paterson, 1979; Dell'Angelo and Tullis, 1988); melt-filled cracks would be oriented relative to the applied stress, resulting in an anisotropic melt distribution. Third, deformation may result in

Deformation-enhanced Fluid Transport in the Earth's Crust and Mantle. Edited by M.B. Holness. Published in 1997 by Chapman & Hall, London. ISBN 0 412 75290 5.

the development of a lattice preferred orientation and/or a fabric defined by the flattening and elongation of grains with respect to the compression axis. A lattice preferred orientation can result in a melt preferred orientation if the effects of crystalline anisotropy are strong; a grain shape fabric produces a melt preferred orientation with melt channels elongated parallel to grain elongation. In this third case, which is strain-controlled, the melt distribution resulting from non-hydrostatic stress conditions should be retained during annealing. In the other cases, which are stress-controlled, given sufficient time the melt network should recover its hydrostatic geometry once the stress is removed and deformation ceases.

Clearly, the distribution of melt in a deforming partially molten rock is determined not only by the interfacial energies of the minerals and melt, but also by the stress state, deformation mechanism, and deformation history. Application of stress and subsequent deformation can cause changes in the connectivity, geometry, and anisotropy of the melt network, and thereby affect both the permeability and rheology of the partially molten rock. This paper presents a review of melt microstructures generated during laboratory deformation of partially molten mantle rocks and examines the implications of these microstructures for melt migration in the upper mantle.

3.2 Experimental observations

3.2.1 Hydrostatic stress conditions

Much work has focused on characterizing melt distribution in partially molten peridotites and olivine-basalt aggregates under hydrostatic stress conditions (Waff and Bulau, 1979, 1982; Cooper and Kohlstedt, 1982; Toramaru and Fujii, 1986; von Bargen and Waff, 1988). A wide range of compositions, pressure and temperature conditions, and experimental durations have been examined; in all experiments, the measured effective dihedral angles for basaltic melt in contact with olivine range between 20° and 50°, suggesting that for all melt fractions, melt forms an interconnected network (Waff and Bulau, 1979, 1982; Cooper and Kohlstedt, 1982; Toramaru and Fujii, 1986; von Bargen and Waff, 1988). This conclusion has been confirmed experimentally (Daines and Richter, 1988). Observations of grain boundaries using scanning electron microscopy, transmission electron microscopy, and high resolution analytic electron microscopy have further demonstrated that at low melt fractions most grain boundaries are free of melt (Cooper and Kohlstedt, 1982; Vaughan *et al.*, 1982; Kohlstedt, 1990), indicating melt in olivine aggregates is primarily confined to tubules and pockets at three and four grain junctions and does not form a continuous thin film along all grain boundaries.

In detail, however, the geometry of the melt network in olivine aggregates is complicated by the anisotropy of minerals with respect to interfacial energies. Although the triple junction geometry described by many models dominates the melt distribution at very low melt fractions, many olivine–melt interfaces are planar rather than smoothly curved indicating that interfacial energies are indeed anisotropic in this system (Cooper and Kohlstedt, 1982). Furthermore, Waff and Faul (1992) determined that (010) crystal faces of olivine are preferentially wetted by basaltic melt. The reflected light micrograph in Figure 3.1(a) shows a typical melt distribution in an olivine-basalt aggregate under static conditions. Image analysis of melt pocket geometry in olivine-basalt aggregates indicates that the effect of crystalline anisotropy on the melt network geometry is significant, leading to a melt network composed of relatively large penny-

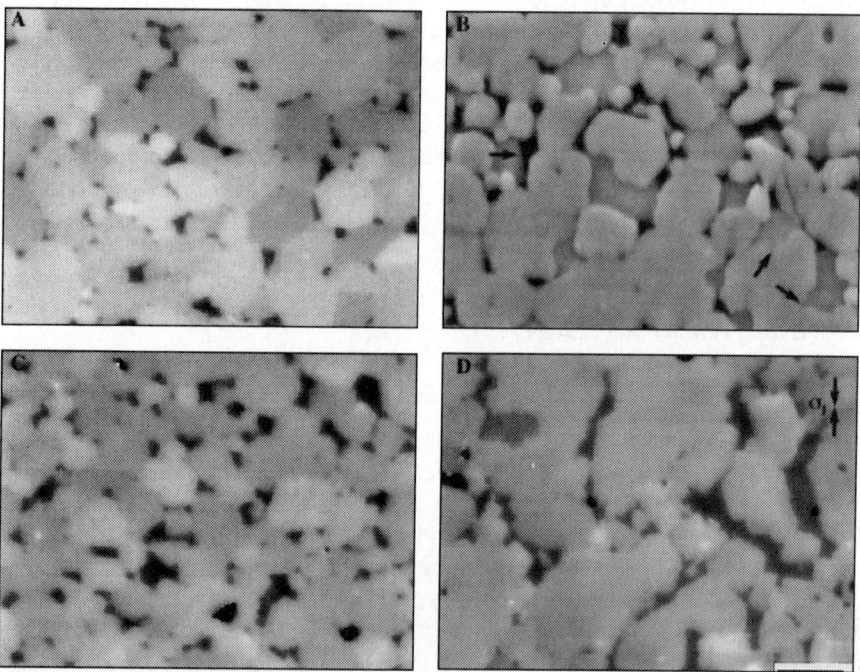

Figure 3.1 Reflected light images of (a) an undeformed olivine-basalt aggregate, (b) an undeformed lherzolite, (c) an olivine-basalt aggregate deformed by diffusion creep at a differential stress of 30 MPa, and (d) an olivine basalt aggregate deformed by dislocation creep at a differential stress of 150 MPa. All samples were fabricated at 1250 °C and 300 MPa confining pressure in a gas-medium pressure apparatus. In all images olivine is light grey, pyroxene is medium grey, quenched melt is dark grey and angular, and pores are black and circular. Scale bar is 10 μm for (a), (b) and (c), but 20 μm for (d). In all four images melt is in triangular triple junctions as well as in irregular features along some two-grain boundaries. Planar mineral–melt interfaces in all images demonstrate the effects of crystalline anisotropy with respect to interfacial energies. In (b) some triple junctions containing pyroxene appear to be melt-free at the resolution of the optical image. The melt distribution in (c) is virtually indistinguishable from that shown in (a) indicating that deformation in the diffusion creep regime does not affect significantly melt distribution. However, (d) exhibits a pronounced preferred orientation of melt pockets at an angle of 15–20° to the maximum principal stress σ_1 which is oriented as shown.

shaped melt pockets connected by smaller cigar-shaped melt channels (Faul *et al.*, 1994; Daines and Kohlstedt, 1996). In the study of Daines and Kohlstedt (1996), secondary electron and reflected light images of hot-pressed olivine-basalt aggregates with two different grain sizes (<10 μm and <40 μm) and melt fractions between 0.01 and 0.12, were analysed to determine the area, perimeter, length, and orientation of melt pockets. In these samples increasing melt fraction and increasing grain size resulted in an increase in the size of melt pockets; this increase was accompanied by a change in shape of the melt pocket, with pockets becoming more elongated and irregular as the melt content and grain size increased. These changes, shown in Figure 3.2, were interpreted to be due to the breakdown of the triangular triple junction morphology as melt fraction increases and melt pockets begin to extend across grain boundaries, linking triple junction tubules by narrow penny-shaped melt channels and resulting in fewer, larger melt pockets which are also much more elongated. Three-dimensional reconstructions of the melt network recently completed using serial polishing techniques support this view of the geometry of the melt network (Daines *et al.*, 1995).

In partially molten peridotites in which pyroxene is present, the effects of crystalline anisotropy are even more obvious; as shown in Figure 3.1(b), not only are many pyroxene–melt interfaces planar, but also some triple junctions containing pyroxene are melt-free (Toramaru and Fujii, 1986; von Bargen and Waff, 1988; Daines and Kohlstedt, 1993).

3.2.2 Diffusion creep regime

Deformation of fine-grained partially molten rocks in the diffusion creep regime yields melt distributions that are not significantly different from those observed

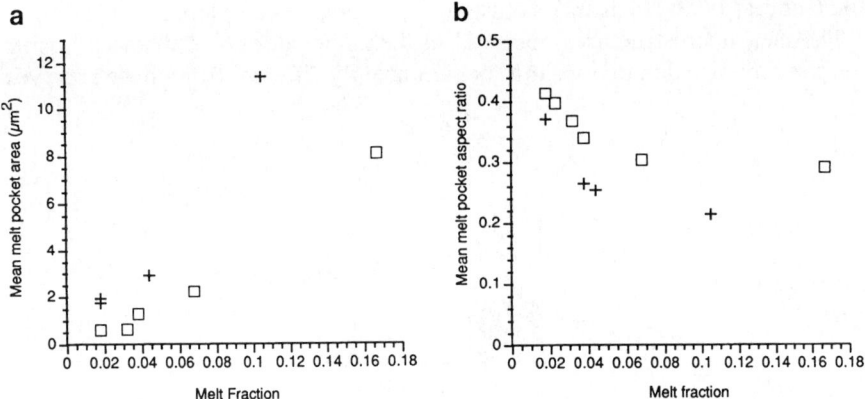

Figure 3.2 Mean area (a) and aspect ratio (b) for undeformed fine and coarse-grained olivine-basalt aggregates with different melt fractions. Fine-grained samples (□) have starting grain sizes less than 10 μm; coarse-grained samples (+) have starting grain sizes less than 40 μm.

in rocks subjected only to hydrostatic conditions (Cooper and Kohlstedt, 1984, 1986; Kohlstedt and Chopra, 1993; Hirth and Kohlstedt, 1995a). Figure 3.1(c) shows a typical melt distribution in an olivine-basalt aggregate deformed in the diffusion creep regime. In the uniaxial hot-pressing experiments of Cooper and Kohlstedt (1984), fine-grained (<15 μm) aggregates of olivine with <7% added natural or synthetic basalt, deformed at differential stresses less than 30 MPa by grain-boundary diffusional creep, were found to be 2–5 times weaker than similar, but melt-free rocks. Melt was predominantly in triple junctions and all triple junctions contained melt. More recent investigations of the deformation of partially molten olivine-basalt rocks at stresses less than 100 MPa, tempera- tures of 1200–1300 °C, and confining pressures of 300 MPa indicate that, in fine- grained rocks with melt fractions greater than about 0.05, the weakening effect of melt is much greater than would be predicted by melt distribution models in which melt is confined to triple junction channels (Hirth and Kohlstedt, 1995a). Observations of laboratory samples containing melt fractions greater than 0.05 indicate that, although a continuous grain boundary film is not present, a signif- icant number of grain boundaries contain thick layers of melt (Kohlstedt and Chopra, 1993; Hirth and Kohlstedt, 1995a). Hirth and Kohlstedt (1995a) concluded that the strain rate enhancement in rocks with melt fractions greater than 0.05 is due to a significant increase in the number of grain boundaries containing melt at higher melt fractions. Comparison of melt pocket character- istics in fine-grained (<10 μm) olivine-basalt aggregates deformed at stresses less than 100 MPa with equivalent rocks subjected only to hydrostatic stress condi- tions, however, indicates that the changes in the geometry of the melt network observed by Hirth and Kohlstedt (1995a) are simply the result of increasing melt fraction and are not due to deformation (Daines and Kohlstedt, 1996). In addi- tion, as shown by Figures 3.3 and 3.1(c), no preferred orientation of melt pockets was observed or could be quantified in the deformed samples (Kohlstedt and Chopra, 1993; Hirth and Kohlstedt, 1995a; Daines and Kohlstedt, 1996).

The melt microstructures observed in lherzolites deformed in the diffusion creep regime also do not appear to be significantly different from those observed

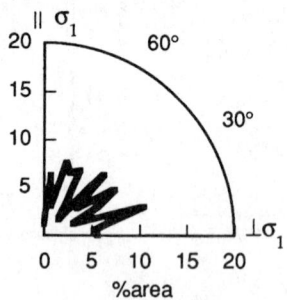

Figure 3.3 Melt pocket orientation by % area for a diffusion creep sample quenched at 93 MPa differential stress. Melt pockets show little preferred orientation.

in undeformed lherzolites (M. Zimmerman, pers. comm.). The presence of pyroxene still results in some melt-free olivine–olivine–pyroxene and olivine–pyroxene–pyroxene triple junctions as well as planar pyroxene–melt interfaces; the relative number of these features does not appear to be affected by deformation.

3.2.3 Dislocation creep regime

The melt distribution in partially molten peridotites deformed in dislocation creep is more varied than that observed in rocks deformed in diffusion creep and can be different from that observed in static experiments. Two types of behaviours have been described: (i) dynamic or stress-induced wetting of grain boundaries occurs, so that many if not all grain boundaries contain melt (Jin *et al.*, 1994; Bai, Jin and Green, this volume), and (ii) melt redistribution occurs such that melt is dominantly in melt pockets that are preferentially oriented relative to the maximum compressive stress (σ_1); the magnitude and orientation of this melt redistribution is controlled by experimental conditions such as confining pressure, melt fraction, grain size, and stress (Ave'Lallemant and Carter, 1970; Bussod and Christie, 1991; Hirth and Kohlstedt, 1995b; Daines and Kohlstedt, 1996).

(a) Stress-induced grain boundary wetting
Stress-induced wetting of grain boundaries during deformation of partially molten peridotite has been described by Jin *et al.* (1994) and Bai, Jin and Green (this volume). In the experiments of Jin *et al.* (1994), fine-grained (20–50 µm) synthetic peridotites were deformed at temperatures of 1200–1225 °C, confining pressures of 1.5 GPa, and differential stresses of 10 to 4 MPa. Deformation under these conditions led to extensive grain boundary migration and grain growth. Melt is still present in triangular triple junctions; in addition, the majority of grain boundaries contain melt along all or part of their length. These grain boundary melt films have very low aspect ratios and no apparent preferred orientation with respect to the stress field (see, e.g. Figure 3.1(b); Jin *et al.*, 1994). Annealing for 40 h after deformation resulted in the break-up of the melt films into rolls and droplets similar to those described during crack healing (Wanamaker and Evans, 1985). The mechanism for the formation of the melt films is not currently understood.

(b) Anisotropic melt distribution
The redistribution of melt into melt pockets preferentially oriented with respect to the stress field has been described in partially molten upper mantle rocks by a number of workers. Ave'Lallemant and Carter (1970) noted zones of melt parallel to the maximum compressive stress (σ_1) in a lherzolite deformed at strain rates of 10^{-5} s^{-1} and confining pressures of 2.5 GPa. Bussod and Christie (1991) found melt anisotropically distributed in melt 'slots' along recrystallized

grain boundaries oriented at ~30° to σ_1 in partially molten lherzolite deformed in a solid-medium apparatus under hydrous conditions at confining pressures of 1.5 GPa, temperatures of 900–1100 °C, and differential stresses of 100 to 400 MPa. Hirth and Kohlstedt (1995b) observed melt not only in three- and four-grain intersections, but also along two-grain boundaries in fine-grained (<40 μm) olivine-basalt aggregates deformed at confining pressures of 300 MPa and differential stresses of 100 to 250 MPa. Figure 3.1(d) shows a typical melt distribution in an olivine-basalt aggregate deformed in the dislocation creep regime. Analysis of the melt distribution in these and similarly prepared samples found that in samples deformed at stresses greater than ~100 MPa more melt was in melt pockets oriented at 15–20° to σ_1 than in melt pockets oriented more perpendicular to σ_1 (Daines and Kohlstedt, 1996). Melt pockets oriented sub-parallel to σ_1 also were more than a factor of two larger, longer and more elongated than those pockets oriented more perpendicularly to σ_1. In samples with larger grain sizes and in samples deformed at higher differential stresses, the preferred orientation of melt was more pronounced. As shown in Figure 3.4 for samples containing melt fractions of ~0.02, the fraction of the total melt in melt

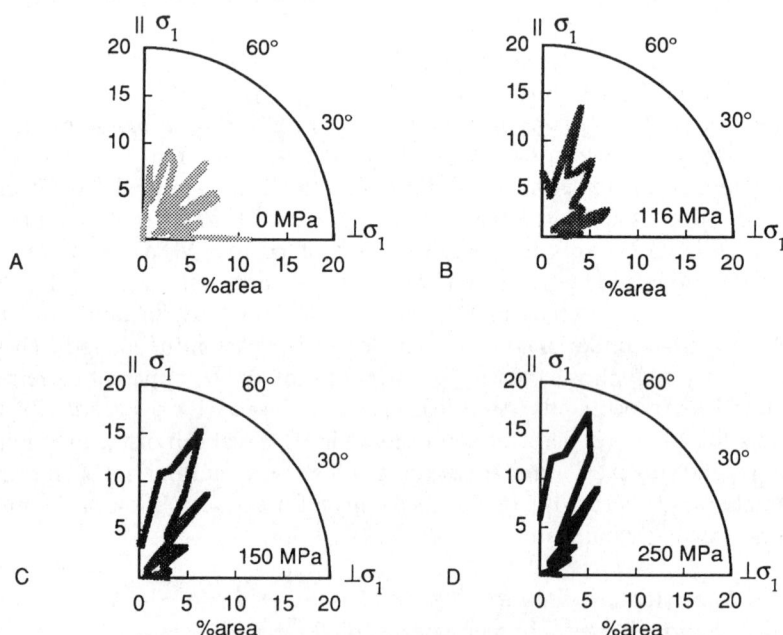

Figure 3.4 Melt pocket orientation by % area for four dislocation creep samples containing melt fractions <0.02. 90° is parallel to the long axis of the samples; in deformed samples the long axis is parallel to the maximum compressive stress, σ_1. With increasing differential stress, more melt is in pockets oriented subparallel to σ_1 (a) undeformed 0 MPa, (b) quenched at 116 MPa, (c) quenched at 150 MPa, and (d) quenched at 250 MPa.

pockets with long axes oriented subparallel to σ_1 doubled as the differential stress was increased from 0 to 250 MPa, suggesting that the magnitude and orientation of the maximum principal stress controls the melt distribution (Daines and Kohlstedt, 1996). Daines and Kohlstedt (1996) also noted that the degree of melt-preferred orientation seemed to be related to the amount of time the stress was applied, rather than to the strain rate or total strain. Similar melt distributions have been described in partially molten lherzolites deformed at similar conditions in compressive creep (Kohlstedt and Zimmerman, 1996).

Shear deformation of olivine-basalt aggregates also results in the development of a preferred orientation of melt relative to the maximum principal stress (Zhang *et al.*, 1995; Kohlstedt and Zimmerman, 1996). In these experiments, olivine-basalt aggregates were deformed in a simple shear geometry under constant load up to high strains. At a differential stress of ~170 MPa and a total strain over 230%, the melt preferred orientation is very well developed with melt aligned in shear bands at about 20° to σ_1.

Daines and Kohlstedt (1996) suggested the melt-preferred orientation exhibited by the samples they analysed might develop during deformation as a result of the concentration of stress at the ends of melt pockets. If melt pockets are treated as elliptical cracks, then in a compressive stress field, melt pockets inclined at an angle γ to σ_1 will yield the greatest tensile stress concentration. γ is given by (Griffiths, 1924)

$$\cos 2\gamma = (\sigma_1' - \sigma_3')/2(\sigma_1' + \sigma_3')$$

where the primes indicate effective stress components ($\sigma_{ij}' = \sigma_{ij} - \alpha P_m \delta_{ij}$; P_m is the melt or pore pressure, δ_{ij} is the Kronecker delta, and α is usually assumed to be equal to 1 (Jaeger and Cook, 1976)). Melt pockets inclined at the angle γ will tend to extend either because the lower mean pressure at the ends of these pockets in comparison to other areas of the melt network causes melt to flow into these regions, or because the lowered effective confining pressure combined with high differential stress leads to microcracking along grain boundaries and extension of these melt pockets. Daines and Kohlstedt (1996) favoured the first of these explanations because melt pockets are not oriented parallel to σ_1 as would be expected if microcracking was the dominant mechanism for developing the melt-preferred orientation. The observation that some minimum interval of time is required for a measurable amount of melt redistribution to occur in response to the applied stress is also consistent with a time-lag associated with melt movement driven by pressure gradients.

Differences in the angle of melt pocket orientation between the samples described by Daines and Kohlstedt (1996) (20°) and those described by Bussod and Christie (1991) (30°) have been interpreted as the result of a difference in pore/melt pressure relative to the confining pressure between these two sets of experiments (Daines and Kohlstedt, 1996). Under undrained conditions such as those imposed in all of these laboratory experiments, the pressure of the melt is

69

between the confining pressure σ_3 and the mean pressure $(\sigma_1 + 2\sigma_3)/3$. In rocks in which the melt is able to freely respond to the imposed stress, the pressure of the melt will be closer to the confining pressure. In rocks in which there is more resistance to the flow of melt, the pressure of the melt will be more equal to the mean pressure. The larger grain size used by Bussod and Christie (1991) results in experimental samples with higher permeability than the fine-grained samples described by Daines and Kohlstedt (1996); higher permeability implies less resistance to flow and hence melt pressures closer to the confining pressure.

Static annealing of rocks deformed in compression or shear results in a new preferred orientation of melt (Daines and Kohlstedt, 1996; Kohlstedt and Zimmerman, 1996). Figure 3.5 shows a diagram comparing the amount of melt in pockets oriented relative to the maximum compressive stress for an olivine-basalt aggregate deformed in compression prior to and after static annealing. This sample was deformed to ~45% strain and quenched at a differential stress of ~111 MPa. Whereas melt is predominantly in pockets subparallel to σ_1 after deformation, melt is preferentially in pockets subperpendicular to σ_1 after annealing for six hours. In sheared samples deformed to ~200% strain, melt is preferentially in melt pockets oriented subparallel to the shear boundary after annealing for 10 h (Kohlstedt and Zimmerman, 1996). Crystalline anisotropy appears to control the melt distribution in the annealed samples as measurements of grain orientations indicate that samples deformed in both geometries have well-developed lattice preferred orientation of olivine. Samples deformed in compression have b axes of olivine preferentially oriented parallel to σ_1 so that (010) planes of olivine are perpendicular to σ_1; samples deformed in shear have b axes of olivine rotated perpendicular to the shear boundary so that (010) planes of olivine are parallel to the shear boundary. (010) planes of olivine are preferentially wetted by melt (Waff and Faul, 1992).

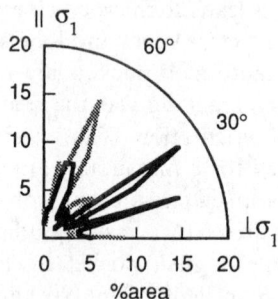

Figure 3.5 Melt pocket orientation by % area of a dislocation creep sample (pi-47) before (black line) and after annealing for 6 h (light grey line). pi-47 was deformed to <45% strain and quenched at a stress of 112 MPa. After annealing melt is preferentially in pockets oriented more nearly perpendicular to the maximum compressive stress.

3.3 Permeability

Permeability is a measure of the resistance of a porous media to fluid flow. Expressions for permeability are usually constructed by invoking a model of the porosity network through which fluid can flow (see Holness, this volume, for a review). The melt network in partially molten upper mantle rocks is similar to the theoretical unit cell shown in Figure 3.6 with sheet-like channels at two-grain junctions, tubular pores with triangular or circular cross-sections at three-grain junctions, and roughly equidimensional pockets at four-grain junctions. In both undeformed and deformed rocks, increases in melt fraction result in increased wetting of two-grain boundaries so that more sheet-like structures are present connecting triple junction tubules and four-grain junction pockets. The effect of this wetting is to limit the size of the tubules and/or pockets; that is, when grain boundary films are present, tubules and/or pockets will be smaller than when grain boundary films are absent. If the sheets or films are very thin and have large lateral dimensions with respect to the radius of the triple junction tubes, their contribution to the hydraulic flow will be negligible (Bernabé, 1995). This observation suggests that permeability may not increase as rapidly with increasing melt fraction as would be predicted based on models in which melt is confined to three- and four-grain junctions (e.g. von Bargen and Waff, 1986).

Figure 3.7 shows the change in local conductance of a unit cell melt network similar to that shown in Figure 3.6 as the aspect ratio of grain boundary films is increased. The hydraulic conductance of the circular tube is $\pi r^4/8L$, where r is the tube radius and L is the tube length; the hydraulic conductance of the grain boundary films is $\pi d b^3/12L$, where b is the film width and d is the grain size (e.g. Bear, 1972; Scheidegger, 1974). Total porosity or melt fraction is held constant so that as the grain boundary film aspect ratio increases and a larger fraction of melt is found in grain boundary films, the radius of the triple junction tube

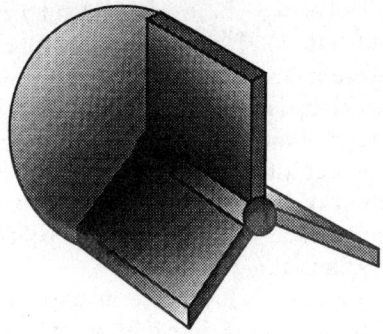

Figure 3.6   A sketch of a unit cell of a typical theoretical melt or fluid network. Sheet-like structures are at two-grain boundaries; tubular features are along three-grain edges; and nodal pores are found at four-grain junctions.

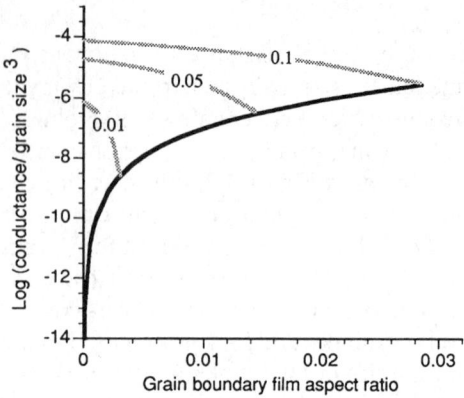

Figure 3.7 Hydraulic conductance of porosity or melt networks composed of a circular tube and three grain boundary films with given aspect ratios. Grey curves, labelled by melt fraction, show the combined conductance of grain boundary films and triple junction tubes. The solid black line indicates the conductance of a network composed only of grain boundary films with a given aspect ratio. As the aspect ratio of grain boundary films increases, more of the available porosity/melt is in these films, resulting in a decrease in size of the melt tube. The decrease in tube size results in a decrease in the hydraulic conductance of the network. The decrease in conductance is more profound at lower melt/porosity fractions.

decreases. It is assumed for this calculation that no nodal pores are present. The solid black line indicates the conductance of a network composed only of grain boundary films with a given aspect ratio. From Figure 3.7 it is clear that as more of the total porosity or melt is present as grain boundary films the conductance of the network decreases. At low porosity or melt fraction, local conductance is almost 1000 times smaller when porosity is confined to grain boundary films as opposed to triple junction tubes. At higher porosity or melt fraction the effect is less marked. It is important to note that the maximum aspect ratios of grain boundary films in this model are dictated by the melt fraction chosen and are lower than the melt pocket aspect ratios reported by either Faul *et al.* (1994) or Daines and Kohlstedt (1996). This difference is due to the fact that current analyses based on two-dimensional images do not separately measure the characteristics of different components of the melt network. Instead, the properties reported are the average of measurements of all components. For example, the mean melt pocket aspect ratio measured by Daines and Kohlstedt (1996) includes a combination of low aspect ratio grain boundary films and higher aspect ratio tubes and pockets and so is likely to be larger than the aspect ratio of the grain boundary films alone.

The decrease in hydraulic conductance in undeformed rocks and in rocks deformed in the diffusion creep regime due to the presence of melt along some two-grain boundaries will probably not be as extensive as that shown in Figure 3.7 because that calculation assumes that all grain boundaries contain films regardless of melt fraction. It is only at melt fractions greater than 0.05 that a high percentage of grain boundaries contain melt (Hirth and Kohlstedt, 1995b). However, the decrease in hydraulic conductance of the magnitude shown in Figure 3.7 could be applicable to rocks in which stress-induced wetting of a majority of grain boundaries occurs. Annealing of rocks in which stress-induced wetting occurs should result in an eventual return to the melt distribution and hence permeability of an undeformed rock with similar melt fraction and grain size provided that a lattice preferred orientation of olivine has not developed.

In contrast to the decrease in permeability that occurs as the number of wetted grain boundaries increases, the presence of melt-free triple junctions in rocks containing pyroxene could result in an increase in permeability, provided the melt network retains connectivity. This increase would be due to the increase in size of melt channels which must occur if melt fraction is to remain constant when the number of melt channels is reduced. In other words, at the same melt fraction and grain size, rocks with pyroxene have fewer melt channels than rocks without pyroxene; hence, the melt channels must be larger. Larger channels mean less resistance to flow. Enhanced melt migration rates in pyroxene-bearing rocks have been observed (Daines and Kohlstedt, 1993), suggesting that this interpretation of the effect of pyroxene on permeability of partially molten upper mantle rocks could be valid. However, in addition to the effects of melt channel size, connectivity of the melt network is an important determinant of the network's permeability. The presence of melt-free triple junctions implies that melt channels will be less well connected in rocks containing pyroxene. Lower connectivity implies an increased path length for flow or increased tortuosity and hence, decreased permeability. Toramaru and Fujii (1986), using a percolation model of permeability, suggested that connectivity and hence permeability will be dependent on the modal amount and relative grain sizes of olivine and pyroxene in the mantle. For example, abundant small pyroxene grains disseminated throughout an olivine matrix could reduce melt connectivity so much that the corresponding increase in the size of melt channels is not enough to maintain permeability. Currently not enough is known about the effects of pyroxene grain size, shape, distribution, and orientation on the grain-scale distribution of melt, nor about the effects of deformation on these factors, to predict precisely the effects of pyroxene on the permeability of deforming upper mantle rocks.

The redistribution of melt into melt pockets preferentially oriented with respect to the stress field could greatly enhance permeability in partially molten upper mantle rocks parallel to the maximum compressive stress (σ_1). If, as is often modelled (see Holness, this volume, for a review), permeability k is proportional to melt fraction ϕ squared or cubed, the ratio of the permeability subparallel to σ_1, $k_{||}$, to the permeability perpendicular to this stress, $k_\perp$, is as shown in Table 3.1 for the samples whose melt distribution is given in Figure 3.4. From this very simplistic calculation, assuming $k \propto \phi^2$, the sample deformed at the highest stress (250 MPa) is almost 50 times more permeable subparallel (0° to 30°) to σ_1 as it is subperpendicular. This value is likely to be an over-

Table 3.1 Anisotropy in melt distribution and permeability with stress

Differential stress (MPa)	0	115	150	250	Annealed
$\phi_\perp$ to σ_1	0.006	0.0041	0.0027	0.0016	0.0275
$\phi_\parallel$ to σ_1	0.006	0.0068	0.0085	0.0113	0.0114
$k_\parallel$ to $k_\perp$	1.0	2.7	9.8	49.0	0.2

estimate, however, because the shape of melt pockets also changes with orientation and stress. Melt pockets subparallel to σ_1 are more elongated than those subperpendicular, suggesting that sheet-like features along two-grain boundaries are more well-developed subparallel to σ_1. At the same melt fraction, these features have lower conductance than tubular channels.

The redistribution of melt that occurs on removing the applied stress and annealing of the partially molten rock also results in an anisotropic permeability structure in samples in which a lattice preferred orientation of olivine is present. Table 3.1 shows the ratio of $k_\parallel$ to $k_\perp$ for the annealed sample whose melt distribution is shown in Figure 3.5. Permeability after annealing is about five times greater perpendicular to the original axis of the applied stress than it is parallel. In other words, permeability is five times greater subparallel to the approximate orientation of (010) planes of olivine. Rocks deformed in shear exhibit similar anisotropy in melt distribution and hence permeability after annealing, except that in this deformation geometry (010) planes of olivine and melt pockets are aligned parallel to the shear boundary or at 45° to σ_1 (Kohlstedt and Zimmerman, 1996).

Melt migration experiments can also provide insight into permeability. In these experiments melt migrates from one portion of a laboratory sample to another driven by capillary forces. The resulting melt distribution profiles are analysed using modified porous flow equations to extract various physical properties including permeability and the permeability–porosity relationship (Riley et al., 1990; Riley and Kohlstedt, 1991; Daines and Kohlstedt, 1993). Permeability of laboratory samples calculated using this method range from $\sim 1 \times 10^{-15} \, \text{m}^2$ for olivine aggregates infiltrated by potassium–aluminum silicate melts (Riley and Kohlstedt, 1991) to $1 \times 10^{-18} \, \text{m}^2$ for olivine aggregates infiltrated by basaltic melts (Daines and Kohlstedt, 1993). The differences in permeability may be explained by differences in melt composition and uncertainties in the value of the melt viscosity (Daines and Kohlstedt, 1993). These values compare to a permeability of $\sim 5 \times 10^{-16} \, \text{m}^2$ for a sample with similar grain size and melt fraction calculated assuming $k = \phi A / 24\pi$, where A is the measured mean melt pocket area. This permeability relationship is the result of assuming the melt network is simply a bundle of tubes with cross-sectional areas equal to the measured distribution of melt pocket areas (Bear, 1972; Scheidegger, 1974).

One surprising result of the melt migration experiments in which the melt sink portion of the sample initially contained no melt was that permeability was linearly dependent on melt fraction (Riley et al., 1990; Riley and Kohlstedt,

1991). Most theoretical models suggest quadratic or cubic dependence (see Holness, this volume). In samples in which the sink originally contained some melt, however, linear, quadratic and cubic relationships provided equally good fits to the experimental data (Daines and Kohlstedt, 1993). The linear relationship determined by Riley *et al.* (1990) may be due to the design of these experiments which results in very high local stresses at the boundary between the melt source and the initially melt-free sink (Daines and Kohlstedt, 1993). A linear porosity–permeability relationship would reduce the strength of the anisotropy in permeability calculated for rocks undergoing deformation and for those in which a lattice preferred orientation of olivine is present.

3.4 Implications for melt migration

One mode of melt migration in the upper mantle is porous flow of melt through a permeable, deformable mantle matrix. This process has been extensively modelled, resulting in a set of equations governing melt flow and several non-dimensional parameters which can be used to examine the effect of changing physical properties on the velocity, time and length scales of melt migration (e.g. McKenzie, 1984, 1985b; Richter and McKenzie, 1984). The permeability k and viscosity η of the mantle matrix are two of the physical properties which figure prominently in the scaling parameters. The length scale or compaction length δ_c is proportional to $(k\eta)^{1/2}$; the time scale τ_0 is proportional to η; and the velocity of the solid W_0 is proportional to $(k\eta)^{1/2}$ (Richter and McKenzie, 1984). Permeability and viscosity are themselves interrelated. If permeability is low, increases in melt fraction due to an increasing degree of melting could cause a marked decrease in viscosity. Conversely, if permeability is sufficiently high such that melt fraction remains low, melt will have little effect on the rheological behaviour of the partially molten rock.

In mantle regions deforming primarily by diffusion creep, permeability is only likely to be different from that predicted by current permeability models used in melt migration simulations when a majority of two-grain boundaries contain melt. The melt fraction required for ~50% of grain boundaries to be wetted is ~0.05 (Hirth and Kohlstedt, 1995b). Estimates of melt fractions present in the upper mantle are generally low (<0.05) (e.g. McKenzie, 1985a; Johnson *et al.*, 1990; Riley *et al.*, 1990), implying that melt fractions at which significant grain boundary wetting and potential breakdown of current models of permeability occur are not attained in the bulk of the upper mantle.

In mantle regions where stress-induced wetting of grain boundaries occurs, the reduction in permeability associated with the formation of grain boundary films should be more than offset by the reduction in the viscosity of the deforming rock, leading to more rapid melt migration. The experiments of Jin *et al.* (1994) suggest an effective viscosity of about 10^{12} Pa s, a value about a factor of 10^6 lower than generally assumed for the upper mantle (Stevenson, 1994). Even though permeability may be reduced by as much as a factor of 10^3, migration velocities, as indicated by calculation of W_0, will be greater than 30 times larger than those of rocks in which melt is confined to three- and four-grain junctions. This increase in melt migration velocity could limit chemical equilibration between migrating melt and residual matrix (e.g. Spiegelman and Kenyon, 1992), potentially leading to the trace element and isotopic signatures such as those observed in mid-oceanic ridge basalts and abyssal peridotites (Salters and Hart, 1989; Johnson et al., 1990). The application of this analysis to the upper mantle, however, is made difficult by the current state of uncertainty about the mechanism which controls the phenomenon of stress-induced wetting. This uncertainty leads to difficulty in scaling this phenomenon to mantle time scales and conditions, a scaling that Stevenson (1994) pointed out must be completed before it can be assumed that stress-induced wetting occurs in the mantle.

In mantle regions where an anisotropic melt distribution is present due to high stress or significant lattice preferred orientation of olivine, the resulting anisotropic permeability structure could lead to melt focusing. Melt focusing is particularly important at mid-ocean ridges where the width of the zone of melt emplacement, the volume of melt emplaced, and the degree of partial melting of the mantle require a mechanism for concentrating or focusing melt flow into the ridge. Several mechanisms for melt focusing have been suggested (e.g. Spiegelman and McKenzie, 1987; Rabinowitz et al., 1987; Buck and Su, 1989); in particular, Phipps Morgan (1987) suggested that the development of a strain-induced anisotropic permeability focuses melt toward mid-ocean ridges. In Phipps Morgan's (1987) model an anisotropy in permeability of three (i.e. $k_x = 3k_y$) is sufficient to strongly affect melt migration beneath the ridge axis. As shown in Table 3.1, the anisotropy in permeability suggested by Daines and Kohlstedt (1996) is close to, or exceeds, this value for samples deformed at differential stresses over 100 MPa and in samples that exhibit a lattice preferred orientation of olivine. This observation implies that the melt distribution and permeability anisotropy developed during dislocation creep could cause the focusing of melt flow depending on the orientation of the stress field and the lattice preferred orientation.

Figure 3.8 shows the flow patterns, orientation of the maximum compressive stress and the orientation of high permeability paths that may develop during passive mantle rifting assuming that the melt distribution is controlled by stress. (010) planes of olivine are oriented parallel to the mantle flow direction during simple shear to high strains (Zhang and Karato, 1995), so high permeability paths are also parallel to the mantle flow direction if the melt distribution is strain-controlled, that is, controlled by the lattice preferred orientation of olivine. In the stress-controlled case at depth below the ridge axis, the highest permeability paths are oriented such that melt flow away from the ridge axis may be enhanced. However, at shallower depths where the upwelling mantle begins to flow away from the ridge axis, melt flow towards the ridge axis may be enhanced due to rotation of the maximum principal stress and the subsequent

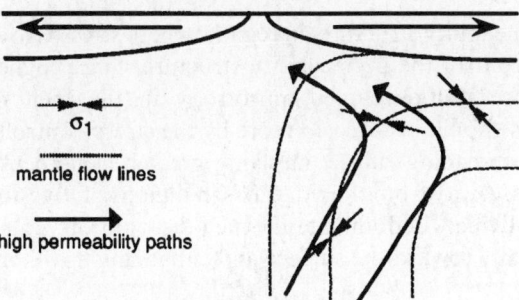

Figure 3.8 Flow patterns, orientation of the maximum compressive stress and the orientation of high permeability paths that may develop during passive mantle rifting. Thin dashed lines indicate mantle flow lines, thick grey arrows show the orientation of the maximum compressive stress, and solid black lines show estimated high permeability paths if the melt distribution is controlled by stress. Because (010) planes become aligned parallel to the mantle flow direction, high permeability paths will be parallel to the mantle flow lines if the melt distribution is controlled by strain.

reorientation of melt. In the strain-controlled case, melt flow will be enhanced parallel to the direction of mantle flow, so that vertical permeability will be enhanced directly beneath the ridge axis, but will be diminished away from the ridge axis relative to the horizontal permeability as (010) planes become oriented parallel to the oceanic plate. As discussed by Waff and Faul (1992), this anisotropy could help focus melt towards the ridge axis and could also limit the width of the zone of melt emplacement.

Measurements of seismic anistropy provide evidence that a lattice preferred orientation of olivine is developed in the upper mantle (e.g. Hess, 1964; Nicolas and Christensen, 1987) and so a lattice preferred orientation of olivine could control the permeability structure in some regions. To determine whether stress-controlled melt distribution and permeability structure is applicable to the mid-ocean ridge environment, it is necessary to scale the processes reponsible for the development of the stress-controlled melt distribution to the grain size of the mantle. To scale the observed melt preferred orientation behaviour from laboratory to geologic conditions, Daines and Kohlstedt (1996) noted that the stress intensity factor near the ends of melt pockets scales with the square root of melt pocket length and hence approximately with the square root of the grain size. In the experiments of Daines and Kohlstedt (1996), the melt preferred orientation developed at differential stresses of ~100 MPa for grain sizes of ~100 μm; therefore, similar melt microstructures should begin to develop at stresses of a few MPa for rocks with grain sizes of 10 mm to 10 MPa for rocks with grain sizes closer to 1 mm. These stresses are appropriate for the upper mantle.

Since both stress- and strain-controlled melt distributions could occur in the mantle, the nature of the anisotropic permeability structure will depend on whether deformation is ongoing and on the preferred orientation of olivine in the rock. In regions where stress is high, the orientation of the stress field will control the permeability structure; in regions where stress is low, the orientation of olivine will control the permeability structure. Since higher levels of stress result in a more well-developed anisotropy in the melt distribution, fast spreading ridges should be affected more by the stress-controlled melt distribution than slow spreading ridges, implying greater melt focusing toward fast-spreading ridges. Away from the ridge axis in regions of low stress, the preferred orientation of olivine will dominate the melt distribution, potentially inhibiting melt flow vertically toward the surface and enhancing flow horizontally toward the ridge axis.

3.5 Summary

The geometry and connectivity of the melt network in partially molten upper mantle rocks are sensitive functions of melt fraction, grain size, composition, and deformation history. Although at low melt fractions melt is mostly confined to three- and four-grain junctions, at higher melt fractions wetting of two-grain boundaries becomes prevalent. Deformation by diffusion creep does not appreciably alter the melt distribution from that observed in the static regime. Deformation by dislocation creep, however, can result in either stress-induced wetting of most grain boundaries or the development of a preferred orientation of melt. Annealing of deformed samples results in an eventual return to the static melt distribution in the case in which stress-induced wetting occurs, but results in a different melt preferred orientation in the case in which a lattice preferred orientation of olivine has developed.

The wetting of two-grain boundaries that occurs with increasing melt fraction could result in reductions in permeability. It is unlikely that this reduction will have much effect on melt migration in a mantle that is deforming by diffusion creep, because melt migration rates are thought to be rapid enough to keep melt fractions below the value at which significant wetting occurs (<0.05). If the mantle is deforming and stress-induced wetting is occurring, however, the reduction in permeability will be more than offset by the large decrease in creep strength and melt migration rates will be enhanced. Finally, the development of a melt preferred orientation controlled either by stress or by strain could result in significant focusing of melt flow in mid-ocean ridge environments due to permeability anisotropy.

References

Ave'Lallemant, H. G. and Carter, N. L. (1970) Syntectonic recrystallization of olivine and modes of flow in the upper mantle. *Geol. Soc. Am. Bull.*, **81**, 2203–20.

Bai, Q., Jin, Z.-M. and Green, II, H.W. (this volume) Experimental investigations of the rheology of partially molten peridotites at upper mantle pressures and temperatures.

Bear, J. (1972) *Dynamics of Fluids in Porous Media*, Elsevier, Amsterdam.

REFERENCES

Bernabé, Y. (1995) The transport properties of networks of cracks and pores. *J. Geophys. Res.*, **100**, 4231–41.

Buck, W. R. and Su, W. (1989) Focused mantle upwelling below mid-ocean ridges due to feedback between viscosity and melting. *Geophys. Res. Lett.*, **16**, 641–4.

Bussod, G.Y. and Christie, J.M. (1991) Textural development and melt topology in spinel lherzolite experimentally deformed at hypersolidus conditions. *J. Petrol.*, Special Lherzolite Issue, 17–39.

Cooper, R.F. and Kohlstedt, D.L. (1982) Interfacial energies in the olivine-basalt system, in *High Pressure Research in Geophysics* (eds S. Akimoto and M.H. Manghnani), Center for Academic Publications, Tokyo, pp. 217–28.

Cooper, R.F. and Kohlstedt, D.L. (1984) Sintering of olivine and olivine-basalt aggregates. *Phys. Chem. Minerals*, **11**, 5-16.

Cooper, R.F. and Kohlstedt, D.L. (1986) Rheology and structure of olivine-basalt partial melts. *J. Geophys. Res.*, **91**, 9315–23.

Daines, M.J. and Kohlstedt, D.L. (1993) A laboratory study of melt migration. *Phil. Trans. R. Soc. Lond.* **A 342**, 43-52.

Daines, M.J. and Kohlstedt, D.L. (1996) Melt distribution in rocks deformed in compressive creep. *J. Geophys. Res.* (in press).

Daines, M.J. and Richter, F.M. (1988) An experimental method of directly determining the interconnectivity of melt in a partially molten system. *Geophys. Res. Lett.*, **15**, 1459–62.

Daines, M.J., Scott, T.J., Morin, P. *et al.* (1995) Three-dimensional reconstructions and two-dimensional analysis of melt networks in partially molten rocks. *EOS AGU Trans.*, **76**, F631.

Dell'Angelo, L.N. and Tullis, J. (1988) Experimental deformation of partially melted granitic rocks. *J. Metamorphic Geol.*, **6**, 495–515.

Drury, M.R. and Van Roermund, H.L.M. (1989) Fluid-assisted recrystallization in uppermost peridotite xenoliths from kimberlites. *J. Petrol.*, **30**, 133–52.

Faul, U.H., Toomey, D.R. and Waff, H.S. (1994) Intergranular basaltic melt is distributed in thin, elongated inclusions. *Geophys. Res. Lett.*, **21**, 29–32.

Griffiths, A.A. (1924) The theory of rupture, in *Proc. 1st Int. Congr. Appl. Mech.* (eds C.B. Biezano and M. Burgers), Tech. Boekhandel en Drukkerij J. Waltman Jr., Delft, pp. 54–63.

Hess, H. (1964) Seismic anisotropy of the uppermost mantle under oceans. *Nature*, **203**, 629–31.

Hirth, J.G. and Kohlstedt, D.L. (1995a) Experimental constraints on the dynamics of the partially molten upper mantle: deformation in the diffusion creep regime. *J. Geophys. Res.*, **100**, 1981–2001.

Hirth, J.G. and Kohlstedt, D.L. (1995b) Experimental constraints on the dynamics of the partially molten upper mantle: deformation in the dislocation creep regime. *J. Geophys. Res.*, **100**, 15441–9.

Holness, M.B. (this volume) The permeability of non-deforming rock.

Jaeger, J.C. and Cook, N.G.W. (1976) *Fundamentals of Rock Mechanics* (2nd edn), Chapman and Hall, London.

Jin, Z., Green, H.W. and Zhou, Y. (1994) Melt topology in partially molten mantle peridotite during ductile deformation. *Nature*, **372**, 164–7.

Johnson, K.T.M., Dick, H.J.B. and Shimizu, N. (1990) Melting in the oceanic upper mantle: an ion microprobe study of diopsides in abyssal peridotites. *J. Geophys. Res.*, **95**, 2661–78.

Kohlstedt, D.L. (1990) Chemical analysis of grain boundaries in an olivine-basalt aggregate using high-resolution, analytical electron microscopy, in *The Brittle-Ductile Transition in Rocks: The Heard Volume* (eds A.G. Duba, W.B. Durham, J.W. Handin, and W.F. Wang), *American Geophysical Union*, Washington, D.C., pp. 211–18.

Kohlstedt, D.L. and Chopra, P.N. (1993) Influence of basaltic melt on the creep of polycrystalline olivine under hydrous conditions, in *Magmatic Systems* (ed. M.P. Ryan), Academic Press, New York.

Kohlstedt, D.L. and Zimmerman, M.E. (1996) Rheology of partially molten mantle rocks. *Ann. Rev. Earth Sci.* (in press).

McKenzie, D. (1984) The generation and compaction of partially molten rocks. *J. Petrol.*, **25**, 713–65.

McKenzie, D. (1985a) 230Th-238U disequilibrium and the melting processes beneath ridge axes. *Earth Planet. Sci. Lett.*, **72**, 149–57.

McKenzie, D. (1985b) The extraction of magma from the crust and mantle. *Earth Planet. Sci. Lett.*, **74**, 81–91.

McKenzie, D. (1989) Some remarks on the movement of small melt fractions in the mantle. *Earth Planet. Sci. Lett.*, **95**, 53–72.

Nicolas, A. and Christensen, N.I. (1987) Formation of anisotropy in upper mantle peridotites: a review, in *Composition, Structure, and Dynamics of the Lithosphere-Asthenosphere System* (eds K. Fuchs and C. Froidevaux), AGU, Washington, D.C., pp. 111–23.

Phipps Morgan, J. (1987) Melt migration beneath mid-ocean spreading centers. *Geophys. Res. Lett.*, **14**, 1238–41.

Rabinowitz, M., Ceuleneer, G., and Nicolas, A. (1987) Melt segregation and flow in mantle diapirs below spreading centers: evidence from the Oman ophiolite. *J. Geophys. Res.*, **92**, 3475–86.

Richter, F.M. and McKenzie, D. (1984) Dynamical models for melt segregation from a deformable matrix. *J. Geol.*, **92**, 729–40.

Riley, G.N.J. and Kohlstedt, D.L. (1991) Kinetics of melt migration in upper mantle type rocks. *Earth Planet. Sci. Lett.*, **105**, 500–21.

Riley, G.N.J., Kohlstedt, D.L. and Richter, F.M. (1990) Melt migration in a silicate liquid-olivine system: an experimental test of compaction theory. *Geophys. Res. Lett.*, **17**, 2101–4.

Salters, V.J.M. and Hart, S.R. (1989) The hafnium paradox and the role of garnet in the source of mid-ocean ridge basalts. *Nature*, **342**, 420–2.

Scheidegger, A.E. (1974) *The Physics of Flow through Porous Media*, University of Toronto Press, Toronto.

Spiegelman, M. and Kenyon, P. (1992) The requirements for chemical disequilibrium during melt migration. *Earth Planet. Sci. Lett.*, **109**, 611–20.

Spiegelman, M. and McKenzie, D.P. (1987) Simple 2-D models for melt extraction at mid-ocean ridges and island arcs. *Earth Planet. Sci. Lett.*, **83**, 137–52.

Stevenson, D.J. (1994) Weakening under stress. *Nature*, **372**, 129–30.

Toramaru, A. and Fujii, N. (1986) Connectivity of melt phase in a partially molten peridotite. *J. Geophys. Res.*, **91**, 9239–52.

Urai, J.L., Means, W.D. and Lister, G.S. (1986) Dynamic recrystallization of minerals, in *Mineral and Rock Deformation: Laboratory Studies. The Paterson Volume* (eds B.E. Hobbs and H.C. Heard), American Geophysical Union, Washington, D.C., pp. 161–200.

Van der Molen, I. and Paterson, M.S. (1979) Experimental deformation of partially molten granite. *Contrib. Mineral. Petrol.*, **70**, 291–318.

Vaughan, P.J., Kohlstedt, D.L. and Waff, H.S. (1982) Distribution of the glass phase in hot-pressed, olivine-basalt aggregates: an electron microscopy study. *Contrib. Mineral. Petrol.*, **81**, 253–61.

REFERENCES

Von Bargen, N. and Waff, H.S. (1986) Permeabilities, interfacial areas and curvatures of partially molten systems: results of numerical computations of equilibrium microstructures. *J. Geophys. Res.*, **91**, 9261–76.

Von Bargen, N. and Waff, H.S. (1988) Wetting of enstatite by basaltic melt at 1350 °C and 1.0- to 2.5-GPa pressure. *J. Geophys. Res.*, **93**, 1153–8.

Waff, H.S. and Bulau, J.R. (1979) Equilibrium fluid distribution in an ultramafic partial melt under hydrostatic stress conditions. *J. Geophys. Res.*, **84**, 6109–14.

Waff, H.S. and Bulau, J.R. (1982) Experimental determination of near-equilibrium textures in partially molten silicates at high pressures, in *High Pressure Research in Geophysics* (eds S. Akimoto and M.H. Manghnani), Center for Academic Publications, Tokyo, pp. 229–36.

Waff, H.S. and Faul, U.H. (1992) Effects of crystalline anisotropy on fluid distribution in ultramafic partial melts. *J. Geophys. Res.*, **97**, 9003–14.

Wanamaker, B.J. and Evans, B. (1985) Experimental diffusional crack healing in olivine, in *Point Defects in Minerals* (ed. R.N. Schock), AGU, Washington, D.C., pp. 194–210.

Zhang, S. and Karato, S.-I. (1995) Lattice preferred orientation of olivine aggregates deformed in simple shear. *Nature*, **375**, 774–7.

Zhang, S., Zimmerman, M., Daines, M.J. *et al.* (1995) Lattice preferred orientation and melt distribution in experimentally sheared olivine-basalt rocks. *EOS AGU Trans.*, **76**, S281.

CHAPTER FOUR

The influence of deformation on the extraction of crustal melts: a consideration of the role of melt-assisted granular flow

E.H. Rutter

4.1 Introduction

The separation of the melt phase from a partially molten rock is an essential first step in the formation of plutons and, ultimately, in the eruption of lavas on the Earth's surface. The past 18 years have seen considerable advances in our understanding of the melt segregation process, especially when it is driven by the density difference between the melt and the matrix of solid grains (e.g. Turcotte and Ahern, 1978; Richter and McKenzie, 1984; McKenzie, 1984, 1985; Fowler, 1985; Scott, 1986; Sleep, 1988; Stevenson, 1989; Fountain *et al.*, 1989; Riley *et al.*, 1990; Spiegelman, 1993; Sparks and Parmentier, 1994). Separation of the melt requires a corresponding reduction in the pore volume of the remaining matrix, which therefore must suffer a volumetric strain (compaction). The compaction may be isotropic, or be accompanied by a distortional strain (a uniaxial shortening in the case of gravity-induced compaction), or the compaction may accompany a more general strain if the rock mass is also undergoing tectonic deformation.

Fluid can flow through a rigid porous matrix in response to an externally applied pressure gradient on the fluid. If the solid matrix is itself deformable, that part of the deformation of the matrix that is compactional or dilatational can impart momentum to the pore fluid and cause flow of the fluid relative to the matrix. Some deformations of the solid matrix can take place at constant volume, in which case the pore fluid is carried about with the distorting matrix. If the permeability of the matrix or its surroundings is effectively zero, then

Deformation-enhanced Fluid Transport in the Earth's Crust and Mantle. Edited by M.B. Holness.
Published in 1997 by Chapman & Hall, London. ISBN 0 412 75290 5.

permanent deformations will be isovolumetric (except for elastic strain changes) and the conditions are described as **undrained**.

The rate at which the melt may be squeezed out under **drained** conditions depends on the melt viscosity, the mechanical properties of the matrix, the matrix permeability and the stress state. Low viscosity ($\sim 10^2$ Pa s) melts, such as basalts, can separate under gravity alone on a time-scale (<1 year) that is short by geological standards. Granitic melts may be $\sim 10^6$ times more viscous, and require geologically unrealistic times for the assembly of melt volumes commensurate with the size of plutons under the influence of gravity alone. Isovolumetric strain paths, in which melt would not be squeezed out, are feasible, but such occurrences may testify to the difficulty of separating granitic melt from its matrix. In high-grade metamorphic complexes in mountain belts it is not unusual to find migmatite suites that appear to have been extensively deformed whilst they contained substantial melt fractions (Brown *et al.*, 1995; Brown and Rushmer, this volume). On the other hand, granitic plutons are often emplaced under conditions of contemporaneous tectonic activity (e.g. Hollister and Crawford, 1986; Hutton *et al.*, 1990; Davidson *et al.*, 1992; D'Lemos *et al.*, 1992; Vigneresse, 1995a, 1995b; Brown and Rushmer, this volume; Grocott and Wilson, this volume). It follows that the extraction of those melts from their protoliths also takes place in a tectonically active environment (Vigneresse, 1995a). It is therefore reasonable to infer that the extraction kinetics of melt from its protolith may be enhanced through deformation-induced melt pressure gradients that are much higher than can be induced through gravity alone.

The induction of large melt pressure gradients is favoured by a strong matrix, but high mechanical strength is not a property to be expected of a partially molten rock. Thus there is a limit to the extent to which tectonic stress gradients can accentuate melt separation kinetics by porous flow through the matrix of solid grains. It has been argued by Clemens and Mawer (1992) and Clemens *et al.* (this volume) that the rapid separation of granitic magma from its source region requires transport in dykes. Rubin (1995) has considered the ability of such dykes to survive large transport distances before freezing. A network of dykes and veins draining a source region effectively increases the permeability of the rock mass by a factor of $\sim 10^6$, and will allow gravity-driven compaction and melt extraction even of viscous granitic liquids. Rutter and Neumann (1995) argued that tectonic stresses would be sufficient to feed granitic melt through the intergranular pore spaces and into a network of veins at a rate comparable to the rate of gravity-driven drainage of the veins. Thus such a two-stage model can provide a solution to the problem of the extraction of viscous melts to form plutons.

Understanding the melt extraction problem requires knowledge of the mechanical properties of the partially molten rock. One of several deformation mechanisms may dominate: (a) dislocation creep of the grain matrix, (b) diffusion creep, possibly facilitated by the presence of the high diffusivity melt, or (c) granular flow of the matrix of grains, with or without fracturing of the grains. Experimental data suggest that any of these mechanisms can be dominant under particular conditions.

It is the purpose of this paper to review the extraction of melts when aided by tectonic deformation, with particular emphasis on granitic rocks. Available experimental data on the behaviour of partially molten granite are limited, therefore special attention will be given to the application of a theoretical constitutive law for diffusion accommodated granular flow of the partially molten rock.

4.2 Experimental deformation of partially molten rocks

With a few exceptions, relatively little is known about deformation mechanisms of partially molten rocks. The transition from dislocation creep to diffusive mass transfer creep in hot-pressed olivine aggregates containing a few vol.% of basaltic melt at 300 MPa confining pressure or under unconfined conditions has been demonstrated in a number of studies (Cooper and Kohlstedt, 1984; Cooper, 1990; Kohlstedt, 1992; Hirth and Kohlstedt, 1995a, 1995b). They showed that when local microstructural equilibrium at solid–solid–melt junctions was attained, melt did not wet the grain boundaries, thus diffusion creep is only accelerated through the presence of the high-diffusivity melt filling the pores. On the other hand, based on similar experiments but at much higher confining pressures (1500 MPa), Jin et al. (1994) and Bai, Jin and Green (this volume) have suggested that grain boundaries may become wetted dynamically during deformation, potentially leading to an acceleration of diffusion creep.

Dell'Angelo and Tullis (1988) deformed partially molten granitic rocks in a solid medium apparatus at 900 °C and 1500 MPa confining pressure, in which the matrix deformed by dislocation creep. Dell'Angelo et al. (1987) also used microstructural evidence from fine-grained, partially molten synthetic aggregates to infer that the flow involved diffusion creep, provided a certain minimum a mount of melt was present.

Other experimental studies on partially molten granitic rocks have used natural rocks deformed in gas confining medium apparatus at lower confining pressure (c. 300 MPa) (e.g. Arzi, 1978; van der Molen and Paterson, 1979; Paquet and François, 1980; Paquet et al., 1981; Auer et al., 1981; Rutter and Neumann, 1995). In all of these studies, deformation of the matrix of solid grains was by some combination of grain fracturing and sliding between grains or fragments, with some possible but unknown amount of strain by diffusive mass transfer creep. Confining pressures were not sufficiently high to inhibit fracture or to favour hydrolytic enhancement of plasticity in quartz.

The amount of melt in these experimental studies on granitic aggregates was controlled either by the addition of different amounts of water (Arzi, 1978; van der Molen and Paterson, 1979; Dell'Angelo et al., 1987; Dell'Angelo and Tullis,

1988) or by temperature, according to the water released from phyllosilicate mineral breakdown reactions (Paquet and François, 1980; Paquet *et al.*, 1981; Rutter and Neumann, 1995). It is either difficult or impossible to maintain constant viscosity of the melt phase through a series of experiments with different melt fractions. Either the melt viscosity decreases at constant temperature with increasing water content, or at constant water content the viscosity decreases with increasing temperature (Figure 4.1(a)). The range of viscosities encountered as water content is varied is, however, typically much greater than the effect of large variations of temperature. These effects must be borne in mind when evaluating the results of experiments involving a range of melt fractions. In contrast to granites, the melt fraction in experiments on partially molten olivine rocks could be held constant over a range of temperatures because it was produced by the addition of a measured amount of basalt glass to the olivine aggregate (e.g. Hirth and Kohlstedt, 1995a, 1995b).

Melt fractions produced from the relatively coarse-grained granitic rocks used for rock mechanics experiments are not chemical equilibrium fractions, such as would be produced from experiments on finely granulated powders (Figure 4.1(b)), and except in the case of fine-grained samples or for wet, low-viscosity melts, it is unlikely that microstructural equilibrium is attained between the melt and the grains in typical experimental durations. However, Laporte (1994) has shown that equilibrium dihedral angles between quartz and hydrous silicic melts are low (12° to 18° at 900 °C) but that many grain boundaries are wetted by melt films. Contacts between unlike solid phases and crystalline anisotropy leads to further variability in the connectivity of the melt

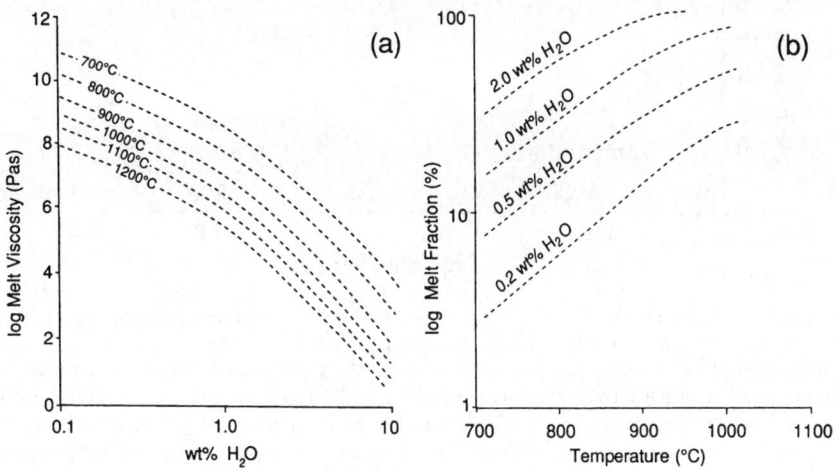

Figure 4.1 Variation of viscosity of granitic melt with temperature and water content (after Shaw, 1965, and Persikov, 1991). (b) Calculated melt fractions versus temperature for different water contents for a fertile quartzo-feldspathic protolith at 500 MPa melt pressure (after Clemens and Vielzeuf, 1987).

phase (Laporte and Watson, 1995). From a high-resolution transmission electron microscope study, Drury and Fitzgerald (1996) have reported very thin (1.5 nm or less) silicic melt films in an olivine-pyroxene rock synthesized by hot-pressing. These films apparently coexist stably with the bulk of the melt that resides in melt tubes at three-grain junctions. The effects of intergranular displacements during deformation are additionally likely to ensure that most grain boundaries are covered dynamically with melt films.

Out of experiments on granitic rocks emerged the concept of a **rheologically critical melt percentage** (Figure 4.2) (e.g. Arzi, 1978; van der Molen and Paterson, 1979). At low melt fractions the interlocking of the grains of the remaining solid leaves the rock reasonably strong. At melt fractions above a few tens per cent, it is supposed that the grain interlocking is destroyed, so that they become melt-supported and are able to slide over one another with little resistance. A great deal of petrogenetic theory has been founded upon this premise, yet the direct experimental evidence is not strong or unequivocal. Part of the effect in experiments is presumably attributable to the aforementioned decrease of viscosity of the melt as melt fraction is increased, either by wetting or by temperature increase. At least for relatively viscous, dry granitic melt, log

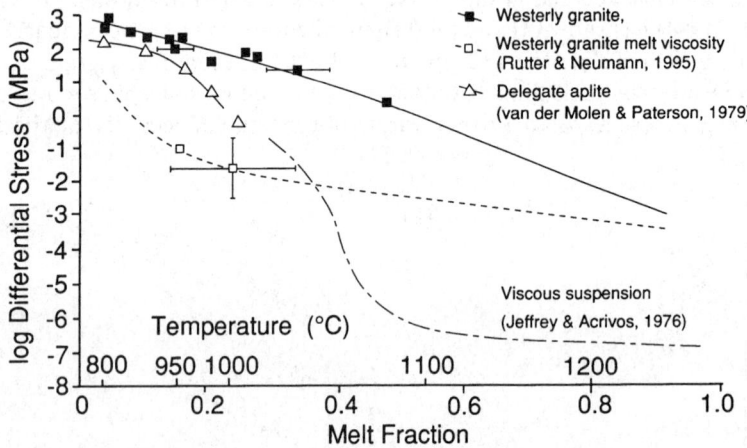

Figure 4.2 Plot of log peak strength versus melt percentage and temperature (solid squares) for the failure of partially molten Westerly granite at a constant strain rate of 8×10^{-5} s^{-1} and constant total water content of 0.3 wt% (after Rutter and Neumann, 1995). Open squares show stresses supported at the same strain rate from melt viscosity measurements at 950 and 1000 °C, and the dashed curve shows extrapolation of these stresses to higher and lower temperatures using the measured activation enthalpy for flow. Horizontal error bars show uncertainty in melt percentage, not temperature. Open triangles show strength data for partially molten Delegate aplite at a similar strain rate, in which the melt fraction was varied by changing water content at a constant temperature of 800 °C (after van der Molen and Paterson, 1979). Above 20 vol.% melt, strength begins to fall rapidly towards the strength of a wet melt, leading to the inference of a rheologically critical melt percentage. In contrast, Westerly granite appears to show no such effect.

'viscosity' of the partially molten rock appears to decrease smoothly with melt fraction (Figure 4.2) up to complete melting (Rutter and Neumann, 1995), without the sudden drop at about 30 vol.% melt for wet (i.e. much less viscous) granitic melt reported by van der Molen and Paterson (1979). It is not unreasonable to anticipate a rheologically critical melt percentage, perhaps if melt viscosity is low or if flow of the matrix involves diffusive mass transfer, when the resistance to grain boundary sliding may be lowered.

At a given strain rate, the strength of granitic rock decreases dramatically over the melting interval (Figure 4.3). Westerly granite (Rutter and Neumann, 1995) shows a strength drop of more than $\times 500$, but it should also be borne in mind that at any given melt fraction the flow stress of the melt itself may be 100 to 1000 times weaker again than the partially molten rock (Figure 4.2). From experiments in which the amount of melt at a given temperature was reduced by partial drainage into a 'sink' of porous quartzite, it is clear that melt fraction has a greater influence on the strength of the rock than temperature alone. No data are available to show whether the strength of partially molten granite is affected by variations in confining pressure. No substantial effect would be expected if the rock follows the law of effective stress, or if deformation of the solid matrix is by intracrystalline plasticity or by diffusive mass transfer.

Van der Molen and Paterson (1979) and Rutter and Neumann (1995) presented experimental data that show that the strength of partially molten

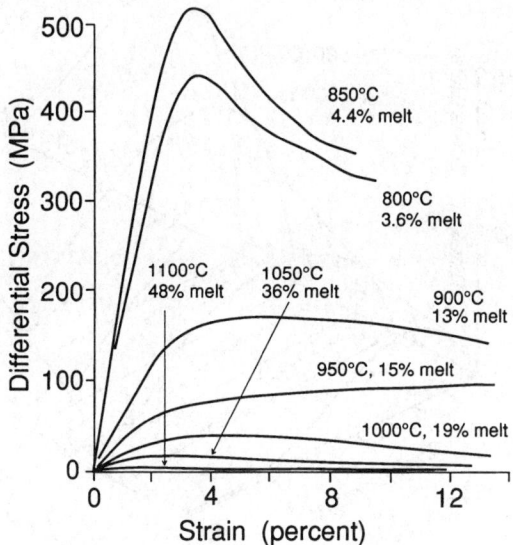

Figure 4.3 Stress–strain curves at a constant strain rate of 8×10^{-5} s^{-1}, 250 MPa confining pressure and after 2.5 h of pre-heating, to show the effects of temperature on the strength and mode of failure of Westerly granite (after Rutter and Neumann, 1995). At 900 °C and below, samples failed ultimately with a brittle fault. At higher temperatures, failure was macroscopically ductile. Over the temperature range shown, strength decreases by a factor of $\times 500$.

87

granite decreases with decreasing strain rate. Rutter and Neumann (1995) presented a 'flow law' for partially molten Westerly granite:

$$\dot{\varepsilon} = 10^{-5.3}\sigma^{2.92}\exp(-510000/RT) \tag{4.1}$$

in which $\dot{\varepsilon}$ is strain rate, σ is the differential stress causing flow, T is temperature in K and R is the gas constant. Deformation involved a combination of grain fracturing and granular flow, but there was no evidence for intracrystalline plasticity of any solid phase. Equation 4.1 must be regarded as a purely empirical relationship that may not tolerate much extrapolation outside the range of experimental conditions. The flow stress is the average flow stress between 10% and 15% shortening. It is not a 'steady state' flow stress because the microstructure is still evolving to produce weakening with increasing strain (Figures 4.3 and 4.4). It also includes the effects of the variation of melt fraction with changing temperature and is therefore particular to Westerly granite. It is likely that with decreasing strain rate or decreasing grain size there will be a transition to grain-size sensitive flow by melt-assisted diffusive mass transfer (diffusion creep) (Dell'Angelo and Tullis, 1988; Hirth and Kohlstedt, 1995a, 1995b). Extrapolation of the 'flow law' given by equation 4.1 to low strain rates suggests that at high temperatures the flow stress will be less than that required to acti-

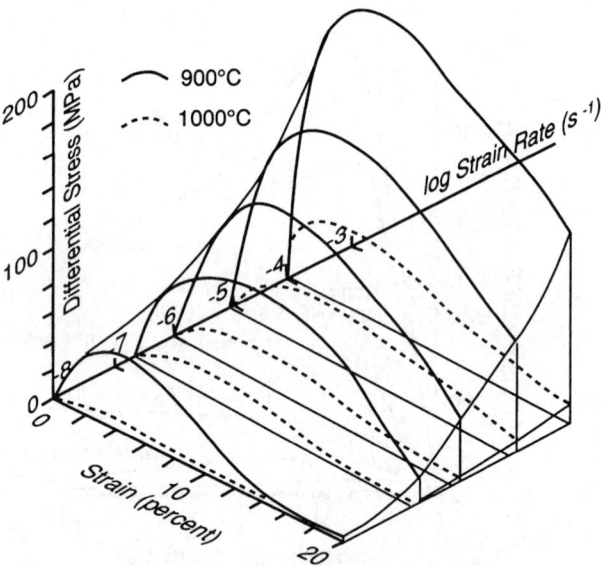

Figure 4.4 Isometric plot to illustrate the shape of stress/strain curves and their variation with log strain rate at 900 and 1000 °C for experimentally deformed Westerly granite. An increase in temperature or a decrease in strain rate markedly reduces strength, and with increasing strain the curves tend to decay after attaining a peak stress early in the deformation history (after Rutter and Neumann, 1995).

vate intracrystalline plastic flow, but at lower temperatures (*c.* 700 or 800 °C) at low (geological) strain rates the plastic flow strength of quartz may be even lower. Figure 4.5 compares the above 'flow law' with an extrapolation of experimental data for quartz plasticity to a strain rate of 10^{-12} s^{-1}.

4.3 The melt extraction problem

4.3.1 Generation of pore pressure

In order to separate liquid from a partially molten rock the matrix and its surroundings must be permeable and the matrix must compact. Compaction may involve the change of grain shape by intracrystalline plasticity or diffusive transfer processes, perhaps aided by the high diffusivity pathways provided by

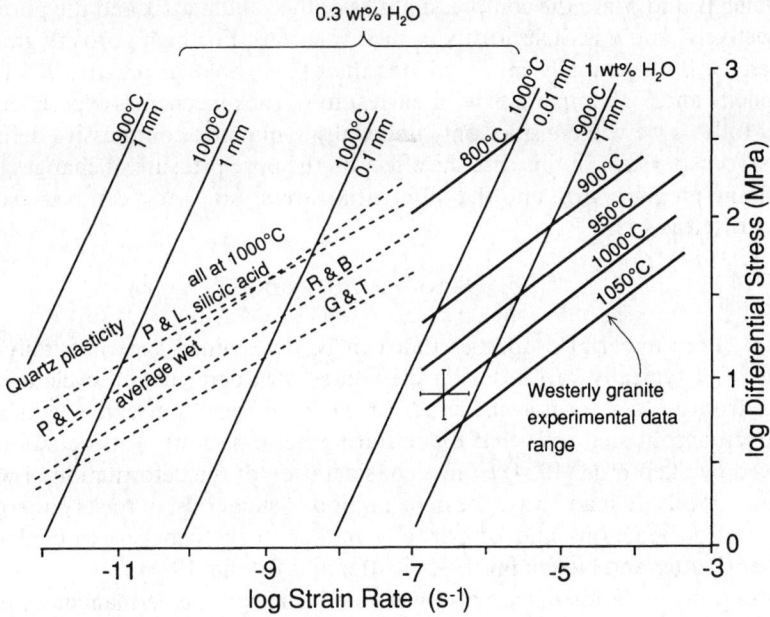

Figure 4.5 Log differential stress versus log strain rate plot to compare the observed mechanical behaviour of partially molten Westerly granite (solid lines at right, error bars are one standard error for the multiple regression fit to the experimental data) with extrapolated plastic flow laws for 'wet' quartz (dashed lines) at 1000 °C. Except at low temperatures, the partially molten rock is expected to be weaker than quartz. P & L = Paterson and Luan (1990) flow laws for synthetic quartzite prepared from silicic acid and an average flow law from previously published data. G & T = flow law of Gleason and Tullis (1995). R & B = flow law of Rutter and Brodie (unpublished data). The steep lines are flow behaviour calculated from the diffusive mass transfer model described in this chapter. For 0.3 wt% water in the melt at 1000 °C the effect of grain size is shown, and lines for 900 °C compare the effects of increasing water content from 0.3 wt% to 1.0 wt% at a constant grain size of 1 mm.

the melt itself. For a given initial melt fraction, a certain amount of compaction may be accomplished by granular flow of the matrix without distortion of individual grains. This will typically be less than 10 vol.% (the volume strain involved in the change from cubic to hexagonal close packing of spheres). The generation of pore pressures by granular flow is well illustrated by the quicksand phenomenon. If, however, melt is being continuously produced as it is expelled, granular flow can account for a significant degree of compaction and shape change of the aggregate without the need for internal distortion of individual grains.

Elastic distortion of a porous aggregate reduces the total volume because Poisson's ratio is generally less than 0.5. Through the compressibility of the pore fluid, a pore pressure change, Δp_e, is generated through a change in the mean stress, $\Delta \bar{\sigma}$, defined as $(\Delta \sigma_1 + \Delta \sigma_2 + \Delta \sigma_3)/3$:

$$\Delta p_e = \Delta \bar{\sigma} \beta_s/(\beta_s + \phi \beta_f) \quad \text{or} \quad \Delta p_e = \Delta \bar{\sigma} B \qquad (4.2)$$

in which β_s and β_f are the compressibilities of the solid matrix and the pore fluid respectively, and ϕ is the porosity (= melt fraction). For high porosity granular materials, β_s is generally much greater than β_f, so that in practice $B \approx 1$. For simplicity and for comparison with the results of rock mechanics experiments, in what follows we will consider only uniaxially symmetric compressive deformations ($\sigma_1 > \sigma_2 = \sigma_3$). In this case the effect on the pore pressure of changes in the confining pressure, $\Delta \sigma_3$, and the differential stress, $\Delta \sigma_1 - \Delta \sigma_3$, can be separated and written as:

$$\Delta p_e = B[\Delta \sigma_3 + A(\Delta \sigma_1 - \Delta \sigma_3)/3] \qquad (4.3)$$

A and B are material properties that can be determined experimentally. The value of A typically varies during the course of a deformation cycle and can range from >1 (e.g. for porous, underconsolidated soils) to negative values (e.g. for overconsolidated soils that dilate during deformation). This equation was derived by Skempton (1954) from a consideration of the deformation of water-saturated soils. In many ways the deformation of such soils, of rocks undergoing dehydration reactions and of partially molten rocks can be expected to be similar (Rutter and Neumann, 1995; Rutter and Brodie, 1995).

Pore pressure gradients are also generated through the permanent deformation of the solid matrix. From the equations of motion of a viscous material (e.g. Jaeger, 1964), an induced hydrostatic pressure gradient is proportional to the volumetric strain rate gradient in the same (x) direction:

$$\partial p/\partial x = (\lambda + 2\eta/3)\, \partial \dot{\varepsilon}_v/\partial x \qquad (4.4)$$

where $\dot{\varepsilon}_v$ is the volumetric strain rate, and λ and η are viscosity moduli defined in an analogous way to Lamé's constants of elasticity. Thus:

$$\sigma_1 = -p + (\lambda + 2\eta)\dot{\varepsilon}_1 + \lambda \dot{\varepsilon}_2 + \lambda \dot{\varepsilon}_3 \qquad (4.5)$$

with corresponding expressions for σ_2 and σ_3. p is a superposed hydrostatic pressure. Defining the bulk modulus of viscosity as $K = \partial p/\partial \dot{\varepsilon}_v$ and the tensile viscosity modulus as $\xi = \partial \sigma_1/\partial \dot{\varepsilon}_1$, it can be shown that:

$$K = (\lambda + 2\eta/3) \quad \text{and} \quad \xi = \eta(3\lambda + 2\eta)/(\lambda + \eta) \tag{4.6}$$

It is implicit here that the matrix flows in a Newtonian viscous manner, a characteristic that will not necessarily apply to geologic materials. If the vertical length scale is sufficiently long, a significant pressure gradient arises from the density contrast, $(\rho_s - \rho_f)$, between the solid matrix and the liquid. Spatial variations in melt fraction, ϕ, can also give rise to a capillary pressure gradient that is a function $f(\phi,\gamma)$ of melt fraction and γ, the fluid–solid interfacial energy (Stevenson, 1989; Cooper, 1990; Riley et al., 1990). The latter contribution is probably of negligible importance in nature, but it can be dominant in experimental configurations (Riley et al., 1990). Thus the total pressure gradient balance becomes:

$$\partial p/\partial x = (\lambda + 2\eta/3)\,\partial \dot{\varepsilon}_v/\partial x - (1 - \phi)(\rho_s - \rho_f)g - f(\phi,\gamma) \tag{4.7}$$

where g is the gravitational acceleration.

This expression shows that local melt pressure gradients can arise as a result of rheological contrasts between different rock types, such as mafic layers and leucosomes in migmatites, leading to a potential for self-amplification of lithologic contrasts (e.g. Robin, 1979; Brown et al., 1995). Depending on the strain plan and the structural configuration of the contrasting lithologies, the rheologically softer material will generally support a lower differential stress and/or deform at a higher rate. Brown et al. (1995) argue that melt will be squeezed from the weaker material into the stronger material, inferring that the mean stress in the stiffer lithology will be lower than in the less stiff material. This is counter to the implications of equation 4.7, which indicates that the melt will be squeezed from the stronger to the weaker material, where the differential stress-induced pore pressure will be lower if they are both deforming at comparable rates. In practice, this will be from the more mafic (melting) phase towards the (more molten and weaker) leucosome phase.

4.3.2 Compaction and melt segregation under gravity

Gravity-driven melt segregation has dominated theoretical work on the problem of melt extraction during the past 18 years. Most attention has been focused on the application of this approach to the segregation of low-viscosity basaltic melt from its parent rock. Recent experimental studies have been made in a high temperature centrifuge to investigate melt segregation under an enhanced gravity field (Dorfman et al., 1994; Bagdassarov et al., 1996). Theoretical models deal with the compaction of the solid matrix against an inferred underlying impermeable base (McKenzie, 1984, 1985; Richter and McKenzie, 1984). Both melt and solid matrix were assumed to be linearly viscous. Above the basal compacting layer, the upward flux of liquid eventually prevents compaction (Figure 4.6), therefore the system is characterized by a length scale, δ_c, over which compaction takes place. This was shown by McKenzie (1984) to be:

$$\delta_c = [k(\lambda + 4\eta/3)/\mu]^{1/2} \tag{4.8}$$

in which μ is the viscosity of the melt and k is the permeability of the matrix. The matrix permeability has usually been estimated by assuming it can be related to grain size, d, and porosity, ϕ, using equations of the form:

$$k = d^2\phi^m/M \tag{4.9}$$

in which exponent m is usually between 2 and 3 and numerical constant M is of order 1000. The relative velocity between the melt and the matrix, w_o, is given by:

$$w_o = k(1 - \phi)(\rho_s - \rho_f)g/\mu\phi \tag{4.10}$$

A characteristic time-scale is given by the compaction time, τ_o, which is the time

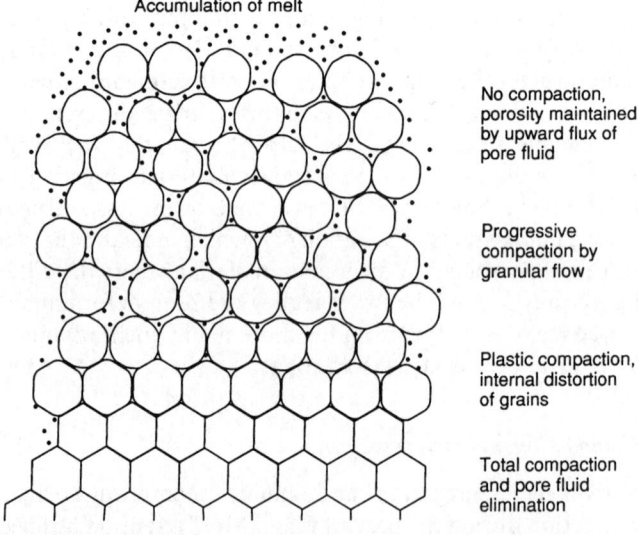

Accumulation of melt

No compaction, porosity maintained by upward flux of pore fluid

Progressive compaction by granular flow

Plastic compaction, internal distortion of grains

Total compaction and pore fluid elimination

Figure 4.6 Illustration of gravity-driven extraction of melt and resultant compaction of the matrix of grains. Compaction is assumed to take place against an undeformable substrate but everywhere the bulk stresses are hydrostatic (the lateral constraint locally equals the vertical pressure). The lowermost grains are permanently deformed to eliminate pore space. Above, there is a transition to pore space reduction by intergranular sliding. Above the compacted zone, the upward flux of melt prevents compaction, and melt eventually accumulates at the top of the section.

required to reduce the melt fraction by a factor e at the base of the compacting layer:

$$\tau_o = \delta_c / [w_o(1 - \phi)] \qquad (4.11)$$

The time, t_h, to reduce the porosity of a layer of thickness h, by a factor e, and the amount of melt extracted in time t_h (equivalent to a layer of thickness h_m) are given by:

$$t_h = \tau_o h / \delta_c \quad \text{and} \quad h_m = h\phi(e - 1)/e \qquad (4.12)$$

Equations 4.8 to 4.12 have been employed widely in studies of the extraction of melts driven by the density difference between melt and solid matrix and resisted by the strength of the matrix and the fluid viscosity. Studies such as that of Fountain *et al.* (1989) have extended the scope of the model to include also heat-flow and the energy balance involved in the melt production.

McKenzie (1985) explored briefly the application of this model to the segregation of granite plutons in the continental crust, and this work was further extended by Wickham (1987). From these studies it is clear that by taking reasonable values for the material characteristics, it is feasible for gravity-driven compaction to extract a substantial fraction of the melt from a column of protolith on a time-scale of *c.* 10 Ma at 800 °C if the melt is wet and of low viscosity (e.g. 10^4 Pa s), but a dry melt at the same temperature ($\mu \sim 10^{11}$ Pa s) would not yield melt over the whole of geological time. The results of such calculations are illustrated graphically in Figure 4.7. Even for a wet melt, the length of required time is only just feasible compared to the time-scale of orogenic processes. On the other hand, independent geological evidence (e.g. Sawyer, 1991) shows that granitic melt segregation can occur in extremely short times.

4.3.3 Tapping of melt by dykes and veins

Several authors have appealed to dykes and vein networks as the mechanism for the transport of granitic magmas (and also basic magmas) away from their source regions (e.g. Wickham, 1987; Sleep, 1988; Clemens and Mawer, 1992; Rubin, 1995; Rutter and Neumann, 1995; Vigneresse, 1995a, 1995b). Clemens and Mawer (1992) show excellent field examples of small vein networks that are common in migmatite terrains. Clemens and Mawer (1992) and Clemens *et al.* (this volume) argue that dyking is the principal method of transport of substantial granitic plutons through a large amount of crustal thickness. The point is reinforced by Vigneresse (1995a, 1995b). Rubin (1995) explores by means of numerical calculations the conditions that must be satisfied to avoid melt freezing before it reaches its emplacement depth, and concludes that the propensity for high-viscosity melts to freeze not long after they leave the source region may mean that dyke transport alone is not necessarily a complete solution to the problem of the mechanism of transport of granitic magma.

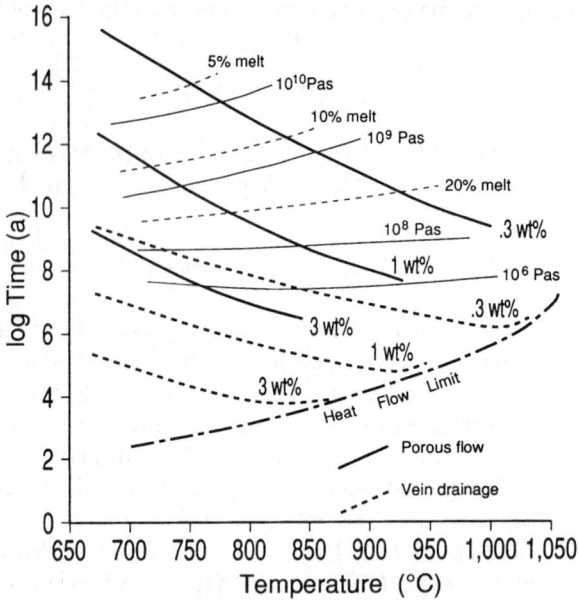

Figure 4.7 Comparison of the time required to extract 10% of the volume of a granitic rock as melt according to (a) the porous flow model for intergranular flow with a grain size of 1 mm (upper solid curves, labelled with melt water contents of 0.3, 1.0 and 3.0 wt%), and (b) the porous flow model applied to a network of veinlets of volume fraction 5% (lower dashed curves for same range of melt water contents). An upper limit to extraction rate is set by the rate that the latent heat of fusion can be supplied to the material (heat flow limit curve). Contours of melt volume fraction and melt viscosity are also shown on the upper curves (based on Rutter and Neumann, 1995).

Rutter and Neumann (1995) argue that once melt has segregated into an interconnected network of veinlets in the source region, gravity-driven compaction can segregate that melt rapidly into a large volume collection point for transport to shallower levels. Isothermal conditions are assumed. The critical point is that the effective permeability of the vein-pervaded rock mass is much higher than the permeability for intergranular porous flow. An effective permeability can be derived for channel (Couette) flow through a network of parallel-sided veinlets. For example, for a parallel array of N channels per metre, and each of width b:

$$k = Nb^3/12 \qquad (4.13)$$

This enhances the effective permeability by a factor on the order of 10^6 relative to intergranular porous flow when $d \sim 1$ mm and $\phi \sim 0.01$, so that even viscous melts can be collected in geologically reasonable periods of time (Figure 4.7). If veinlet networks are essential to the collection of granitic magmas prior to the transportation stage, the question remains of how the melt is transported into the veins.

4.3.4   Role of tectonic deformation in the segregation of melt into dykes and veins

Melt pressure gradients induced by gravity, $(\rho_s - \rho_f)g$, may be only on the order of 3×10^{-3} MPa m^{-1}, for a 10% density contrast between melt and solid matrix. Tectonically induced local gradients in mean and differential stress, and the corresponding induced gradients in melt pressure, may be much larger (c. 0.1 to 1.0 MPa m^{-1}) hence tectonic deformation may play a pivotal role in the segregation of relatively viscous magmatic liquids.

The rate of production of melt, and ultimately the rate of extraction, is limited by the heat flow (Bergantz, 1989). If the rate of melt extraction is limited by the low permeability of the surroundings, distortional flow of the migmatite may occur before all the melt is removed, if indeed it is all removed. Normally, the amount of melt in a partially molten rock may be expected to be controlled by the balance between heat flow, the pressure gradients induced by gravity and regional deformation, the decreasing fertility of the matrix as melt is progressively extracted, and the way in which the permeability is controlled by the melt fraction. Erection of a model for the segregation of a finite amount of melt is complicated if all these effects are taken into account simultaneously, although Fountain *et al.* (1989) and Thompson and Connolly (1995) have gone some way towards the formulation of such a thermomechanical model. This problem is also analogous to that of the production and flow of water in a crustal section undergoing dehydration reactions, that has been partially addressed by Walder and Nur (1984) and more comprehensively by Wong *et al.* (1997). In the present context of the role of deformation in melt extraction, it is useful to take the simplified view of Rutter and Neumann (1995) in order to estimate the rate of melt extraction, by assuming a period of steady-state melt production at all points in the solid matrix. A steady-state condition might be approximated for the extraction of melt corresponding to, say, 10% to 20% of the total rock volume. This will provide a comparative index of the potential of tectonic deformation in controlling the rate of melt extraction.

For simplicity we assume uniaxially symmetric loading of columns of partially molten rock, of lateral dimension $2a$, bounded by an interconnected network of veinlets that intersect in the σ_1 direction (Figure 4.8). Melt is required to flow radially through the matrix pores to the veins. The veins must be held open by the non-hydrostatic stresses in the rock, via the potential for shear failure along surfaces obliquely interlinking the veins (Figure 4.8). This is essential, so that the fluid pressure in the veins, p_v, can be less than σ_3 ($= \sigma_2$) and less than the melt pressure in the pores of the matrix. This is required so that a melt pressure gradient can develop to drive the melt out of the source area (Sleep, 1988). For steady-state radial (r) permeation of melt, against a background of a uniform rate of melt production, $\dot{\phi}$, at all points in the matrix Poisson's equation for the pore pressure, $p(r)$, must be satisfied:

$$\nabla^2 p(r) = - \dot{\phi}\mu/k \qquad (4.14)$$

95

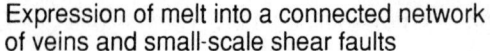

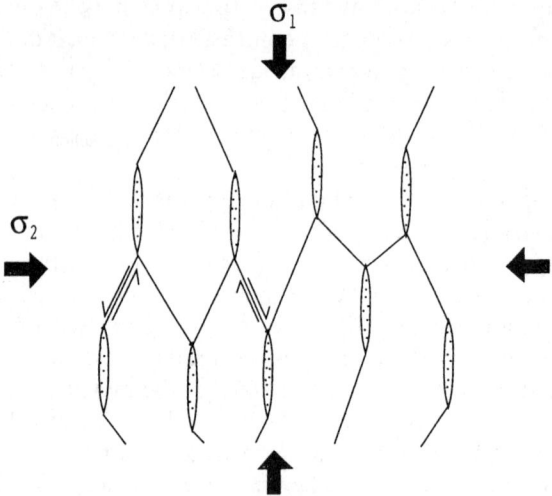

Pore pressure can be less than σ_2

Figure 4.8 Schematic illustration of a network of σ_1-parallel veins, of horizontal separation 2a, linked by shear surfaces. The veins are inferred to be linked in three dimensions, to form a high permeability network of 'pores'. The opening of the veins is partially facilitated by the nonhydrostatic stress state through sliding on the shear surfaces, and the melt pressure in the veins can be slightly less than σ_2. Melt is expressed into the veins by porous flow, driven by deviatoric stress-induced pore pressure gradients.

where ∇^2 is the Laplacian operator. The solution for these boundary conditions is:

$$p(r) - p_v = \phi\mu(a^2 - r^2)/4k \qquad (4.15)$$

Thus a parabolic variation of fluid pressure must develop between the veins, induced by the applied stresses. Rutter and Neumann (1995) assumed that the deformable, melt-saturated matrix of grains would behave like a water-saturated soil. At low effective pressures (applied mean stress minus pore pressures), consolidated soils respond to stress by a combination of distortion and dilatation, which favours localization of the deformation as faults. An increase of effective pressure tends to suppress dilatation, therefore the strength of the material increases. At sufficiently high effective pressures, however, the porous matrix starts to collapse, and the strength falls with further increase in effective pressure, as illustrated in Figure 4.9. At a sufficiently high effective pressure, pore collapse can be induced by the effective pressure alone, without facilitation by shear stresses acting along grain boundaries. The yield surface forms a closed loop, the size of which decreases with increasing porosity. At elevated tempera-

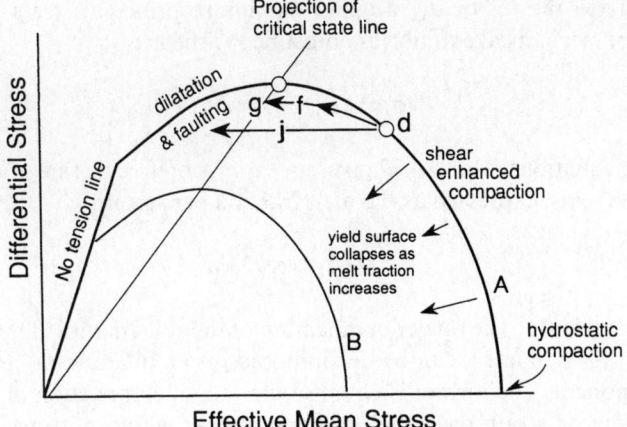

Figure 4.9 Schematic representation of the yield surface for a porous rock. The yield surface forms a closed loop, the size of which shrinks as porosity increases, for example from curve A to curve B. The projection of the critical state line is the projection of the crest of successive yield loops (e.g. small circle at intersection with crest of loop A) at different porosities (Muir Wood, 1990). To the left of this line the deformation is dilatant, and tends to result in localized faulting. To the right it is compactive and tends to result in distributed, ductile flow. Yield is possible even on the abscissa if the effective hydrostatic pressure is sufficient to initiate pore collapse. Variations in pore fluid (melt) pressure cause effective pressure changes and hence behavioural changes. For example, from the small circle labelled d on curve A, progressive compaction during failure can lead to a displacement of effective stress to the left (path f) until isovolumetric failure is achieved at g, or pore pressure rise (path j) at constant differential stress, perhaps from melt influx from beneath, may displace the stress state to the left until dilatant yield (faulting) is produced.

tures, the loop is also expected to shrink with decreasing strain rate. In the region of negatively sloping yield surface, deformation tends to become distributed through the volume of the material because the deformation reduces the porosity and causes local expansion of the yield surface (strain hardening). This is the region of **shear enhanced compaction** (SEC) (Curran and Carroll, 1979). Matrix compaction induces the elevation of pore fluid pressure that is required for deformation to be able to aid melt extraction.

Rutter and Neumann (1995) assumed that $p(r)$ in equation 4.15 was approximated by Skempton's equation (equation 4.3), thereby relating the fluid pressure to the applied stresses:

$$p(r) = (\sigma_1(r) + 2\sigma_3)/3 \qquad (4.16)$$

with the assumption here that $B = 1$ (the pores are fully saturated with melt) and that $A = 1$ (an arbitrary but conservative assumption that means that the effect of deviatoric stress on the pore pressure is equal to the change in the mean stress on the solid framework). The radially varying value of σ_1 that must develop within each column must be in equilibrium with the total force acting along each column due to the regionally applied maximum principal stress, σ_a. Thus a condition of mechanical equilibrium must be satisfied:

$$\sigma_a a^2 = 2\int_o^a \sigma_1(r)\, r\, dr \qquad (4.17)$$

Combining equations 4.14 to 4.17 gives an expression for the rate of melt extraction into the veins, expressed as the melt fraction per second:

$$\dot{\phi} = 8k(\sigma_1 - \sigma_3)/3a^2\mu \qquad (4.18)$$

$\dot{\phi}$ is also equivalent to the rate of compaction of the solid matrix. The total strain rate tensor for the matrix can be decomposed into a dilatational and a distortional component. The former is equal to the rate of compaction of the matrix. An isovolumetric strain path such as simple shear would not give rise to any melt extraction, hence the partially molten rock would simply distort without any melt loss, but more general strain paths will involve melt extraction accompanied by distortion. Rutter and Neumann used their empirical 'flow law' for the deformation of partially molten Westerly granite as an estimator for rate of matrix compaction, arguing that the compaction strain rate was likely to be within an order of magnitude of the distortional strain rate. Thus:

$$\dot{\phi} \approx 10^{-5.3}(\sigma_1 - \sigma_3)^{2.92} \exp(-510000/RT) \qquad (4.19)$$

Equations 4.18 and 4.19 form a simultaneous pair with $\dot{\phi}$ and $(\sigma_1 - \sigma_3)$ as unknowns. Thus rate of melt extraction into veins can be estimated for different values of differential stress and for different temperatures. At combinations of high temperature and lower melt viscosities, the rate of gravity-driven melt extraction from the vein network dominates and veins tend to collapse, therefore the overall rate of the melt extraction is governed by the rate that melt is squeezed into the veins. At lower temperatures the rate of gravity-driven melt extraction from the veins becomes slower, and therefore overall rate-controlling. The results of these calculations are presented in Figure 4.10, in which the relation between melt fraction, temperature and water content (ψ) is given by:

$$\phi = 10^{-3.68}\, \psi^{0.9} \exp(0.00875\, T_c) \qquad (4.20)$$

(after data of Clemens and Vielzeuf, 1987), and T_c is temperature in degrees Celsius. The data of Shaw (1965) were used to infer a relationship between water content, temperature and melt viscosity:

$$\mu = 10^{-6.82}\, \psi^{-3.3} \exp(Q/RT) \qquad (4.21)$$

Activation enthalpy Q was taken as 320 kJ/mol, although a lower value would probably be more appropriate at higher melt water contents (Persikov, 1991). In

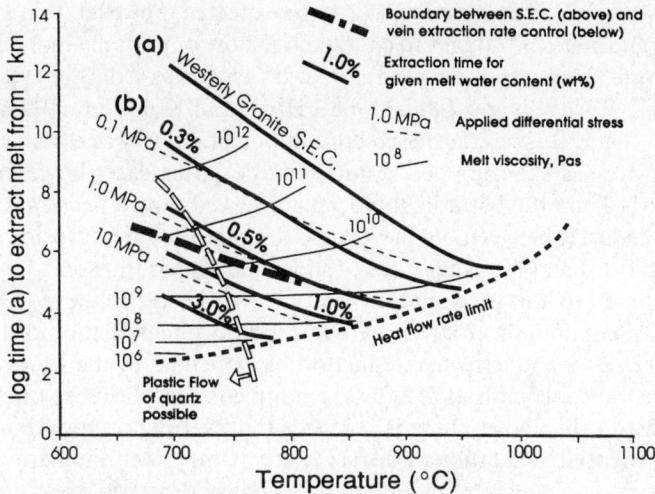

Figure 4.10 Computed time to extract about 10 vol.% of melt to the top of a 1 km, thickness of (a) protolith of the properties of Westerly granite and (b) (remaining curves) for a hypothetical quartzo-feldspathic protolith of grain diameter 2 mm, containing 0.3, 0.5, 1.0 and 3.0 wt% water. Shear enhanced compaction (SEC) is rate controlling above the bold dashed line, and extraction via the vein network below it. At short times, extraction rate is limited by heat-flow. Contours are shown for differential stress (MPa) and melt viscosity (Pa s). Only in the lower left of the figure are combinations of stress and temperature sufficient to activate intracrystalline plasticity in quartz. (After Rutter and Neumann, 1995.)

the above model, the local deformation-induced pressure gradients (for realistic strength values for the partially molten rock) driving the melt into the veins are on the order of 1 MPa m^{-1}, and increases as the strength of the solid matrix increases. This is about 300 times greater than the gravity-induced pressure gradient. Further, the distance over which porous flow has to occur in the solid matrix of grains is on the order of 1 m, rather than 1000 m or more. This obviates the problem of the extraction of very viscous melts from their protoliths, by reducing the required time-scale by the order of 10^5, for a given melt viscosity.

4.3.5 Deformation of solid matrix by diffusion creep

The constitutive law (equation 4.19) describing the deformation of the melt-filled matrix of grains was obtained from axisymmetric compression experiments on partially molten Westerly granite (Rutter and Neumann, 1995). Based on microstructural observations, the flow involved cataclastic deformation of grains, granular flow between grains and viscous flow of the melt. There was no evidence for matrix deformation by intracrystalline plasticity, which would have required higher stresses than the rock was able to support in the partially molten condition. Although strain rates down to 3×10^{-7} s^{-1} were accessed, no evidence

99

was seen of flow by diffusion creep. It is to be expected, however, that at the low strain rates and stresses thought to characterize flow of partially molten granitic rocks in nature, intergranular melt-assisted diffusive mass transfer creep will be important (Dell'Angelo and Tullis, 1988). Hirth and Kohlstedt (1995a, 1995b) have shown that diffusive transfer creep is important at low stresses in partially molten hot-pressed, fine grained dunite. In the latter case, however, it was argued that because the basaltic liquid was observed not to penetrate fully the stressed boundaries between olivine grains, the rate-limiting diffusion step was solid-state diffusion as far as the pores, followed by rapid transport through the melt-filled pores to the precipitation site. The presence of the pore volume, which cannot support shear stress, caused both an intensification of the inter-granular stresses and an effective reduction of the length of the slowest part of the diffusive transport path. Both effects tend to enhance the creep rate.

In contrast to the above, there is evidence that wetting of grain boundaries, and hence potential facilitation of diffusion creep, may occur in many rocks. In single solid phase aggregates under conditions of hydrostatic stress, crystalline anisotropy leads to melt penetration along certain grain to grain contact orientations (e.g. Waff and Faul, 1992). Drury and Fitzgerald (1996) showed that nanoscale films of relatively silicic melt can exist in polymineralic ultramafic rocks for which measured equilibrium dihedral angles are apparently 20–50°, and Hess (1994) argued theoretically that stable thin melt films can exist in grain (particularly interphase) boundaries when the equilibrium dihedral angle is greater than zero. In granitic rocks it appears from experiments that their melts penetrate and wet a large proportion of grain boundaries, owing to the combined effects of intrinsically low equilibrium dihedral angles, a large number of contacts between unlike mineral grains and the effects of crystalline anisotropy (Laporte, 1994; Laporte and Watson, 1995). Adding to this the perturbing effect of deviatoric stress, it seems likely that melt-assisted diffusion creep will be effective across most stressed grain interfaces in such rocks.

Unlike other deformation mechanisms, constitutive laws for diffusive mass transfer processes, with or without intergranular sliding, can be completely derived from purely theoretical considerations. Such flow laws are also commonly characterized by a linear stress/strain rate relationship, i.e. Newtonian viscosity. Diffusive mass transfer flow can occur at constant volume in non-porous materials, or it can accommodate both volume compaction and distortion in porous materials, in which case the bulk and shear viscosities can be separately specified. It is of interest, therefore, to investigate deformation-enhanced melt extraction when the porous matrix is deforming by melt-assisted diffusion creep.

Melt-assisted diffusive mass transfer is thought to be comparable to water-assisted diffusive mass transfer (pressure solution), which is commonly responsible for porosity reduction in sandstones and limestones. Many approaches have been made to the erection of models for porosity reduction by pressure solution, although most do not consider the effects of sliding between grains

(Weyl, 1959; Stocker and Ashby, 1973; Rutter, 1983; Pharr and Ashby, 1983; Spiers and Schutjens, 1990; Lehner, 1990, 1995; Shimizu, 1995). Most recently, Paterson (1995) has developed models to describe granular flow accommodated by material transfer via an intergranular fluid. This is close to the problem of the combined distortional and compactional flow of a partially molten rock, hence the latter approach will be adapted here.

Diffusive transfer creep involves sequential dissipative steps, principally the chemical potential drop that drives the diffusion out of the stressed interface and the potential drop associated with the kinetics of 'dissolution' or precipitation at either end of the diffusion path. The larger step can be said to control the overall kinetics. Because flow laws for the two types of kinetic control have different grain size sensitivities, there will be a transition between them at a particular grain size, with diffusion control being favoured at larger grain sizes. Because rocks undergoing partial melting tend to be rather coarse-grained, we will assume that diffusion kinetics are the principal rate control. Of the cases considered by Paterson (1995), that of removal of material from stressed interfaces followed by precipitation into pores is most pertinent here. For uniaxially symmetric shortening strain, he derived for the deviatoric axial and volumetric strain rates respectively:

$$\dot{\varepsilon}_1 = F_1 V^2 cD\phi^m [(\sigma_3 - p) + 3(\sigma_1 - \sigma_3)]/RTd^2 \qquad (4.22)$$

$$\dot{\phi} = F_2 V^2 cD\phi^m (\sigma_1 + 2\sigma_3 - 3p)/3RTd^2 \qquad (4.23)$$

in which V is the molar volume of the diffusing component in the solid phase, D is the grain boundary diffusion coefficient, which contains an exponential temperature dependence, p is the pressure of the pore fluid and d is the grain size. F_1 and F_2 are geometric factors of order unity and c is the equilibrium molar concentration of the diffusing component in the pore fluid under hydrostatic pressure conditions. Exponent m has a value between 1 and 2 and corresponds to the exponent of porosity in Archie's law, that links diffusive transport through pore fluids to porosity. These equations contain a number of approximations and estimates of values for material-dependent parameters, and are expected only to provide order of magnitude estimates.

Equations 4.22 and 4.23 were derived to describe 'pressure solution' creep, when water is the pore fluid and the hydrostatic solubility of the solid phase is low, so that ideal solution behaviour obtains. However, diffusion of silica in a silica-rich melt cannot be compared to diffusion of ions in a weak aqueous solution. Some idea of the equilibrium concentration of silica at the the interface between a dissolving quartz crystal and a melt of different composition (e.g. albite) can be obtained by extrapolating silica concentrations determined by electron probe at different distances from the solid surface back to the solid surface (e.g. Chekhmir and Epel'baum, 1991). This gives concentrations in the range 80 to 90 wt% SiO_2 over a wide range of temperatures. It is probably more realistic to compare such diffusive transfer to intrinsic point defect diffusion in crystalline solids, in which an equilibrium point defect concentration is determined from consideration of the configurational entropy of the material. In such a case, the flow law for diffusion creep would not contain a solubility term, c, but the geometric constant is different and the apparent activation enthalpy for creep contains a term corresponding to the energy of formation of a point defect by thermal fluctuations.

In the model of Paterson (1995), the porosity appears as a device to modify the effective diffusion coefficient in the same way as porosity scales other transport properties, such as electrical conductivity. Thus the creep rate goes to zero as porosity goes to zero, reflecting the absence of a high diffusivity phase in a zero porosity material. Porosity can also be introduced into the model to take account of the varying radius of the diffusion path and the way in which stress concentrations arise at grain contacts as porosity increases. Paterson (1995) chose to use numerical constants to account for these effects. For approximately cubic close packing the interfacial area varies with porosity approximately as $(1 - 0.4\phi^{2/3})$, and this factor has been incorporated below.

Diffusivities of components in magmatic melts are often estimated from viscosity data via the Stokes-Einstein relation, and the predicted inverse linear relation has repeatedly been verified experimentally (Chekhmir and Epel'baum, 1991). For SiO_2 diffusion these authors report the relation:

$$\log D_{SiO_2} = -4.58 \log \mu - 2.633 \tag{4.24}$$

This may be combined with equation 4.21 above to provide a useful empirical generalization for the silica diffusivity as a function of temperature and water content in granitoid melts.

Modification of equations 4.22 and 4.23 as per above yields the following forms for the flow laws:

$$\dot{\varepsilon}_1 = C_1 C_2 C_3 V D \phi^m [(\sigma_3 - p) + 3(\sigma_1 - \sigma_3)]/RTd^2 \tag{4.25}$$

$$\dot{\phi} = C'_1 C'_2 C'_3 V D \phi^m (\sigma_1 + 2\sigma_3 - 3p)/3RTd^2 \tag{4.26}$$

in which $C'_1 \approx 6$, $C'_2 \approx \frac{1}{4}[\pi/(1 - 0.4\phi^{2/3})]^{1/2}$ and $C'_3 \approx (1 - 0.4\phi^{2/3})^{-1}$ and m is taken to be 2. It is convenient to assume $C_1 C_2 C_3 \approx C'_1 C'_2 C'_3$. Note that $C'_1 C'_2 C'_3 V D \phi^m /RTd^2$ corresponds to modulus $1/\xi$ and $3C'_1 C'_2 C'_3 V D \phi^m /RTd^2$ corresponds to modulus $1/K$, thus $\xi \approx K/3$, which leads to $8\eta \approx \lambda$ and $\dot{\varepsilon}_1 \approx 4\dot{\varepsilon}_3$ when $\sigma_3 = p$.

It is assumed that the overall deformation rate is controlled by the diffusion of silica in the melt, and that D can be replaced by the function of melt viscosity given in equation 4.24. The viscosity is in turn assumed to be given by equation 4.21. The activation enthalpy for viscosity, which contains the main temperature dependence in the flow law, was derived from the summary of data on the influence of water content on activation enthalpy for granitic melts given by Persikov (1991). A power function fits this data quite well:

$$H \text{ (kJ/mol)} = 10^{2.38} \, \psi^{-0.155} \tag{4.27}$$

The melt fraction as a function of temperature and water content (ψ) is calculated from equation 4.20. The resulting flow rates have been calculated and presented in Figures 4.5 and 4.11. The form of equation 4.18 for the rate of melt extraction is not affected by the form of the flow law for the deformation of the solid matrix of grains because the effect of the matrix stress on the melt pressure is assumed to arise simply through an effective Skempton pore pressure coefficient, A. However, the overall rate of melt extraction cannot proceed faster than the rate at which the solid compacts, as described by equation 4.26. Both equations 4.18 and 4.26 describe flows of a linear viscous character, thus the compaction rate described by one will always be faster than that described by the other, as Figure 4.11 shows, but the overall rate of melt extraction will be controlled by the interactions between the two processes. From equation 4.26 the compaction rate varies with the stress state according to:

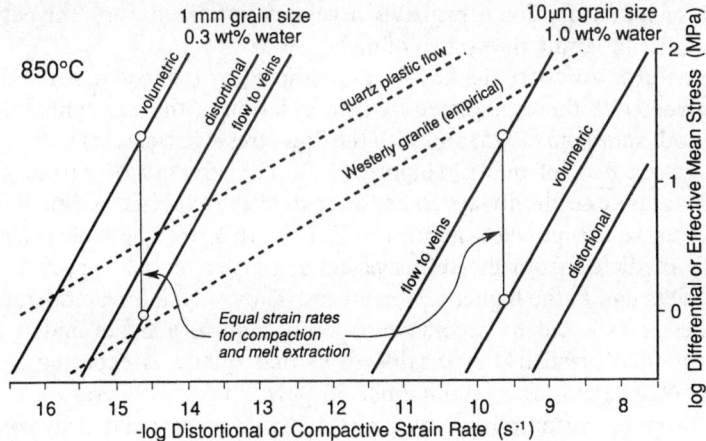

Figure 4.11 Computed rates of compactional or distortional strain rate (as appropriate) for deformation of partially molten granite (solid lines) by melt-aided diffusive mass transport creep, compared with extrapolation of flow laws for quartz plasticity (Rutter and Brodie, unpublished data) and for partially molten Westerly granite (after Rutter and Neumann, 1995), (dashed lines), all at 850 °C. Diffusive transfer creep laws are shown for 1 mm grain size with 0.3 wt% water in the melt and 10 mm grain size with 1.0 wt% water. For the former case, the rate of melt removal driven by shear-enhanced compaction (flow to veins line, representing equation 4.18) is potentially faster than the rate of matrix compaction, and vice versa for the latter case. The constant vertical separation between the lines corresponds to a fixed ratio of effective mean stress to differential stress in order for the compaction and melt extraction rates to become equal. In either case, the differential stress probably adjusts by distortional flow to make these rates equal, as indicated.

103

$$\dot{\phi} \propto [(\sigma_1 - \sigma_3) + 3(\sigma_3 - p)] \tag{4.28}$$

If a change in differential stress alters p by $(\sigma_1 - \sigma_3)/3$ through the poroelastic effect, then nominally the driving force for compaction of the matrix is unaffected by changes in differential stress, whereas the rates of matrix distortion and melt flux to veins are directly proportional to differential stress. From Figure 4.11, equal rates of compaction and melt extraction only occur (for given grain size, temperature and water content) for a fixed ratio of effective mean stress to differential stress. This corresponds to fixed points along a line passing through the origin and of constant slope on the yield diagram (Figure 4.9). It seems likely that distortional flow of the matrix will cause the necessary adjustments to differential stress such that the rate of compaction matches the melt extraction rate.

The model for granular flow accommodated by melt-enhanced diffusion presumes that deformation is always compactive and stable, but the condition of deformation of a weak matrix with a tendency for the melt pressure to rise can conceivably reduce the effective pressure so much that granular flow is displaced out of the shear-enhanced compaction regime and becomes dilatant (Figure 4.9). The result may be deformation of the partially molten rock at constant volume (critical state), or the localization of deformation into faults or veins. Indeed, such deformation is probably necessary initially to form the network of veins required to drain the system of melt.

It is useful to compare the flow rates predicted by the melt-assisted diffusion creep model with those predicted by extrapolation of the experimental data of Rutter and Neumann (1995) and with the flow stresses associated with intracrystalline plastic flow of quartz (Figure 4.11). The strength of partially molten Westerly granite seems always to be lower than the expected plastic flow stress for wet quartz, except below about 750 °C (Figure 4.10). The same is true of the strength predicted from the diffusive mass transfer model, especially for the wetter melts and at the higher temperatures. Thus only at low temperatures are flow stresses expected to become sufficiently high to activate matrix flow by intracrystalline plasticity, especially given that quartz is expected to be the weakest phase for intracrystalline plastic flow. Higher-temperature magmatic flow is probably accomplished by granular flow accommodated by melt-aided diffusive mass transfer. Unfortunately, it is probably generally true that syntectonic plutons have had microstructural evidence of synmagmatic deformation mechanisms removed by annealing.

The predicted high strain rates for diffusive transfer accommodated granular flow of very fine-grained partially molten granite (Figure 4.11) are reasonably consistent with the results of experiments on very fine-grained (10 μm) sintered aggregates (1 wt% water added) reported by Dell'Angelo and Tullis (1988), and for which they presented microstructural evidence of diffusive transfer creep. They reported, for a strain rate of $10^{-6}\,s^{-1}$ at 900 °C and with a melt fraction of about 4 vol.%, a flow stress less than 100 MPa (below the differential load measurement accuracy for the solid-medium testing machine used). Under these conditions, from the model a flow stress in the range 1 to 100 MPa is expected to result in flow at about $10^{-7}\,s^{-1}$. Experiments such as these should now be repeated, but taking advantage of the greater differential load resolution of a gas-medium testing machine, or the molten salt medium technique for higher confining pressures in the GPa range.

4.4 Summary and conclusions

A great deal of field evidence supports the view that the extraction of granitic melt from crustal protoliths requires assistance by deformation (Brown and Rushmer, this volume). Unlike low-viscosity basaltic melts, the separation of granitic melts by porous flow under the influence of its density contrast with the solid matrix alone seems to require geologically unrealistic time periods, except under the most favourable circumstances (low melt viscosities, favoured by high water content and high temperature). Tectonic deformation, however, can provide substantial enhancement of the pressure gradient in the melt. Rutter and Neumann (1995) described a simple, two-stage model of granitic melt separation, based on deformation-enhanced compaction of the matrix, leading to initial segregation of the melt into veinlets. The latter are inferred to provide a high permeability network for gravity-driven segregation and transport of large volumes of melt. The model is too simplistic to give anything more than a comparative snapshot of the rates of deformation-enhanced versus gravity-driven melt separation. Ultimately, the rate of melt extraction is limited by heat flow, and proper account of this will have to be taken in more sophisticated modelling.

The results of deformation experiments on partially molten Westerly granite were used initially to provide a basis for estimating the 'viscosity' of the solid matrix, but the model of Paterson (1995) for granular flow assisted by diffusion through an intergranular fluid has been adapted here to provide an estimator of the shear and bulk viscosities of the solid matrix. Such a flow mechanism is expected to dominate the behaviour of partially molten rocks at low strain rates. Although Dell'Angelo and Tullis (1988) presented microstructural evidence from experiments in support of this contention, there is as yet no unequivocal experimental evidence of the mechanical behaviour of granitic rocks deforming in this way. On the other hand, Hirth and Kohlstedt (1995a, 1995b) have demonstrated diffusion creep in synthetic dunite with pores filled with basalt melt. They argued that because the basalt did not wet stressed grain boundaries, solid-state grain boundary diffusion controlled the overall rate of deformation, a conclusion apparently in conflict with the observations of Bai, Jin and Green (this volume). Available experimental evidence for granitic melts suggests that the melt will wet many or most grain boundaries, through a combination of the effects of crystalline anisotropy, dissimilar contacting solid phase compositions or the production of textural disequilibrium under non-hydrostatic stress.

105

Table 4.1 List of symbols used

σ_i	Principal stress component (Pa), $i = 1, 2$ or 3.
ε_j	Conventional strain component, $j = x$ (linear) or v (volumetric).
p	Pore fluid (melt) pressure (Pa).
p_v	Melt pressure in veins.
β_s	Solid matrix compressibility (Pa^{-1}).
β_f	Pore fluid compressibility (Pa^{-1}).
A	Skempton pore pressure coefficient.
B	Skempton pore pressure coefficient.
x	Distance along coordinate axis (m).
r	Radial distance (m).
λ, η	Solid matrix viscosity moduli, analogous to Lamé parameters (Pa s).
K	Bulk modulus of viscosity (Pa s).
ξ	Tensile viscosity modulus of solid matrix (Pa s).
k	Matrix permeability (m^2).
ϕ	Porosity (= melt fraction).
δ_c	Compaction distance (m).
w_o	Melt velocity ($m\,s^{-1}$).
τ_o	Compaction time (s).
h	Layer thickness (m).
μ	Melt viscosity (Pa s).
ρ_s	Solid density ($kg\,m^{-3}$).
ρ_f	Melt density ($kg\,m^{-3}$).
γ	Melt-solid surface energy ($J\,m^{-2}$).
b	Vein width (m).
N	Number of veins per metre (m^{-1}).
$2a$	Distance between veins (m).
R	Gas constant (J/mol).
T	Temperature (K).
T_c	Temperature (°C).
ψ	Water content of melt (wt %).
Q	Activation enthalpy (J/deg.mol).
d	Grain diameter (m).
g	Gravitational acceleration ($m\,s^{-2}$).
D	Diffusion coefficient ($m^2\,s^{-1}$).
V	Solid phase molar volume (m^3).
c	Equilibrium molar concentration.
n	Stress exponent in flow law.
m	Melt fraction exponent.
F_i, C_i, C'_i	Empirical constants in flow laws, $i = 1, 2$ or 3

Hence in the modified Paterson (1995) model, rate control by diffusion through the melt film is assumed. Experiments on synthetic granitic aggregates, similar to those of Hirth and Kohlstedt (1995a, 1995b) on ultramafic aggregates, will be required to address this question. Flow stresses for partially molten granite protoliths are generally expected to be very small, on the order of 1 MPa or even less.

Subject to the limitations on the extrapolation of laboratory rock mechanics data to nature and the uncertainties in crude theoretical flow models, a comparison has also been made between the expected strength of the solid matrix deforming by intergranular melt diffusion, granular flow as observed in the experiments of Rutter and Neumann (1995) and the plastic flow strength of quartz. This shows that intracrystalline plastic flow of the solid matrix is only expected to be rate competitive with the other mechanisms at low temperatures and high stress levels.

It is suggested that the distortional and compactive flow of partially molten rocks is in many ways comparable to the deformation of porous, water-saturated soils, and that the principles of critical-state soil mechanics can be used to help understand the behaviour. The behaviour is complicated, however, by strain rate dependency in the strength of the solid matrix at elevated temperatures. Soil mechanics methods for the probing of the yield surface will be required to explore the phenomenon of shear-enhanced compaction. This will require the development of techniques for the independent control of the pressure of the melt phase.

Acknowledgements

The work on the flow behaviour of partially molten Westerly granite was developed in collaboration with Ditta Neumann. Careful and constructive reviews by Gayle Gleason and Brian Evans led to many improvements and are greatly appreciated.

References

Arzi, A.A. (1978) Critical phenomena in the rheology of partially melted rock. *Tectonophysics*, **44**, 173–84.

Auer, F., Berkhemer, H. and Oehlschlegel, G. (1981) Steady-state creep of fine grain granite at partial melting. *J. Geophys.*, **49**, 89–92.

Bagdassarov, N.S., Dorfman, A.M. and Dingwell, D.B. (1996) Modelling of melt segregation processes by high temperature centrifuging of partially molten granites: 1. Melt extraction by compaction and deformation. *Geophys. J. Int.*, **127**, 616–26.

Bai, Q., Jin, Z.-M. and Green, II, H.W. (this volume) Experimental investigation of the rheology of partially molten peridotite at upper mantle pressures and temperatures.

Bergantz, G.W. (1989) Underplating and partial melting: implications for melt generation and extraction. *Science*, **245**, 1093–5.

Brown, M. and Rushmer, T. (this volume) The role of deformation in the movement of granitic melt: views from the laboratory and the field.

Brown, M., Averkin, Y.A., McLellan, E.L. and Sawyer, E.W. (1995) Melt segregation in migmatites. *J. Geophys. Res.*, **100**, 15665–79.

Chekhmir, A.S. and Epel'baum, M.B. (1991) Diffusion in magmatic melts: New Study, in *Physical Chemistry of Magmas* (eds L.L. Perchuk and I. Kushiro), *Advances in Physical Geochemistry*, **9**, 99–119, Springer-Verlag, New York.

Clemens, J.D. and Mawer, C.K. (1992) Granitic magma transport by fracture propagation. *Tectonophysics*, **204**, 339–60.

107

Clemens, J.D. and Vielzeuf, D. (1987) Constraints on melting and magma production in the crust. *Earth Planet. Sci. Lett.*, **86**, 287–306.

Clemens, J.D., Petford, N. and Mawer, C.K. (this volume) Ascent mechanisms of granitic magmas: causes and consequences.

Cooper, R.F., (1990) Differential stress induced melt migration: an experimental approach. *J. Geophys. Res.*, **95**, 6979–92.

Cooper, R.F. and Kohlstedt, D.L. (1984) Solution-precipitation enhanced diffusional creep of partially molten olivine-basalt aggregates during hot-pressing. *Tectonophysics*, **107**, 207–33.

Curran, J. H. and Carroll, M. M. (1979) Shear stress enhancement of void compaction. *J. Geophys. Res.*, **84**, 1105–12.

Davidson, C., Hollister, L.S. and Schmid, S.M. (1992) Role of melt in the formation of a deep crustal compressive shear zone: The McLaren Glacier metamorphic belt, south central Alaska. *Tectonics*, **11**, 348–59.

Dell'Angelo, L.N. and Tullis, J. (1988) Experimental deformation of partially melted granitic aggregates. *J. Metamorph. Geol.*, **6**, 495–516.

Dell'Angelo, L.N., Tullis, J. and Yund, R.A. (1987) Transition from dislocation creep to melt-enhanced diffusion creep in fine-grained granitic aggregates. *Tectonophysics*, **139**, 325–32.

D'Lemos, R.S., Brown, M. and Strachan, R.A. (1992) The relationship between granite and shear zones: magma generation, ascent and emplacement in a transpressional orogen. *J. Geol. Soc. Lond.*, **149**, 487–90.

Dorfman, A.M., Bagdassarov, N.S. and Dingwell, D.B. (1994) Experimental centrifuge studies of separation processes in partially molten granitic rocks: compaction versus filter pressing. *EOS, Trans. Am. Geophys. Union, Fall Meeting Abstracts*, **586**.

Drury, M.R. and Fitzgerald, J.D. (1996) Grain boundary melt films in an experimentally deformed olivine-orthopyroxene rock: implications for melt distribution in upper mantle rocks. *Geophys. Res. Lett.*, **23**, 701–4.

Fountain, J.C., Hodge, D.S. and Shaw, R.P. (1989) Melt segregation in anatectic granites: a thermomechanical model. *J. Volc. Geotherm. Res.*, **39**, 279–96.

Fowler, A.C. (1985) A mathematical model of magma transport in the asthenosphere. *J. Geophys. Astrophys. Fluid Dyn.*, **33**, 63–9.

Gleason, G.C. and Tullis, J. (1995) A flow law for dislocation creep of quartz aggregates determined with the molten salt cell. *Tectonophysics*, **247**, 1–23.

Grocott, J. and Wilson, J. (this volume) Ascent and emplacement of granitic plutonic complexes in subduction-related extensional environments.

Hess, P.C. (1994) Thermodynamics of thin films. *J. Geophys. Res.*, **99**, 7219–29.

Hirth, G. and Kohlstedt, D.L. (1995a) Experimental constraints on the dynamics of the partially molten upper mantle, 1. Deformation in the diffusion creep regime. *J. Geophys. Res.*, **100**, 1981–2001.

Hirth, G. and Kohlstedt, D.L. (1995b) Experimental constraints on the dynamics of the partially molten upper mantle, 2. Deformation in the dislocation creep regime. *J. Geophys. Res.*, **100**, 15441–9.

Hollister, L.S. and Crawford, M.L. (1986) Melt-enhanced deformation; a major tectonic process. *Geology*, **14**, 558–61.

Hutton, D.H.W., Dempster, T.J., Brown, P.E. and Becker, S.M. (1990) A new mechanism of granite emplacement: intrusion in active extensional shear zones. *Nature*, **343**, 452–5.

Jaeger, J.C. (1964) *Elasticity, Fracture and Flow*. Methuen, London.

Jin, Z.-M., Green, H.W. and Zhou, Y. (1994) Melt topology during dynamic partial melting of mantle peridotite. *Nature*, **372**, 164–7.

Kohlstedt, D.L. (1992) Structure, rheology and permeability of partially molten rocks at low melt fractions, in *Mantle Flow and Melt Generation at Mid-ocean Ridges* (eds J. Phipps-Morgan, D.K. Blackman and J.M. Sinton), *Geophysical Monograph*, **71**, 103–21.

Laporte, D. (1994) Wetting behavior of partial melts during crustal anatexis: the distribution of hydrous silicic melts in polycrystalline aggregates of quartz. *Contrib. Mineral. Petrol.*, **116**, 486–99.

Laporte, D. and Watson, E.B. (1995) Experimental and theoretical constraints on melt distribution in crustal sources: the effect of crystalline anisotropy on melt interconnectivity. *Chem. Geol.*, **124**, 161–84.

Lehner, F.K. (1990) Thermodynamics of rock deformation by pressure solution, in *Deformation Processes in Minerals, Ceramics and Rocks* (eds D.J. Barber and P.G. Meredith), Unwin Hyman, London, pp. 296–333.

Lehner, F.K. (1995) A model for intergranular pressure solution in open systems. *Tectonophysics*, **245**, 153–70.

McKenzie, D. (1984) The generation and compaction of partially molten rock. *J. Petrol.*, **25**, 713–65.

McKenzie, D. (1985) The extraction of magma from the crust and mantle. *Earth Planet. Sci. Lett.*, **74**, 81–91.

Muir Wood, D. (1990) *Soil Behaviour and Critical State Soil Mechanics*, Cambridge University Press, New York.

Paquet, J. and François, P. (1980) Experimental deformation of partially melted granitic rocks at 600 to 900 °C and 250 MPa confining pressure. *Tectonophysics*, **68**, 131–46.

Paquet, J., François, P. and Nedelec, A. (1981) Effect of partial melting on rock deformation: experimental and natural evidences for rocks of granitic compositions. *Tectonophysics*, **78**, 545–65.

Paterson, M.S. (1995) A theory for granular flow accommodated by material transfer via an intergranular fluid. *Tectonophysics*, **245**, 135–52.

Paterson, M.S. and Luan, F.-C. (1990) Quartzite rheology under geological conditions, in *Deformation Mechanisms, Rheology and Tectonics* (eds R.J. Knipe and E.H. Rutter), Geol. Soc. Lond. Spec. Publ., **54**, 299–308.

Persikov, E.S. (1991) The viscosity of magmatic liquids: Experiment, Generalized Patterns. A model for calculation and prediction. Applications, in *Physical Chemistry of Magmas* (eds L.L. Perchuk and I. Kushiro), *Advances in Physical Geochemistry*, **9**, 1-40.

Pharr, G.M. and Ashby M.F. (1983) On creep enhanced by a liquid phase. *Acta Metall.*, **31**, 129–38.

Richter, F.M. and McKenzie, D. (1984) Dynamical models for melt segregation from a deformable matrix. *J. Geol.*, **92**, 729–40.

Riley, G.N. Jr., Kohlstedt, D.L. and Richter, F.M. (1990) Melt migration in a silicate liquid-olivine system: an experimental test of compaction theory. *Geophys. Res. Lett.*, **17**, 2101–4.

Robin, P.-Y.F. (1979) Theory of metamorphic segregation and related processes. *Geochim. Cosmochim. Acta*, **43**, 1587–600.

Rubin, A.M. (1995) Getting granite dykes out of the source region. *J. Geophys. Res.*, **100**, 5911–29.

Rutter, E.H. (1983) Pressure solution in nature, theory and experiment. *J. Geol. Soc. Lond.*, **140**, 725–40.

Rutter, E.H. and Brodie, K.H. (1995) Mechanistic interactions between deformation and metamorphism. *Geological Journal*, **30**, 227–39.

Rutter, E.H. and Neumann, D.H.K. (1995) Experimental deformation of partially molten Westerly granite, with implications for the extraction of granitic magmas. *J. Geophys. Res.*, **100**, 15697–715.

Sawyer, E.W. (1991) Disequilibrium melting and the rate of melt-residuum separation during migmatization of mafic rocks from the Grenville front, Quebec. *J. Petrol.*, **32**, 701–38.

Scott, D.R. (1986) Magma ascent by porous flow. *J. Geophys. Res.*, **91**, 9283–6.

Shaw, H. R. (1965) Comments on viscosity, crystal settling and convection in granitic magmas. *Am. J. Sci.*, **265**, 120–52.

Shimizu, I. (1995) Kinetics of pressure solution creep in quartz: theoretical considerations. *Tectonophysics*, **245**, 121–34.

Skempton, A.W. (1954) The pore pressure coefficients A and B. *Géotechnique*, **4**, 143–7.

Sleep, N.H. (1988) Tapping of melt by veins and dykes. *J. Geophys. Res.*, **93**, 10255–72.

Sparks, D.W. and Parmentier, E.M. (1994) The generation and migration of partial melt beneath oceanic spreading centers, in *Magmatic Systems* (ed. M.P. Ryan), Academic Press, San Diego, pp. 55-76.

Spiegelman, M. (1993) Physics of melt extraction: Theory, implications and applications. *Phil. Trans. Roy. Soc. Lond.*, **A342**, 23–41.

Spiers, C.J. and Schutjens, P.M.T.M. (1990) Densification of polycrystaline aggregates by fluid-phase diffusional creep, in *Deformation Processes in Minerals, Ceramics and Rocks* (eds D.J. Barber and P.G. Meredith), Unwin Hyman, London, pp. 334-53.

Stevenson, D. J. (1989) Spontaneous small-scale melt segregation in partial melts undergoing deformation. *Geophys. Res. Lett.*, **16**, 1067–70.

Stocker, R.L. and Ashby, M.F. (1973) On the rheology of the upper mantle. *Rev. Geophys. Space Phys.*, **11**, 391–426.

Thompson, A.B. and Connolly, J.A.D. (1995) Melting of the continental crust: some thermal and petrological constraints on anatexis in continental collision zones and other tectonic settings. *J. Geophys. Res.*, **100**, 15565–79.

Turcotte, D.L. and Ahern, J.L. (1978) A porous flow model for magma migration in the asthenosphere. *J. Geophys. Res.*, **83**, 767–72.

Van der Molen, I. and Paterson, M.S. (1979) Experimental deformation of partially melted granite. *Contrib. Mineral. Petrol.*, **70**, 299–318.

Vigneresse, J.L. (1995a) Control of granite emplacement by regional deformation. *Tectonophysics*, **249**, 173–86.

Vigneresse, J.L. (1995b) Crustal regime of deformation and ascent of granitic magma. *Tectonophysics*, **249**, 187–202.

Waff, H.S. and Faul, U.H. (1992) Effects of crystalline anisotropy on fluid distribution in ultramafic partial melts. *J. Geophys. Res.*, **97**, 9003–14.

Walder, J. and Nur, A. (1984) Porosity reduction and crustal pore pressure development. *J. Geophys. Res.*, **89**, 11539–48.

Weyl, P.K. (1959) Pressure solution and force of crystallization – a phenomenological theory. *J. Geophys. Res.*, **64**, 2001–25.

Wickham, S.M. (1987) The segregation and emplacement of granitic magmas. *J. Geol. Soc. London.*, **144**, 281–97.

Wong, T.-F., Ko, S.C. and Olgaard, D.L. (1997) Generation and maintenance of pore pressure excess in a dehydrating system. *J. Geophys. Res.*, **102**, 841–52.

Further reading

Jeffrey, D.J. and Akrivos, A. (1976) The rheological properties of suspensions of rigid particles. *J. Am. Inst. Chem. Eng.*, **22**, 417–32.

Murase, T. and McBirney, A.R. (1973) Properties of some common igneous rocks and their melts at high temperatures. *Geol. Soc. Am. Bull.*, **84**, 3563–92.

CHAPTER FIVE

The role of deformation in the movement of granitic melt: views from the laboratory and the field

Michael Brown and Tracy Rushmer

The reader should be warned that [we are] not presenting an accepted or even a complete theory but [our] view of fragments of a subject to which many are contributing and about which ideas are rapidly changing and developing. If it is conceded that much of this is speculation, then it should also be added that many of the accepted ideas have in fact been speculations also.

J. Tuzo Wilson (1963)

5.1 Introduction

Granitic magmatism is believed to be the most important mechanism by which the continental crust, which is of andesitic bulk composition, has differentiated into a dominantly mafic lower part and a more quartzofeldspathic upper part (Rudnick, 1995). It follows that an appreciation of the chemical and mechanical behaviour of crustal materials during anatexis and melt movement is important to understanding overall crustal evolution. Specifically, we require knowledge of potential source regions for granites and melting behaviour of protolith rock types, modes and rates of melt segregation and magma movement, and processes of pluton emplacement, all of which may vary with geodynamic setting (e.g. active circumoceanic continental margin or collisional settings) and tectonic environment (contractional vs. extensional) (e.g. Clemens and Mawer, 1992; D'Lemos *et al.*, 1992; Brown, 1994; Grocott *et al.*, 1994; Paterson *et al.*, 1996; Petford, 1996; Weinberg, 1996; Clemens, Petford and Mawer, this volume; Grocott and Wilson, this volume). Although not complete, based on several decades of work we have an understanding of the first and last steps in this crustal differentiation process. In contrast, the modes and rates of melt segregation and magma movement during crustal anatexis remain poorly understood. This is partly because research has concentrated on the perme-

Deformation-enhanced Fluid Transport in the Earth's Crust and Mantle. Edited by M.B. Holness. Published in 1997 by Chapman & Hall, London. ISBN 0 412 75290 5.

ability of non-deforming rocks (reviewed by Holness, this volume), but also it reflects the limited data available on the physical and mechanical properties of melts, magmas, and partially molten rocks (e.g. Brown *et al.*, 1995a,b; Rutter, this volume).

Exposed segments of the middle and lower continental crust typically are migmatitic, characterized by layers, pods or veins of granitic material (leucosome) in a plastically deformed metamorphic host. Study of migmatites that are products of melting may provide information about melt migration and magma transfer pathways in the crust. A primarily anatectic origin for many migmatites, particularly those produced by high a_{H_2O} volatile phase-present melting, is evidenced by their igneous microstructures, indicated by geochemical data, permitted by estimates of P-T conditions and suggested by accumulation of granite in dilatant sites formed during deformation of the protolith. However, there are also examples of migmatites in which the mineral assemblages and geochemical data suggest that melt loss has occurred, and some leucosomes have been interpreted as cumulate or residual (Weber *et al.*, 1985; Powell and Downes, 1990; Ellis and Obata, 1992; Nyman *et al.*, 1995; Hartel and Pattison, 1996; Carson *et al.*, 1997).

Although leucosomes in migmatites are believed to represent congealed magma, its cumulate product or its residue lodged in the plumbing during the processes of melt segregation, migration and accumulation, any connection between migmatites and upper crustal granites remains a contentious issue (e.g. McLellan, 1988; Vernon and Collins, 1988; White and Chappell, 1990; Brown, 1994; Sawyer, 1996). In addition, it has been proposed that migmatites could represent either stillborn granites formed when melt could not be expelled before congealing in the source, or the remnant traces of granite passage through the crust. Thus, there is no agreement on whether the drainage network implied by interconnected migmatite leucosomes represents the pathways by which granite melt percolated through the source to major transfer sites, such as dykes or shear zones.

In anisotropic protoliths, plumbing identified by granite stuck in the system defines a connected network of diverse channels and conduits (veins, inter-boudin partitions, fold axial surfaces, shear zones, etc.) that helped melt migration and magma transfer through the crust. This suggests that melt segregation is not driven by gravity alone, but by deformation of a source subject to applied differential stress – melt flows down pressure gradients or is squeezed out of the deforming matrix (McLellan, 1988; Sawyer, 1991, 1994; Brown, 1994; Brown *et al.*, 1995a,b; Williams *et al.*, 1995; Collins and Sawyer, 1996; Oliver and Barr, 1997). Collection of granite magma into vein networks is the likely mode of segregation and migration in isotropic protoliths, such as orthogneiss source materials (Clemens and Mawer, 1992; Rutter and Neumann, 1995; Hartel and Pattison, 1996; Rutter, this volume). We suggest that movement of crustal melts generally requires synanatectic deformation. In examples where melting post-dated deformation, melt does not appear to segregate from residual minerals and patch migmatites are the result (e.g. McLellan, 1989). An additional driving force for segregation may come from the volume increase associated with low a_{H_2O} volatile phase-absent melting (e.g. Clemens and Mawer, 1992; Brown, *et al.*, 1995b).

In this chapter recent laboratory data and field observations are used to illustrate the changes in our thinking about melt segregation and magma movement in the crust. There are concerns with the direct application of any experimental or outcrop study to active geological processes, in particular how they can be scaled and applied to understanding the crustal differentiation process. We are aware that strain rates on rock samples deformed in laboratory experiments under dynamic conditions are much higher than those expected to apply in nature. In addition, field outcrops provide only a snapshot of a continuous process frozen in time. We do not know whether the snapshot represents a picture captured at fast shutter speed, so that the processes of melting and extraction have been frozen in progress at an instant in geological time, or whether the picture is the equivalent of a longer exposure that has captured a sequence of superimposed stages as melt generation declined and the extraction process decayed. Would the resulting pictures look any different?

5.2   Insights from experiments: melt segregation at the grain scale

Two different kinds of experimental studies are used to investigate the nature of melt distribution on the grain scale. Hydrostatic experiments (experiments done so that no externally applied differential stress is placed on the sample) focus on characterizing equilibrium melt distribution among different residual minerals. Dynamic experiments (experiments performed with externally applied differential stress) provide information on the rheological changes that occur as a function of increasing melt fraction during active deformation. The main questions addressed in both types of studies are the following. What is the melt distribution at the grain scale and how much melt is needed before connectivity can be achieved, either by wetting grain boundaries or using brittle fracture systems or in cataclastic shear zones? Is it possible to predict the permeability, or the point at which melt segregation is possible given the composition of the melt and the types of residual minerals and/or the deformation environment? What are the effects of strain rate, melt fraction, melt composition and the types of rheological changes that accompany progressive melting on deformation mechanisms? Observed modes of failure include brittle fracture, melt-enhanced cataclastic flow and homogeneous ductile flow of the material, which in turn change the dominant mode of melt migration.

5.2.1   Hydrostatic experiments

Hydrostatic experiments on aggregates of individual mafic and felsic mineral phases with felsic melts have been used to describe melt distribution in polymin-

113

eralic non-deforming rocks (e.g. Jurewicz and Watson, 1984, 1985; Laporte, 1994; Laporte and Watson, 1995; Lupulescu and Watson, 1995). Most of these experimental studies first document the presence of an equilibrium texture, and then use the data to estimate connectivity of the melt phase as a function of pressure, temperature, bulk composition and sometimes water content of the melt phase (summarized by Holness, this volume). These results have given us an approach by which to estimate both the vol.% melt at which connectivity is achieved, and the permeability of the partially molten rock on a grain scale. This information in turn has formed the basis of melt segregation models, based on compaction by gravity-driven two-phase flow (McKenzie, 1984; Rutter, this volume). However, the models are applicable only to a non-deforming crustal melting environment and cannot be applied directly to active tectonic settings in which synanatectic deformation normally occurs.

Besides surface energy controls on melt distribution, melting reactions themselves may have a major influence on melt distribution at the grain scale. Recent experiments on muscovite–quartzite rock cores (Rushmer et al., 1995; Connolly et al., 1997) have been used to investigate the development of a microfracture network in response to the positive ΔV_r of the melting reactions. These experiments provide transient permeability data at the grain scale at the onset of reaction due to the development of significant melt overpressure and consequent hydraulic fracturing. Melting reactions can produce transient periods of increased permeability on the grain scale and this small-scale process may be the most important first step in the development of a large-scale fracture network necessary for efficient melt segregation (e.g. Clemens, Petford and Mawer, this volume; Rutter, this volume). The results are applicable to environments that undergo fast rates of heating in which rapid rates of melt production might be expected, for example during magmatic underplating in arcs, or ridge subduction at trenches, and in contact metamorphism.

The results from hydrostatic experiments form the basis for suggesting that the lower crust may have significant permeability at the grain scale during melting and that episodes of melt fracturing observed at the outcrop scale may be initiated by these grain-scale processes. Thus, melt distribution in the crust when it is not actively deforming may be envisioned as a combination of melt-filled tubes and planar pores between grains and melt-filled cracks. Connectivity is suggested at low melt fractions in the lower crust (<10 vol.%) for both mafic and felsic rock-types.

5.2.2 Dynamic experiments

Over the past two decades, the experimental results from several different deformation studies on partially molten systems (e.g. synthetic systems by Arzi (1978), and granite by van der Molen and Paterson (1979)) have been used to argue that in the range c. 25–40 vol.% melt fraction there is a major change in the rheology of partially molten granite. Researchers have used this result to suggest that granitic melt cannot segregate from its source below this critical melt fraction (*c.* 25–40 vol.%) because the viscosity of water-undersaturated granitic melt is thought to render it stagnant until the source is weakened. While the basic idea of a major rheological transition present in a partially molten system is robust, application of the experimental results to melt segregation in granite source regions is probably invalid (Rutter, this volume).

The results from dynamic experiments suggest that generally there are three rheological regimes corresponding to different melt fractions (Rutter and Neumann, 1995; Rushmer, 1995, 1996; Rutter, this volume). At low melt fractions, a brittle regime is present where melt-filled fractures develop parallel or subparallel to the shortening direction. This regime appears even at higher pressures where rocks deform plastically without the presence of melt. With increasing melt fractions, melt-enhanced cataclastic flow is observed within shear zones (Figure 5.1). In this regime, extensive grain fracturing by the presence of melt occurs in the sample. This confirms earlier experimental observations of Paquet and François (1980) and Paquet *et al.* (1981), among others, who suggested that the mechanical behaviour of partially molten granite might be controlled by deformation concentrated in dilatant shear bands. Melt migration in this regime is likely to be focused along these cataclastic shear zones because of their higher permeability (cf. Paquet *et al.*, 1981). These cataclastic shear zones may be inferred to have occurred at the outcrop scale, although they are likely to be transient features and microstructural evidence of them may be lost during subsequent deformation/recrystallization events. However, field evidence documented by Schmid (1992) and Davidson *et al.* (1994) shows that features produced by melt-enhanced embrittlement are preserved in some terrains. At the highest melt fractions investigated (>20 vol.% in amphibolite to >45 vol.% in granite), grains are separated from one another by the melt phase and steady-state deformation is by homogeneous ductile flow (Figure 5.2). No through-going fractures occur and the deformation is homogeneously distributed (Figure 5.2a). Grains are separated from one another, but are not fractured, and in experiments on amphibolite, for example, clinopyroxene (as pseudomorphs of hornblende), garnet and new plagioclase are found in a matrix of glass (Figure 5.2b). For partially molten rocks in this rheological regime, the mode of melt segregation changes from fracture-dominated to a shear-enhanced compaction process in which the melt is squeezed out of the deforming matrix (Sawyer, 1991, 1994; Brown, 1994; Brown *et al.*, 1995a,b; Rutter and Neumann, 1995). Thus, there are rheological transitions related to increasing melt fraction, but the 'critical melt fraction' model for melt segregation (Arzi, 1978; van der Molen and Paterson, 1979; Wickham, 1987), which requires the melt to remain in the source until melt fractions reach *c.* 25–40 vol.%, is without a firm foundation and should be abandoned.

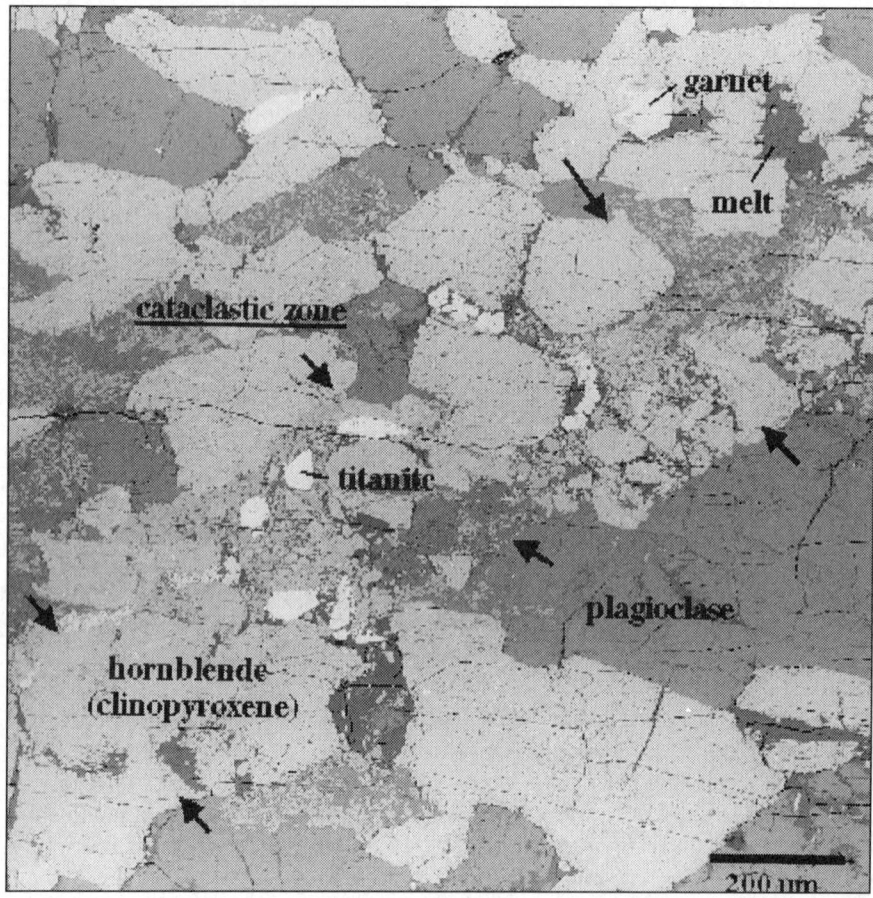

Figure 5.1 SEM backscatter photomicrograph of a longitudinal section of an amphibolite core deformed with 10–15 vol.% melt (935 °C, 1.8 GPa confining pressure, $10^{-5}\,s^{-1}$ strain rate). Photograph is oriented so the externally applied differential stress is from the top downwards. Deformation is localized and is dominantly cataclastic (melt-enhanced cataclastic regime). One of the shear zones is indicated by the sets of arrows. Width of the shear zone ranges from 300 to 500 μm Reaction is more extensive and garnet, clinopyroxene psuedomorphing hornblende and new plagioclase are found in the shear zone.

5.2.3   Application of experimental data to the crust

The application of experimental data to the middle and lower crust may require knowledge of the geodynamic setting during crustal differentiation, since rates of heating and deformation are expected to vary with geodynamic setting. Field evidence shows that melt-enhanced cataclasis and fracturing most likely does occur in the earth's middle and lower crust, but not during all stages of orogenic events. Crustal differentiation during periods of fast melting, such as during

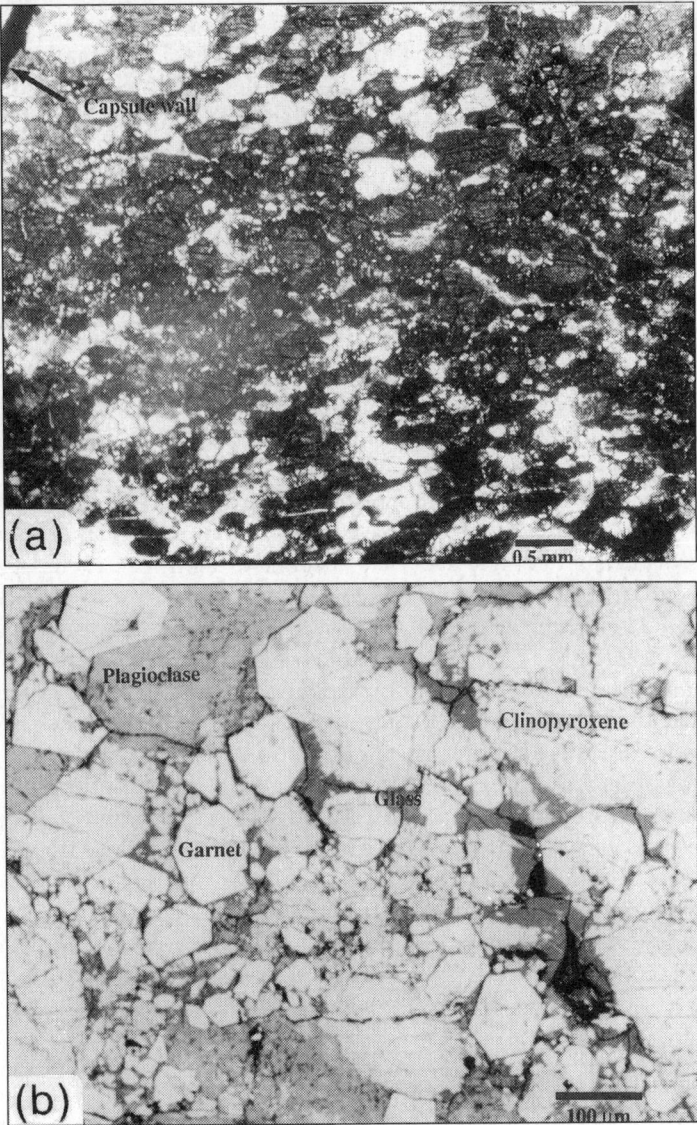

Figure 5.2 (a) Photomicrograph of the hottest region in an amphibolite core deformed at 1000 °C, 1.8 GPa confining pressure and 10^{-5} s^{-1} strain rate. Sample shows more homogeneous deformation at this melt fraction; the melt fraction varies but may exceed 20 vol.%. Deformation is accommodated by homogeneous ductile flow. There are no through-going fracture zones observed at these higher melt fractions. (b) SEM photomicrograph of the same amphibolite sample showing details at higher magnification. Garnet is up to and sometimes greater than 100 μm in size, and hornblende is almost completely pseudomorphed by clinopyroxene. Glass, representing former melt (dark grey; the black areas are where glass has been plucked from the sample during preparation), is observed in pools and lenses surrounding garnet and pseudomorphed hornblende grains. Relict plagioclase is still present; new albite grains are found associated with the glass.

117

underplating by basalt, or during convergence at geologically fast strain rates, may be controlled by fracture-dominated modes of melt segregation. In environments where the build-up of melt is slow or deformation is slow, or absent, melt distribution at the grain scale may be controlled by interfacial energies, reaction volume changes and/or melt movement along grain boundaries; at high melt fractions a matrix compaction/melt extraction process may take over. Figure 5.3 summarizes schematically the changes in deformation behaviour as melt fraction increases in amphibolite and granite, based on Rushmer (1996) and Rutter and Neumann (1995). Although changes are observed from fracture- to mineral-dominated microstructures at low melt fractions, to melt-dominated structures at high melt fractions, melt segregation is not controlled by this rheological transition. The transitions between the rheological regimes occur at different melt fractions in different protoliths, and in the crust may occur at melt fractions different from those in the experiments, according to rates of heating and strain.

In the experiments on amphibolite, pyroxene as the solid product of the incongruent melting reaction first appears as pseudomorphs after hornblende at the edges of through-going cracks. This occurs at temperatures just above the solidus, concurrent with a change in behaviour from macroscopically homogeneous ductile behaviour to fracturing caused by melting. At higher melt fractions, pyroxene is most abundant in the wider shear zones formed during melt-enhanced cataclastic flow. Our present interpretation of these experimental data is that the presence of melt causes the nucleation of the zones of failure. Furthermore, melting does not occur throughout the bulk of the sample and melt is not observed to migrate into the cataclastic shear zones in the experiments. One interpretation of these observations is that the presence of the initial melt nucleates the shear zones, and positive feedback related to cataclastic deformation during shearing promotes additional melting within the shear zones. Small variations in differential stress that produce sites of higher or lower pressure may determine sites of initial melting, ideally combined with multiphase grain contacts between phases required for the particular melt-producing reaction.

5.3 Insights from the field: melt segregation at the outcrop scale

The primary driving force for melt segregation in the crust is physical: either related to buoyancy of the melt, driven by gravity; or related to differential stress, driven by pressure gradients. During orogenesis variations in differential stress must be reflected in effective pressure gradients over the same length scale. In most high-grade middle and lower crustal rocks the differential stress may be tens of MPa; it is this that helps to drive the physical process of melt movement. Also, there will be thermodynamic implications for diffusion. Chemical potentials in a system under applied differential stress are inhomogeneous and drive diffusive mass transfer (Robin, 1979). At the outcrop scale, melt migration may be enhanced not only by an increase in the pressure gradient from one part to another, but also by melt fraction because this corresponds to an increase in porosity (which in turn enhances the possibility of melt migration since permeability varies with porosity (Holness, this volume)). The structural control on sites of melt accumulation at low melt fractions, before significant migration of melt occurred, is prima facie evidence that sites of initial melting may be controlled by local variations in differential stress. In effect, once the equilibrium for a particular melting reaction has been overstepped, melting may be initiated at sites of different pressure, assuming the appropriate phase assemblage occurs at these sites. For most volatile phase-present solidi with negative dP/dT, melting may begin at sites of increased pressure, whereas for most volatile phase-absent solidi with positive dP/dT, melting may begin at sites of decreased pressure (cf. Ord, 1990; Mancktelow, 1995). Thus, in certain circumstances, melting may occur in zones of higher pressure, in which case melt may not flow to zones of low pressure since these will be zones of lower porosity (Ord, 1990). However, because deformation may enhance or inhibit permeability, it is difficult to evaluate the effect of increasing the porosity due to the increasing melt fraction. Other variables that may affect the permeability include the rates of strain, melt production and volume change of the system.

In this section we are concerned primarily with evidence from leucosome (where mobility is implied to be at outcrop scale), whether congealed melt or magma, cumulus product or compacted residue, and intrusive granite (the implied scale of mobility is larger than outcrop scale but smaller than crustal scale) lodged in the plumbing to infer the mesoscopic and macroscopic flow pathways used to carry melt or magma through the crust.

5.3.1 Melt accumulation sites

(a) In layers: stromatic migmatites
Perhaps the most obvious structural feature in many plastically deformed metamorphic rocks is relict compositional layering and/or a tectonite fabric. Layer-parallel leucosomes are the definitive feature of stromatic migmatites, also termed metatexites (Brown, 1973). Such migmatites are characterized by symmetrical structures in which leucosome is bordered by residual minerals that may form continuous selvedges; commonly in many sedimentary protoliths the selvedges are composed predominantly of biotite and may be termed melanosomes. Frequently, in metasedimentary protoliths leucosomes develop on pre-existing layering (Johannes, 1988; Brown et al., 1995b). This suggests that some general set of processes may be responsible for migration of granite melt to form leucosome layers. Their formation has been discussed in detail by Brown et al. (1995b). Once melt has segregated into stromae parallel to the relict compositional layering, it appears able to migrate parallel to the layering (Brown, 1994; Brown et al., 1995b) into low pressure sites such as shear zones (Figure 5.4).

119

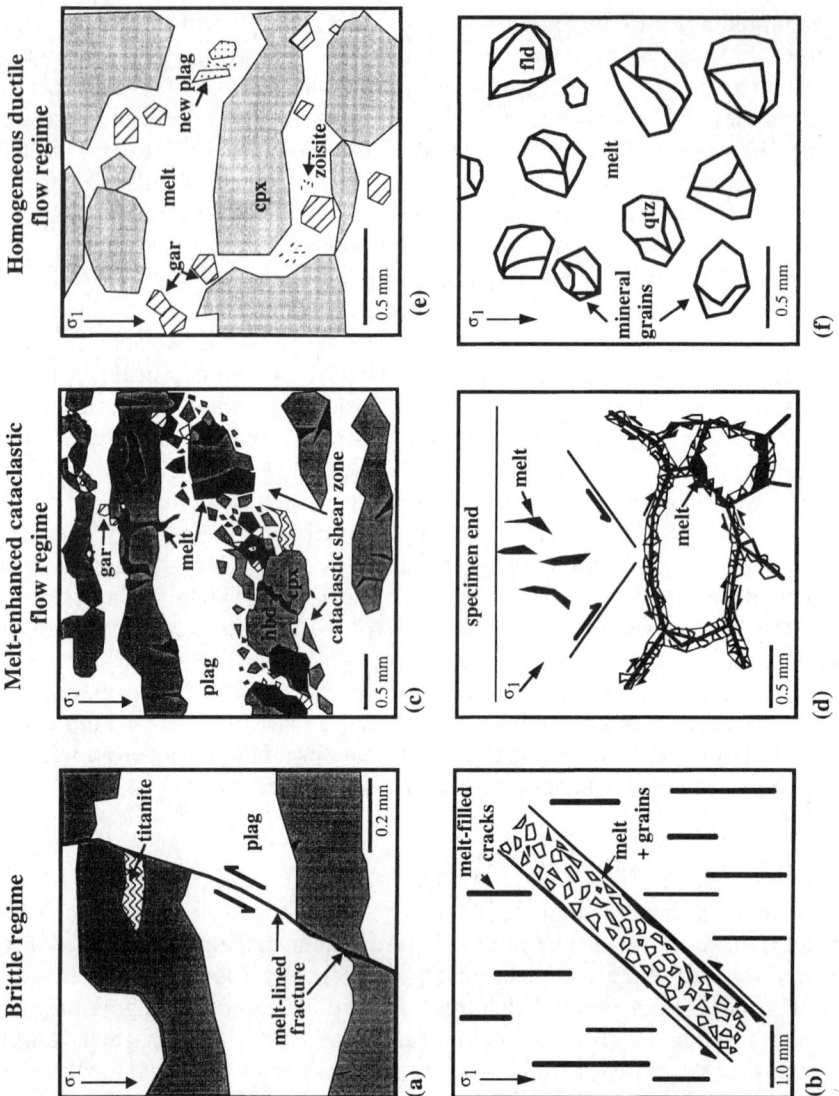

Figure 5.3 Schematic illustration of the microstructures observed at high confining pressure (amphibolite: (a), (c) and (e)) and low confining pressure (granite: (b), (d) and (f)) as a function of increasing melt fraction. The deformation experiments on amphibolite were done at 1.8 GPa and 800–1000 °C, whereas those on granite were done at 0.25 GPa and 800–1100 °C; the strain rate for both experimental sets is 10^{-5} s^{-1}. (a) In the amphibolite, deformation is macroscopically plastic with no through-going faults below the solidus (650 and 750 °C) but as melting begins (800–850 °C), micrometre-wide fractures deform the sample (brittle regime) and melt is found along some of these cracks (plag, plagioclase; hbd, hornblende). (c) In the amphibolite, melt is found in dilatant cracks in the hornblende (partially reacted to clinopyroxene, cpx) and in cataclastic shear zones that transect the sample (melt-enhanced cataclastic flow regime, 5–20 vol.% melt). Product phases such as garnet

120

One interesting outcome of the experimental work by Laporte and Watson (1995) that relates to stromatic migmatites is the fact that a combination of a pronounced crystalline anisotropy and a marked preferred orientation of mica flakes leads to a very low permeability normal to mica-rich layers, such as biotite-rich melanosomes in stromatic migmatites. Such a barrier should impede interactions between neighbouring leucosomes and mesosomes. This may promote proliferation of increasingly closely spaced leucosome–melanosome sets with progressive melting and the elimination of mesosome by anatectic erosion, to produce a structure described as 'high-grade metatexite' by Brown (1974) and 'banded diatexite' by Brun and Martin (1978). The barrier to layer-normal flow imposed by the biotite-rich selvedges in metapelitic migmatites in combination with the negative ΔV_r expected during high a_{H_2O} volatile phase-present anatexis will help melt retention in the system, unless deformation encourages magma mobility in the manner envisioned by Brown (1973, 1979), Brown et al. (1995b) and Sawyer (1996).

Although primary layering of different bulk compositions may control melt fraction, deformation controls the formation of sites for melt accumulation, as shown by Greenfield et al. (1996). In the migmatites they describe from the Mt Stafford area of the Arunta Inlier, Australia, 'leucosome networks may fill extensional structures in metapelite layers . . .', while at higher grades, 'the metapsammite layers had a lower proportion of melt and behaved as more rigid blocks that were broken and rotated in a mobile, metapelite-sourced diatexite matrix'. However, the layered structure preserved by the different types of migmatite reflects the original compositional layering in the sedimentary protolith (cf. Johannes, 1988).

Alternative models for the formation of stromatic metatexite migmatites have been proposed by Lindh and Wahlgren (1985) and Maaløe (1992). Lindh and Wahlgren propose a model for the formation of stromatic migmatites by 'diffusion . . . along grain boundaries enhanced by small amounts of "intergranular fluid"' and driven by gradients in mean pressure. This model does not require

(gar), new albitic plagioclase, titanite and zoisite are also present. (e) In the amphibolite, at ≥20 vol.% melt (in the centre of the sample at 1000 °C), deformation is by homogeneous ductile flow and no through-going fractures are observed (homogeneous ductile flow regime). Garnet grains have increased in size, new albitic plagioclase (new plag) and zoisite prisms are present in melt pools. (b), (d) and (f) Interpretation by Rutter and Neumann (1995; modified from their Figure 9) of the evolution of microstructures observed in experiments on partially molten Westerly granite. (b) In the granite, at the lower confining pressure, deformation is brittle below and above the solidus (brittle regime, dilatant fault zone, 0–10 vol.% melt). As melt fraction increases, dilatant melt-filled cracks, which lead to faulting, are observed. (d) In the granite, cataclastic deformation dominates the sample in the melt-enhanced cataclastic flow regime (1–45 vol.% melt). Formation and collapse of melt-filled pores leads to the shear-enhanced compaction model for melt segregation and migration in this regime (Rutter and Neumann, 1995). (f) In the granite, a transition between cataclasis and ductile flow is observed in these experiments at melt fractions ≥45 vol.% (homogeneous ductile flow regime), which is a higher melt fraction than observed in the experiments on amphibolite at higher pressures (≥20 vol.%).

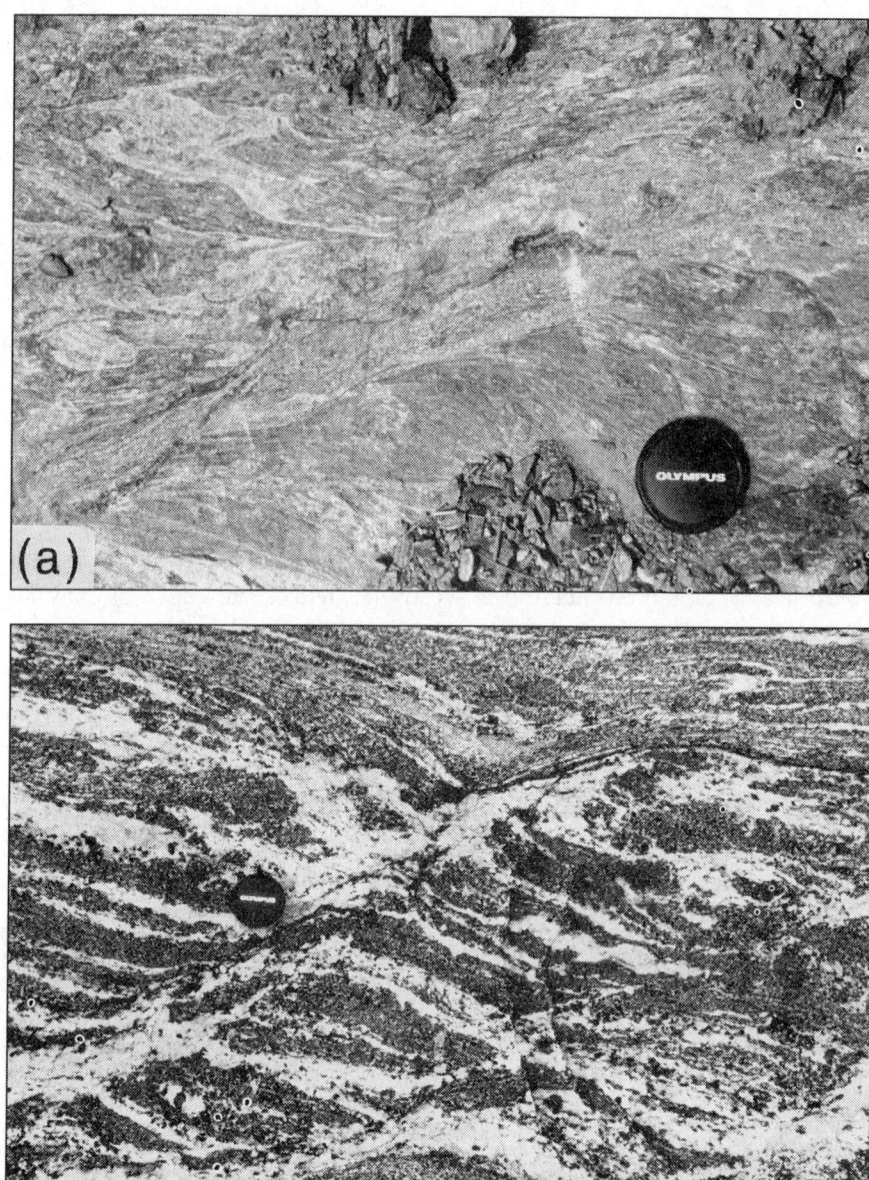

Figure 5.4 (a) Granite-filled shear zone in migmatitic host; Ryoke Metamorphic Belt, Japan. (b) Leucosome within amphibolite protolith interpreted to show migration of anatectic melt into extensional (from the top down to the left) shear zones; Tolstik Peninsula, Karelia, Russia.

heterogeneities in the protolith and may apply to isotropic rocks such as orthogneisses and granulites. Further, the model can explain two common features in migmatites: non-minimum melt composition of some leucosomes; and similar mineral compositions in both mesosomes and leucosomes (cf. the equilibration volume of Powell and Downes, 1990). Maaløe (1992) argues that leucosomes develop along shear fractures by diffusion-controlled melting, and biotite selvedges develop by liquid-phase sintering. In the model, formation of the biotite selvedges is based on the expansion of the leucosome and the contraction of the melting mesosome. This model is based on an assumption that once formed, melt remains at grain corners and edges; subsequent experiments suggest that low dihedral angles are to be expected in all common crustal protoliths, which invalidates the assumption of stagnant melt upon which the model was based.

(b) Boudinage structure
Interboudin partitions have long been recognized as low-pressure sites to which material may be transferred by diffusive or advective mass flux (e.g. Stromgard, 1973; van der Molen, 1985; Mancktelow, 1995). The strong anisotropy characteristic of many supracrustal migmatite terrains leads to boudinage at all scales and accumulation of leucosome between boudins, spectacularly illustrated in several papers (e.g. Pattison and Harte, 1988; Brown, 1994; Williams *et al.*, 1995; Greenfield *et al.*, 1996). Furthermore, this process may be scale-invariant. Gapais *et al.* (1995) observed that the extensional deformation pattern of many thickened orogenic belts may result from pervasive thinning of the ductile crust, accompanied by décollement along the weakest units to allow emplacement of granite sheets, and by boudinage or pinch and swell of the strongest units to allow local uplift of migmatitic lower crust. Such a crustal-scale process may generate new melt during decompression and may create structural sites for granite ascent and emplacement.

In the example shown in Figure 5.5, a discordant leucosome patch occurs in an interboudin partition within foliated amphibolite. The leucosome patch contains large orthopyroxene crystals as solid products of the incongruent melting reaction that involved hornblende breakdown under low a_{H_2O} volatile phase-absent conditions. Within the amphibolite boudins, orthopyroxene ± garnet were the peritectic products of the incongruent melting reaction, and orthopyroxene commonly nucleated on hornblende; these reaction products are spatially related to thin foliation-parallel leucosomes. Our interpretation of these observations is counter-intuitive. Although, superficially, it appears that melt migrated parallel to the foliation in the amphibolite boudin to accumulate in the interboudin partitions, this may not be what happened. As Mancktelow (1995) pointed out, extension along strong layers may lead to underpressure. We suggest that the incipient sites of interboudin partitions also represented sites of initial melting. The large size of the peritectic orthopyroxene crystals in the interboudin partition, up to an order of magnitude larger than the grain size of

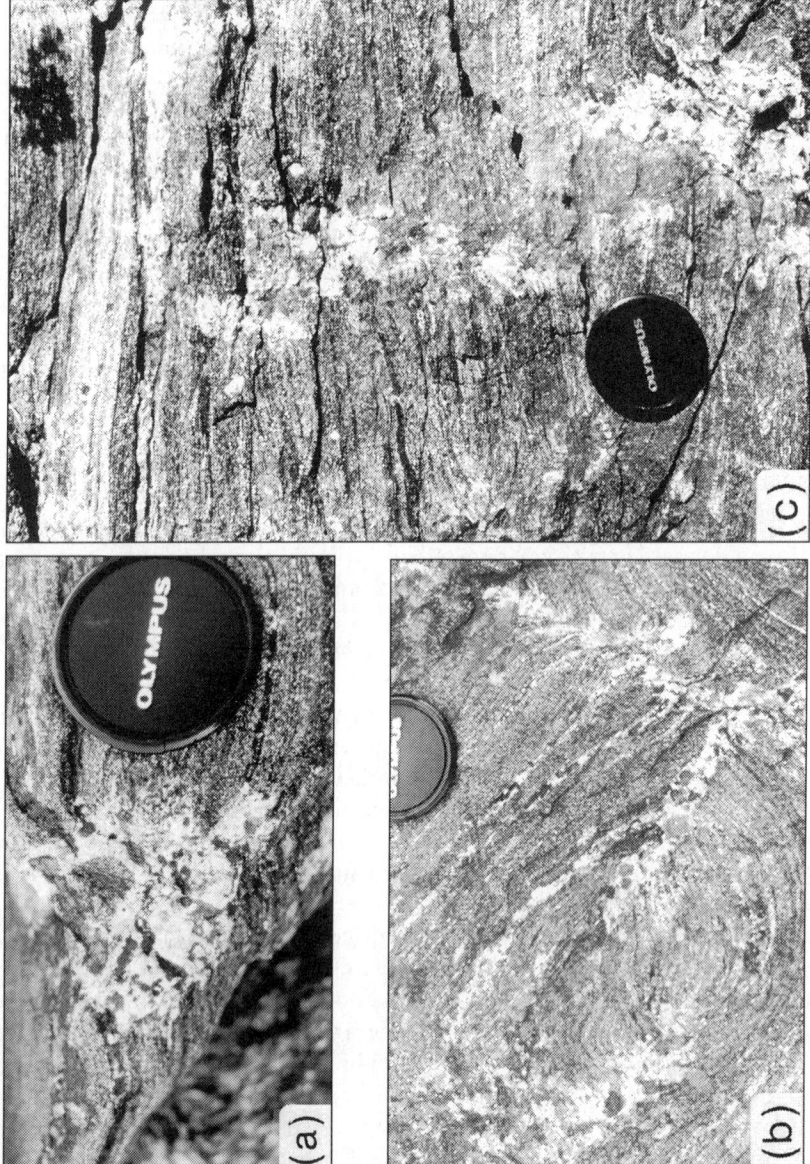

Figure 5.5 (a) Discordant volume of leucosome that contains large orthopyroxene crystals located in a partition within boudinaged fine-scale layered amphibolite. Boudinage amphibolite layers occur within predominantly metasedimentary migmatites of the Southern Brittany Metamorphic Belt, France. The origin of this texture/structure is discussed further in the text. (b) Veined amphibolite in which leucosome veins contains large orthopyroxene crystals, a reaction product of the incongruent melting reaction involving breakdown of hornblende; Southern Brittany Metamorphic Belt, France. (c) Vein array of leucosome within foliated amphibolite; Southern Brittany Metamorphic Belt, France.

124

the host amphibolite, suggests that a diffusion process was involved in the initial development of the texture, because continued growth of first-formed orthopyroxene was energetically favoured over nucleation of additional orthopyroxene crystals (cf. Powell and Downs, 1990). It is only when the melt fraction has exceeded some critical value that it can migrate along interboudin partition in the manner suggested for fluid flow by Oliver *et al.* (1990).

(c) Folds

The accumulation of melt in surfaces parallel to the axial planes of folds is illustrated in Figure 5.6 and has been widely reported from migmatite terrains (e.g. Edleman, 1973; Allibone and Norris, 1992; Hand and Dirks, 1992; Brown, 1994; Johnson *et al.*, 1994; Williams *et al.*, 1995; Collins and Sawyer, 1996). Commonly, melt associated with folds concentrates in the hinge zone, and occurs in greater abundance in antiformal rather than synformal hinges (Allibone and Norris, 1992; Collins and Sawyer, 1996). During progressive tightening of flexural flow folds, strain is partitioned preferentially into the limbs, so that melt is expected to migrate from limbs to hinges in response to deformation. In the example described by Allibone and Norris (1992), melt accumulation sites change from leucosome filled extensional shears and fractures that intersect millimetre-thick stromatic leucosomes on fold limbs to thicker and more abundant vein- and dyke-like leucosomes in the hinge region of the fold. In the case of folds formed at heterogeneities that disrupt the flow pattern during general non-coaxial progressive deformation, the axial plane essentially is parallel to a principal plane of strain that includes the principal elongation. At low effective confining pressure induced by high melt pressure during anatexis, this fabric may be the preferred migration pathway for the melt, which may explain the common location of leucosomes in the axial plane direction (cf. Brown and Solar, 1998).

The leucosomes described by Hand and Dirks (1992) are metre-scale bodies of granite that occur along the axial planes of macroscopic crenulation folds. They contain coarse-grained garnet and oxide phases as solid products of incongruent melting involving biotite breakdown. The absence of these peritectic phases outside the axial-planar leucosomes led Hand and Dirks to suggest that the reaction activation energy had overstepped sufficiently only in the high strain zones to allow melting to start in these sites. Once the peritectic products of the incongruent melting had nucleated, continued melting concentrated in the deformation zones because growth of the peritectic products at the initial sites of nucleation was energetically favoured over nucleating these phases elsewhere. Hand and Dirks (1992) argued that initial reaction probably produced grain-supported volumes of higher melt content in these high strain zones than in the surrounding matrix. In response to imposed strain rates and fluid pressures, diffusion caused elongation of the melt bodies in the plane normal to the maximum principal compressive stress, parallel to the axial planes of the folds. According to these authors, strain partitioning determined where melting

125

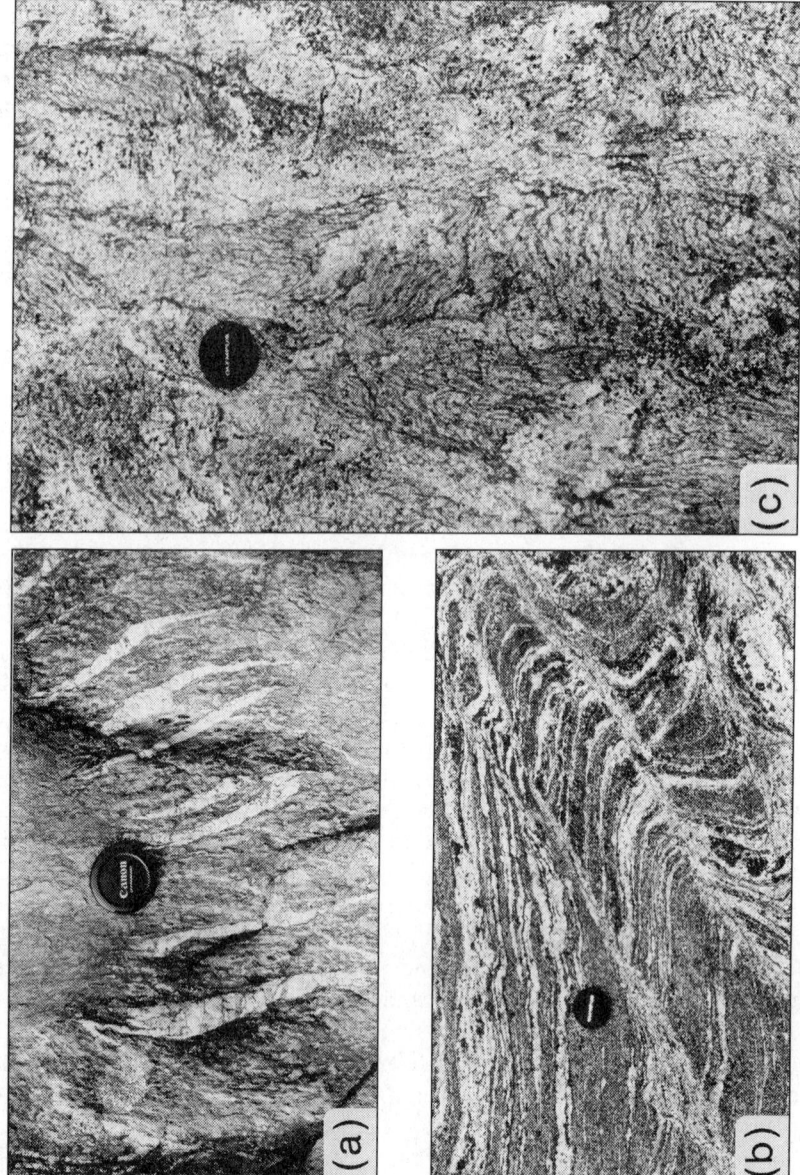

Figure 5.6 (a) Hinge of a mesoscale fold around which leucosomes occur approximately parallel to the crenulation cleavage in the metapelitic layer. In detail, leucosome tails are seen to have rotated into parallelism with this cleavage while some centres are transected by the cleavage, observations that suggest formation of the veins prior to or early during the folding event. Photograph provided by Scott Johnson, School of Earth Sciences, Macquarie University; Cooma Complex, south-eastern Australia. (b) Leucosome in shear zones parallel to axial surfaces of asymmetric folds of stromatic migmatites; Tolstik Peninsula, Karelia, Russia. (c) Sheets of fugitive leucosome parallel to fold axial surfaces; Omeo Complex, Southern Australia.

126

occurred and how melt dispersed along planar deformation bands to create a connected melt network. Hand and Dirks (1992, p. 601) describe a process of melt migration 'through a matrix by continuously dissolving, diffusing and redepositing the solid phase among which it moves', which they term 'diffusion-driven melt migration'. Once connected, axial planar leucosomes may permit melt transfer out of the system.

(d) Veins

Veins of leucosome are common in migmatites formed in isotropic protoliths, such as some orthogneisses and granulites (Figure 5.7). Some of these may represent melt-filled fractures. During extensional fracturing, local pressure gradients promote melt flow towards the fracture, providing a mechanism for segregation of melt (Shaw, 1980; Sleep, 1988; Rutter and Neumann, 1995; Rutter, this volume). Leucosome-filled fractures and vein networks also appear characteristic of melting processes in the innermost zones of contact metamorphic aureoles (e.g. Pattison and Harte, 1988; Symmes and Ferry, 1995), particularly in strongly anisotropic compositionally layered protoliths. In the Onawa aureole, Maine, USA, Symmes and Ferry (1995) describe connected networks of leucocratic veins that surround highly deformed blocks of hornfels. These vein networks are similar to amphibolitic migmatite described by Maaløe (1992), and to an example from the high-temperature migmatitic core to the southern Brittany metamorphic belt, France, which is shown in Figure 5.5(b). We suggest that such vein networks may form during the stepwise increase in melt fraction and increase in volume associated with hydrate breakdown melting reactions. Initially, melt accumulation may have been diffusion-controlled, as suggested by the large orthopyroxene crystals that represent a peritectic product of the incongruent hornblende breakdown melting reaction (Figure 5.5(b)). Additional factors that might be involved are high rates of strain and rapid melting (e.g. Connolly et al., 1997).

Stevenson (1989) showed theoretically that a partially molten rock subjected to applied differential stress is unstable. Melt may migrate by porous flow on the metre scale down small gradients in pressure to form veins parallel to the maximum principal compressive stress. The orientation of these veins is not controlled by original layering in the protolith, and can form in initially isotropic source materials. Initial variations in stress distribution determine their location, so that initial melting occurs at sites of reduced pressure, for an appropriate phase assemblage involving a hydrate breakdown melting reaction. Subsequently, these areas of greater melt fraction become loci of melt accumulation to produce veins. Given the expectation that melting reactions leading to large volume magma production will involve hydrate breakdown at low a_{H_2O}, this mode of vein formation may be common in geodynamic settings such as active circumoceanic continental margins where isotropic orthogneiss and granulite protoliths may be expected to occur (e.g. Figure 5.7).

Clemens and Mawer (1992), Rutter and Neumann (1995), Rutter (this

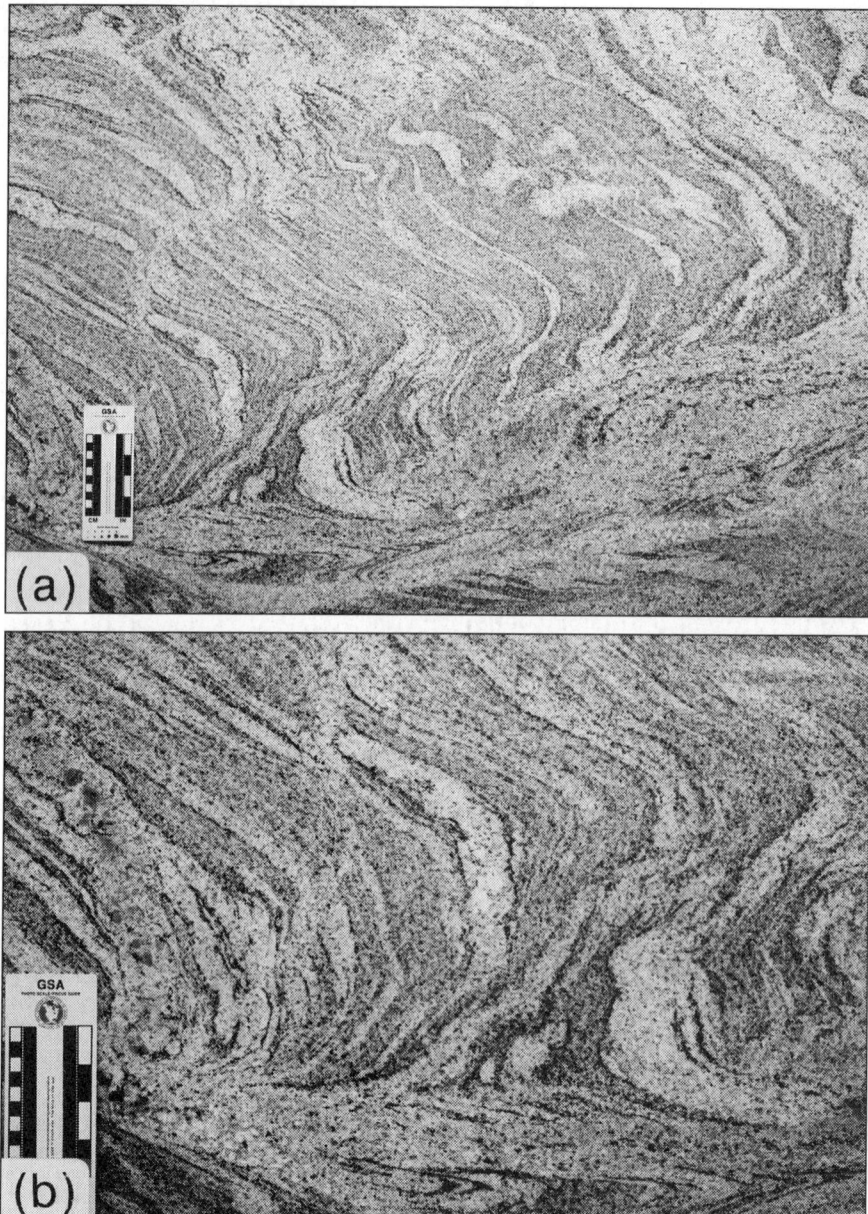

Figure 5.7 Stromatic metatexite migmatite formed from a relatively homogenous gneiss; Sri Lanka. (a) Folded leucosomes with thin marginal melanosomes are cut by discordant patches and veins of leucosome, and in the lower part of the photograph by diatexite in a shear zone. (b) Enlargement of (a) to show the relationship between the folded stroma and the diatexite at the edge of the shear zone. Photographs were provided by Richard S. Fiske and James F. Luhr, Department of Mineral Sciences, Smithsonian Institution/National Museum of Natural History.

volume) and Clemens, Petford and Mawer (this volume) appeal to collection of granite melt into vein networks and ascent to higher levels in the crust through dykes. In this vein network and dyke ascent model, melt is assumed to migrate short distances (hundreds of millimetres to several metres) by porous flow through intergrain pores to the nearest veins, from which it is drained by porous flow. As a general magma transfer mechanism, such a model is attractive.

(e) Shear zones

Commonly, at outcrop-scale, granite is associated with localized deformation in periodically spaced shear bands. At map-scale, some granite plutons are associated with localized deformation in crustal-scale shear zone systems; in such systems, plutons are located in low strain zones enclosed by anastomosing high strain zones (Brown and Solar, 1998). Furthermore, there is plenty of evidence that dilatancy is a common feature of naturally occurring shear zones in high-grade metamorphic belts (reviewed in Hobbs et al., 1990), and other evidence of dilatancy in shear zones comes from experiments (Fisher and Patterson, 1989; Ree, 1994). On the basis of these observations, shear zone systems may represent an important structural control on the movement and accumulation of granitic melt in the crust.

Shear zones in migmatite terrains are commonly sites of melt concentration (Figure 5.4). This implies enhanced permeability along the shear zone and a driving force for melt concentration in the shear zone. Permeability enhancement may occur simply as a function of increasing melt fraction during reaction and/or it may depend on active creation of syntectonic permeability. Cataclasis necessarily involves dilatancy, so that failure by melt-enhanced cataclastic flow involves enhanced permeability. The permeability may reflect a balance between porosity-creation (e.g. by volume change due to melting, intergranular sliding and melt-induced cataclasis) and porosity-reduction (e.g. by shear-induced compaction through intergranular sliding and melt-assisted diffusive mass transfer). Under high-temperature conditions close to the solidus, deforming rocks that contain small amounts of a volatile phase, or into which small amounts of a volatile phase infiltrate, are in a critical state for failure. In this condition the start of melting will cause local transient reduction in strength and may lead to shear instabilities (Casey, 1980). Even when a_{H_2O} is low and/or the rocks are 'dry', deforming rocks become critical and fail as the first melt is generated during hydrate phase breakdown. Under these circumstances, failure along shear instabilities, particularly by melt-enhanced cataclastic flow, leads to grain-size reduction and dilation, both of which result in a small but significant drop in pressure that encourages melting and starts a positive feedback loop between deformation and melting. A positive feedback mechanism in which melting lets the shear instabilities grow implies that the rate of deformation is controlled largely by the rate of melt migration along the shear zone to the propagating tip.

Sometimes shear zones appear to have been developed in already partially molten protoliths. Whether shear zones develop depends on the relative rates of solid matrix strain vs. melt segregation and migration (Rutter, 1995). If the solid matrix deforms more slowly than the rate of melt segregation, then the melt segregation rate will fall to compensate. In other words, the rate of deformation is controlled by the rate of matrix flow. This may be the common case for low a_{H_2O} volatile phase-absent melting, in which case ΔV_r generally is positive. This, in combination with deformation, may generate excess melt pore pressure to facilitate melt escape by mechanisms that involve melt-enhanced embrittlement. On the other hand, if the solid matrix deforms more quickly than the rate of melt segregation, then volumetric strain is limited by melt segregation rate and a greater distortional strain will result. Under this condition, melt-lubricated shear zones or more distributed flow may be expected to develop. Volumetric strain may be accommodated by melt loss through an anastomosing shear zone system. This may be the common situation for high a_{H_2O} volatile phase-present melting during which negative ΔV_r allows melt accumulation, although high synanatectic strain rates may lead to melt-enhanced embrittlement.

5.3.2   Initial leucosome segregation

In the presence of both a concentration and a pressure gradient, both diffusion and advection contribute to the mass flux which may result in the formation of segregated leucosomes. In this section we evaluate in a simple way the relative importance of advective and diffusive mass fluxes in partially molten protoliths. This evaluation is important because of the popular tendency to interpret granitic leucosomes located in dilatant sites as having migrated to the site rather than having accumulated there during melting at the site (e.g. Brown, 1994). Leucosomes that occur around the peritectic products of an incongruent melting reaction have been discussed by Powell and Downes (1990) as a diffusion-controlled texture. These authors suggested that such leucosomes form by a chemical rather than a physical process. In contrast, Williams et al. (1995) described leucosome formation in interboudin partitions associated with melt loss from strain shadow sites spatially associated with garnet crystals in amphibolite boudins adjacent to the partition, although garnets further away from the partition still retain melt frozen as leucosome in pressure shadow sites around garnet. Two examples are shown in Figure 5.8, in which there is no way to distinguish from simple field observations between flow of melt to the dilatant sites and leucosome growth in the dilatant site by diffusion. Although a full discussion of this issue is beyond the scope of this chapter, we wish to emphasize the point that melt flow must be shown and not assumed.

For segregation at low volume fraction of melt, Brown et al. (1995b) develop a mesoscopic-scale two-dimensional two-phase viscous flow cell model, with phase changes. This model predicts either expansion or contraction convection (depending on whether ΔV_r is positive or negative). Convection may lead to melt segregation to form a stromatic structure on a geologically reasonable time-scale. However, Brown et al. (1995b) did not consider fully the relative

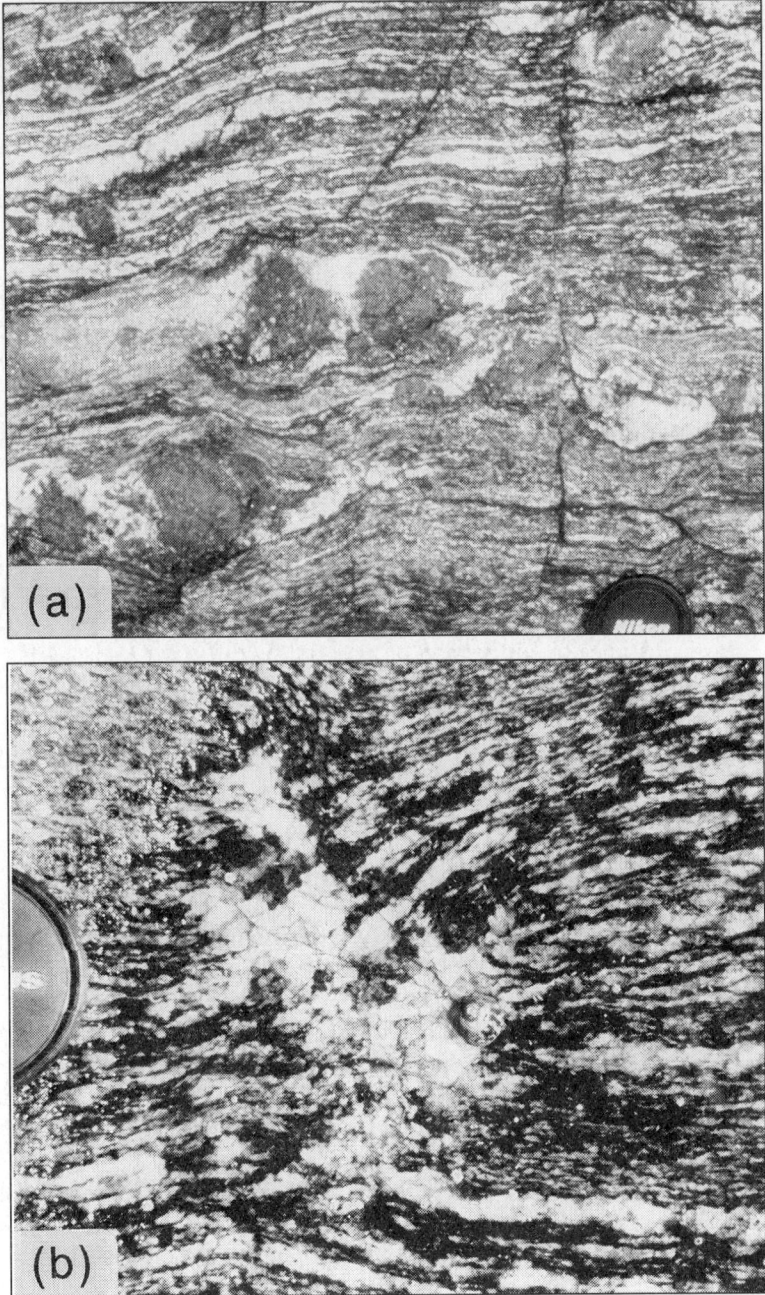

Figure 5.8 (a) Garnet-bearing leucosomes in stromatic migmatite, migmatitic core to the Southern Brittany Metamorphic Belt. (b) Vein of leucosome accumulated in a dilatant site due to shear failure during boudinage; migmatitic core to the Southern Brittany Metamorphic Belt, France.

roles of melt-enhanced diffusive mass transfer and advective mass transfer at these low melt fractions. During more rapid melting that leads to moderate volume fraction of melt, Brown *et al.* (1995b) propose that advective mass transfer may occur by filter pressing in compositionally layered rocks, in response to applied differential stress. Melt migration is by porous flow, driven by differences in mean normal stress between the layers, because of rheological differences, and shear-enhanced matrix collapse. Rates of segregation calculated by the authors are faster, and the model yields an adequate volume of leucosome.

Diffusion may dominate transport processes when convection is suppressed (e.g. when melt occurs as thin films in grain boundaries during early stages of melting; see Chakraborty, 1996). At very low melt fractions and T of $\sim 800\,°C$, assuming mass transfer occurs by two-phase flow driven by the ΔV_r of the melt-producing reactions (Brown *et al.*, 1995b) and diffusion (Chakraborty, 1996), the mass flux perpendicular to the flow direction is the sum of the advective and diffusive mass fluxes. The advective mass flux J_A is vC where v is the velocity vector of the melt phase and C is the concentration, and the diffusive mass flux J_D is $-DAC$, where D is the diffusion coefficient and A is the gradient in concentration C. The scale distance of a system characterized by particular values of D and v is D/v. Flow velocity v transports material in time t a distance L_A, and the characteristic distance for dispersion of material by diffusion L_D is $(Dt)^{0.5}$. The quotient of these two distances $v(t/D)^{0.5}$ shows that for constant velocity flow v and diffusion coefficient D, the distance material is transported by advection can become greater than the distance material is transported by diffusion after sufficient time (Lerman, 1979, pp. 59–60; although strictly a comparison of distances L_A and L_D is a measure of the relative effectiveness of the two processes operating independently). Given diffusivities for Si, the alkalis and Ca in granitic melts with $c.$ 3% H_2O at $800\,°C$ of $c.\,2\times10^{-8}\,m^2\,s^{-1}$ to $c.\,2\times10^{-13}\,m^2\,s^{-1}$ (Baker, 1991), then for melt velocities derived during two-phase flow by Brown *et al.*, (1995b), diffusive mass flux is more effective than advective mass flux at low melt fractions for length and time-scales evaluated by these authors.

At moderate melt fractions, the much faster advective mass flux implied by the higher strain rates involved in filter pressing by porous media flow may result in advective mass transfer overtaking diffusive mass transfer as the more effective process. We evaluate this using the Peclet number Pe, a dimensionless quotient of the total mass flux J and the grain size, and the diffusion coefficient and porosity (Lerman, 1979, p. 65). For fine-grained protoliths and the strain rates derived by Brown *et al.*, (1995b), values for the Peclet number may be <1, and diffusive mass transfer still may be the more effective process. Faster flow rates in coarse-grained protoliths yield values of Peclet number >>1 and advective mass transfer becomes the more effective process.

Other factors may be involved, such as rapid breakdown of coarse-grained hydrate phases during melting (Wolf and Wyllie, 1995). Such rapid breakdown may lead to the formation of low volume water-saturated melts with low viscosities that may promote hydraulic fracturing. Thus, transitional conditions may control the initial stages of melt segregation, and advective mass transfer may become the more effective process at lower melt fractions in such circumstances. As this simple analysis shows, we have a long way to go to quantify the relative efficacies of processes involved in the formation of leucosomes. The distinction is important because of the need to separate structural controls on melt formation from structural controls on melt migration.

5.4   Crustal-scale processes

5.4.1   *Granite generation in relation to tectonic setting*

(a) Continental orogens in collisional settings

Numerical models of the pressure–temperature–time evolution within continental orogens in collisional settings (e.g. Thompson and Connolly, 1995) suggest that temperatures >800 °C within continental crust may be attained only under particular tectonic circumstances, such as extensional collapse of a thickened orogen or by removal of part of the lithosphere by delamination (e.g. Wernicke *et al.*, 1996) or by underplating (Bergantz, 1989). Experimental investigation of low a_{H_2O} volatile phase-absent melting of quartz-saturated muscovite- and biotite-bearing materials at 1.0 GPa suggests that the volatile phase-absent solidus in common lower crustal protoliths is in the range 725–850 °C. Thus, in collisional settings, without volatile phase influx, melt fractions are likely to be low even in fertile metasedimentary protoliths, consistent with conclusions of petrogenetic studies of Himalayan leucogranites (Inger and Harris, 1993; Harris *et al.*, 1995). In collisional settings, the rates of deformation can be rapid and it is likely that synanatectic deformation allows segregation and migration of low melt fractions to occur. Also, such melts (≤10 vol.% melting) are the most likely to transfer to the upper crust due to their low viscosities (e.g. Holtz *et al.*, 1996; Scaillet *et al.*, 1996). However, as Sawyer (1996) points out, such melt compositions are not typical of upper crustal granites, and some process to generate and transfer larger volumes of granitic melt with entrained residual material is required.

(b) Active circumoceanic continental margins

Along active circumoceanic continental margins, magma source compositions, modes of melt segregation and the mechanism of magma ascent may be different from those in collisional settings. In the continental arc setting, anatexis of the lower crust is more likely to be initiated by underplating (Bergantz, 1989) or intraplating (Glazner and Ussler, 1988) of basaltic magmas, and the source materials may be an accretionary complex or pre-existing plutonic rocks. The high temperatures of melting achieved rapidly by advection of thermal energy in basalt may result in melt segregation processes dominated by hydraulic fracture

133

and/or melt-enhanced embrittlement, and pulses of variably mingled, multiply sourced magmas may ascend rapidly through the crust in dykes to be emplaced as plutons during construction of composite batholiths.

5.4.2 Geochemical signatures of segregated melts

Only by understanding the processes that operate during melt segregation, such as those discussed in this chapter, can we begin to unravel the geochemical data that are obtained from granites (Bea, 1991). Rates of melt segregation may determine whether the extracted granitic magma has compositions in equilibrium with its residues or not. Examples of evacuation of disequilibrium melts based on interpretations of geochemical data have been presented by Sawyer (1991), Srogi *et al.* (1993), Watt and Harley (1993) and Watt *et al.* (1996).

In tectonic environments in which the applied differential stress is very low, the ΔV_r of the melting reaction may be a significant control on melt retention or escape and on melt or magma chemistry. Brown *et al.* (1995b) suggested that the generally positive ΔV_r for low a_{H_2O} volatile phase-absent melting reactions under crustal pressures in the absence of deformation normally will promote melt escape. These authors argued that if the rate of melt production exceeds the rate of melt escape, then the increase in melt pressure will lead to melt-enhanced embrittlement and fracture, and promote melt migration out of the system. Deformation will tend to enhance the melt pore pressure due to the positive ΔV_r and enhance further the tendency for melt escape. In contrast, high a_{H_2O} volatile phase-present melt-producing reactions generally will result in melt retention after segregation, unless deformation generates melt escape pathways, because these reactions generally have a negative ΔV_r. The experimental work of Laporte and Watson (1995), discussed above, suggests that the formation of biotite-rich melanosomes along the margins of segregated leucosome may limit the growth of individual leucosome stromae. This supports the suggestion of Brown *et al.* (1995a) that evolution of stromatic structure may occur through anatectic erosion of mesosome by increasing the melt fraction *in situ* through formation of additional leucosome segregations. Further, these authors suggested that the increase of melt fraction *in situ* may lead to breakdown of the solid matrix, which might result in instability due to buoyancy and magma mobility with consequent entrainment of residual material. Of course, in tectonic environments characterized by synanatectic deformation, breakdown of the solid matrix may be deformation-enhanced at lower melt fractions.

High a_{H_2O} melting commonly occurs above the stability of muscovite but below the breakdown of biotite. The implications for magma chemistry are twofold. First, at low melt fraction, leucosome chemistry commonly mimics feldspar, and accessory minerals may be armoured from interaction with melt if they are located inside residual biotite (Watson *et al.*, 1989; Johannes *et al.*,

134

1995). Second, increasing melt fraction and loss of contiguity, whether deformation enhanced or not, results in mixtures of melt and residual phases that may undergo unmixing during ascent (Brown, 1979; Sawyer, 1996). Thus, diatexites mimic better their source chemistry (Brown, 1979; Brown and D'Lemos, 1991; Sawyer, 1996).

The physical process of melt segregation from residual material within ascending magma, and the geochemical consequences of these processes, have been investigated by Sawyer (1996). Sawyer distinguishes between: (i) simple melt segregation in the lower crust, where only the melt fraction migrates; (ii) magma mobility, in which all the melt and residual minerals migrate together; and (iii) magma mobility with melt segregation, in which melt and residual minerals move together as magma, but the melt becomes progressively separated from the residual minerals during flow. In the first case, segregation is so effective that resulting magmas are depleted in rare earth element contents and in FeO, MgO, Rb, Zr and Th contents. In the second case, no segregation occurs and the resulting magma will have the identical composition to its source. According to Sawyer (1996), only the third case produces magmas with chemical compositions comparable to those of granites, and he concludes that flow separation within migrating diatexite magmas represents the principal process occurring in the source regions of granites. However, flow separation without resorption or complete dissolution of entrained phases only represents an end-member in the spectrum of possible behaviour. Carrington and Watt (1995) contrasted the chemistry of restite-free low a_{H_2O} granitic leucosomes in granulite facies migmatites with restite-bearing diatexites and high-level crustally-derived granites. They point out that most amphibolite facies diatexites have REE patterns similar to those of peraluminous granites believed to have been derived by low a_{H_2O} volatile phase-absent melting under granulite facies conditions. Carrington and Watt (1995) suggest that low a_{H_2O} melts undergo modification during ascent, primarily by dissolution of entrained accessory phases and/or peritectic phases, to develop geochemical characteristics closer to granites. This process has been modelled for isentropic adiabatic ascent paths (Holtz and Johannes, 1994; Clemens, Petford and Mawer, this volume) and entrained solid phases can be expected to be resorbed or to dissolve completely during ascent and emplacement. An additional complexity is revealed by detailed investigation of REE patterns in zircons from the Manaslu granite and its potential source the Tibetan Slab migmatites (Barbey et al., 1995). This study confirms previous models in which the Manaslu granite was suggested to be an aggregation of several different batches of magma. Finally we note that Bea (1996) has summarized studies which suggest the REE and, to a lesser extent, U and Th contents of garnet and feldspar from granulite facies metapelites are different from those from amphibolite facies metapelites. Thus, anatexis of metapelites under granulite facies conditions might produce significant differences in the composition of melt in comparison with melt derived from metapelites under amphibolite facies conditions.

5.5 Summary

Crustal anatexis and crustal deformation are both extremely complex processes. They involve many interrelated factors that make it difficult to separate the dominant controls from those that merely influence the outcome. We have emphasized that processes may vary with geodynamic setting. Additionally, there may be significant differences between orogenic belts characterized by clockwise P–T–t paths and those characterized by anticlockwise P–T–t paths (Brown, 1993). The critical point is that the major period of melting in orogenic belts characterized by clockwise P–T–t paths may occur during decompression and will postdate contractional thickening but may be synchronous with tectonic exhumation, whereas in orogenic belts characterized by anticlockwise P–T–t paths major melting may occur synchronously with orogenic thickening close to peak T.

Hydraulic fracturing during melting of the lower crust may be more likely in cases where melt fraction increases rapidly, otherwise volume increase could be accommodated by distributed plastic deformation. In these circumstances, melt movement is largely fracture controlled. Melt fraction may increase rapidly when the melting reactions involve breakdown of a hydrate phase or if the solidus is overstepped significantly, which might occur in contact metamorphic environments or at a convergent plate margin when a ridge becomes subducted. In geodynamic settings that involve basalt underplating, for example along active circumoceanic continental margins, dyking may be the principal mechanism by which granite magma is transferred through the crust. Alternatively, infiltration of a volatile phase into dry rocks already significantly above the temperature of the wet solidus may result in rapid melting. One possible scenario is infiltration of a volatile phase synchronously with shearing. This may induce rapid melting and cataclastic deformation at high strain rates within the shear zone. In these circumstances, melt movement during synanatectic ductile deformation is predominantly by pervasive flow, but evidence of the cataclastic disaggregation of the rock mass may be erased during annealing. In thickened orogens characteristic of collisional geodynamic settings, there is a clearly demonstrated close relationship between magma transfer through the crust and major shear zone systems. We conclude that the process by which magma is transferred through the crust to be emplaced in high-level plutons varies according to geodynamic setting.

Based on observations presented earlier, we conclude that melt segregation at low melt fraction can be surprisingly effective as long as the matrix deforms as a solid-state rock. Doubt has been cast on whether uniformly pervasive, percolative flow of granitic melt can occur on length scales sufficiently large to allow plutonic volumes of granite to accumulate. The emerging view is one in which percolation of granitic melt through the crust might occur over sufficient lengths at short enough time-scales after an efficient, most probably deformation-produced, drainage network has been established. Drainage networks may vary in geometry with geodynamic setting and tectonic style, but the elements within any one setting may be self-similar. These networks may be composed of grain-scale pathways and structurally controlled channels. These deformation-assisted drainage networks are efficient, in the sense that once melt is connected at both the grain scale and the scale of the structurally controlled channels it may be evacuated rapidly, as evidenced by disequilibrium chemical compositions preserved in upper-crustal granites. The rheological behaviour of partially molten crust may be brittle or ductile; however, in both cases, applied differential stress generates pressure gradients that may drive melt extraction. Other controls that impinge upon melt segregation include the P–T–X conditions of melting and the bulk composition.

Dynamic experiments on samples of granite and amphibolite suggest three rheological regimes with increasing melt fraction. Under these circumstances, melt segregation is not controlled by melt fraction, and melt may be tectonically squeezed from the rock under most circumstances if strain rates are high enough. Increased pore pressure due to melting reactions with positive ΔV_r may lead to melt-enhanced embrittlement and formation of fracture and vein networks. These are more likely to occur in homogeneous orthogneiss and granulite protoliths and at faster heating and strain rates, such as might occur more commonly along active circumoceanic continental margins and in contact metamorphic environments. It is outside the scope of this chapter to consider how a vein network might generate the magma pressure sufficient to start a dyke. We are conscious, however, that as melt accumulates within a particular zone in the crust it becomes increasingly unstable gravitationally. Perturbations along the upper surface of a zone of melting within the crust may lead to melt migration into antiformal culminations that might, in turn, nucleate dykes or shear zones to transfer magma upward through the crust (cf. Weinberg, 1996).

There is an important distinction between melt-producing reactions with a negative ΔV_r of reaction and those that involve a positive ΔV_r of reaction. Melting at high a_{H_2O} generally involves a volume reduction, segregation may be by convection and filter pressing, and melt retention is expected, since the system pulls in the melt (Brown et al., 1995b). Melt fraction may increase until the contiguity of the solid framework breaks down and bulk magma mobility with restite entrainment may occur; deformation may drive out the melt or destroy the contiguity at lower melt fractions. Major crustal melting involves volatile phase-absent hydrate breakdown at low a_{H_2O}, which leads to an increase in volume that may promote segregation. Increasing volume may lead to pore fluid pressures high enough to permit melt-enhanced embrittlement, cataclastic flow and fracture; effectively the system is open to melt migration (Brown et al., 1995b). Field evidence is consistent with these observations. Metatexite migmatites evolve to diatexite migmatites through anatectic erosion of the mesosome, and excess melt in leucosomes commonly has congealed while draining through a connected network of structurally created pathways and dilatant sites to a higher structural level. However, we acknowledge that if strain

137

rates are rapid, then the effects of the deformation may overwhelm any effect that relates to positive or negative ΔV_r of reaction. Thus, deformation may serve to disrupt or control the movement of magma that is otherwise influenced by the development of reaction-enhanced permeability and buoyancy. Apparently multigenerational complex structures may develop at outcrop scale during single partial melting events because of the complex interplay between applied differential stress, mechanical heterogeneity, strain rate, ΔV_r of reaction and rock permeability.

Where does melting initiate? Under equilibrium conditions in an isotropic protolith, we expect melting to begin at multiphase grain junctions that include quartz and feldspar (e.g. Mehnert et al., 1973; Busch et al., 1974). The earth's crust is anisotropic and in a state of stress. We may expect slight variations in differential stress within any particular volume of crust. Also, we may expect variations in bulk composition and grain size that may influence the choice of sites where melting is initiated. Our interpretation of the experimental and field data is that the small variations in differential stress give rise to slightly decreased or increased pressures (mean stress) throughout any volume of the crust, which in turn initiate melting once the initial thermal overstep is close to the amount required to overcome the activation energy associated with the melting reaction. For high a_{H_2O} volatile phase-present melting, small increases in pressure may promote the start of reaction, whereas for low a_{H_2O} volatile phase-absent melting, small decreases in pressure may promote the start of reaction. Further, for low a_{H_2O} melting due to mica/amphibole breakdown, the product phases of the reaction may have difficulty nucleating so that once melting is established at a particular site, it is energetically favourable for melting to continue at that site and for the solid products to grow at that site. The process will be diffusion-controlled, at least at low melt fraction. Melt is likely to be connected in all common crustal protoliths in both non-deforming and deforming environments.

We expect that at melt fractions of c. 5–10 vol.% most protoliths will be permeable and that melt flow generally will occur down pressure gradients. A complex interplay between strain rate, rate of melt production, whether the ΔV_r is positive or negative, and the rate of volume change of the system determines how a structurally controlled permeable network may develop. At low strain rates, ΔV_r and rate of volume change may be important, whereas at high strain rates the effect of volume change may be overtaken by the effect of increasing melt pore pressure and melt-enhanced embrittlement. In metamorphic belts representative of the middle crust, evidently structurally linked dilation sites form a connected network through which melt can migrate. The role of shear zones in permitting melt migration, both at the outcrop scale and at the crustal scale, is probably significant and shear zones appear to have been the locus of magma transfer through the crust. We suggest that in zones of high strain and at fast strain rates, shear zones may nucleate once melting begins and that a complex positive feedback mechanism may be established in which increasing volume of melt promotes growth of the shear zone and deformation ensures that melt accumulates in the shear zone system. Although it is hard to separate cause from effect, the main permeable pathways through large volumes of the continental crust in collisional settings may be shear zones; usually these will be effective magma conduits only during active deformation.

The crustal protolith before anatexis may contain existing high strain zones in which grain size may have been reduced during earlier deformation. Activation energies associated with melting reactions are likely to be lower at finer grain sizes for kinetic reasons, and melting may commence preferentially in preexisting shear zones. Thus, there may be no general model in which either shear zones nucleate at sites of initial melting or melting begins in preexisting zones of high strain.

Acknowledgements

We acknowledge colleagues too many to mention individually with whom we have interacted and from whom we have learned, although remaining misconceptions are ours alone. Our ideas have been clarified by recent discussions with J. Connolly, J.-P. Burg, P.A. Candela, J. Clemens, R. Powell and R.H. Vernon. We thank J.-P. Burg, J.C., M. Holness, N. Oliver, P.M. Piccoli, G.S. Solar, R.H. Vernon and G. Watt for rapid critical reviews of the manuscript. Editing by M. Holness significantly improved the text. Finally, the word processing support of J. Martin is greatly appreciated.

References

Allibone, A.H. and Norris, R.J. (1992) Segregation of leucogranite microplutons during synanatectic deformation: An example from the Taylor Valley, Antarctica. *J. Metamorphic Geol.*, **10**, 589–600.

Arzi, A.A. (1978) Critical phenomena in the rheology of partially-molten rocks. *Tectonophysics*, **44**, 173–84.

Baker, D.R. (1991) Interdiffusion of hydrous dacitic and rhyolitic melts and the efficacy of rhyolite contamination of dacitic enclaves. *Contrib. Mineral. Petrol.*, **106**, 462–73.

Barbey, P., Allé, P., Brouand, M. and Albarède, F. (1995) Rare-earth patterns in zircons from the Manaslu granite and Tibetan Slab migmatites (Himalaya): insights in the origin and evolution of a crustally-derived magma. *Chem. Geol.*, **125**, 1–17.

Bea, F. (1991) Geochemical modeling of low melt-fraction anatexis in a peraluminous system: The Peña Negra Complex (central Spain). *Geochim. Cosmochim. Acta*, **55**, 1859–74.

Bea, F. (1996) Controls on the trace-element composition of crustal melts. *Trans. Roy. Soc. Edinburgh: Earth Sciences*, **87**, 33–41.

Bergantz, G.W. (1989) Underplating and partial melting: Implications for melt generation and extraction. *Science*, **245**, 1093–5.

Brown, M. (1973) The definition of metatexis, diatexis and migmatite. *Proceedings of the Geologists' Association*, **84**, 371–82.

Brown, M. (1974) The petrogenesis of the St. Malo Migmatite Belt, North-Eastern Brittany, France. Unpublished Ph.D. thesis, University of Keele, UK.

Brown, M. (1979) The petrogenesis of the St. Malo migmatite belt, Armorican Massif, France, with particular reference to the diatexites. *Neues Jahrb. Mineral. Abh.*, **135**, 48–74.

Brown, M. (1993) *P-T-t* evolution of orogenic belts and the causes of regional metamorphism. *J. Geol. Soc., London*, **150**, 227–41.

Brown, M. (1994) The generation, segregation, ascent and emplacement of granite magma: The migmatite-to-crustally-derived granite connection in thickened orogens. *Earth-Sci. Rev.*, **36**, 83–130.

Brown, M. and D'Lemos, R.S. (1991) The Cadomian granites of Mancellia, north-east Armorican Massif of France: Relationship to the St. Malo migmatite belt, petrogenesis and tectonic setting. *Precambrian Res.*, **51**, 393–427.

Brown, M. and Solar, G.S. (1998) Shear zones and melts: positive feedback in orogenic belts. *J. Struct. Geol.* (in review).

Brown, M., Rushmer, T. and Sawyer, E. (1995a) Introduction to special section: mechanisms and consequences of melt segregation from crustal protoliths. *J. Geophys. Res.*, **100**, 15551–63.

Brown, M., Averkin, Y., McLellan, E. and Sawyer, E. (1995b) Melt segregation in migmatites. *J. Geophys. Res.*, **100**, 15655–79.

Brun, J.-P. and Martin, H. (1978) Relations métamorphisme-déformation au cours de l'evolution. dynamique d'un dôme migmatique: le massif de Saint-Malo (France). *Bull. Soc. géol. France*, 7(xx), 91–101.

Büsch, W., Schneider, G. and Mehnert, K.R. (1974) Initial melting at grain boundaries. Part II: Melting in rocks of granodioritic, quartzdioritic and tonalitic composition. *Neues Jahrb. Mineral. Abh.*, **8**, 345–70.

Carrington, D.P. and Watt, G.R. (1995) A geochemical and experimental study of the role of K-feldspar during water-undersaturated melting of metapelites. *Chem. Geol.* **122**, 56–79.

Carson, C.J., Powell, R., Wilson, C.J.L. and Dirks, P.H.H.M. (1997). Partial melting during tectonic exhumation of a granulite terrane: an example from the Larsemann Hills, East Antarctica. *J. Metamorphic Geol.*, **15**, 105–26.

Casey, M. (1980). Mechanics of shear zones in isotropic dilatant materials. *J. Struct. Geol.*. **2**, 143–7.

Chakraborty, S. (1996) Diffusion in silicate melt, in *Structure, Dynamics and Properties of Silicate Melts* (eds J.F. Stebbins, P.F. McMillan and D.B. Dingwell), Mineralogical Society of America, *Reviews in Mineralogy*, **32**, 411–503.

Clemens, J.D. and Mawer, C.K. (1992) Granitic magma transport by fracture propagation. *Tectonophysics*, **204**, 339–60.

Clemens, J.D., Petford, N. and Mawer, C.K. (this volume) Ascent mechanisms of granitic magmas: causes and consequences.

Collins, W.J. and Sawyer, E.W. (1996) Pervasive magma transfer through the lower-middle crust during non-coaxial compressional deformation: An alternative to diking. *J. Metamorphic Geol.*, **14**, 565–79.

Connolly, J.A.D., Holness, M.B., Rushmer, T. and Rubie, D.C. (1997) Reaction-inducted micro-cracking: An experimental investigation of a mechanism for enhancing anatectic melt extraction. *Geology*, **25** (in press).

Davidson, C., Schmid, S.M. and Hollister, L.S. (1994) Role of melt during deformation in the deep crust, *TERRA Nova*, **6**, 133–42.

D'Lemos, R.S., Brown, M. and Strachan, R.A. (1992) The relationship between granite and shear zones: Magma generation, ascent and emplacement within a transpressional orogen. *J. Geol. Soc., London*, **149**, 487–90.

Edleman, N. (1973) Tension cracks parallel with the axial plane. *Bull. Geol. Soc. Finland*, **45**, 61–5.

Ellis, D.J. and Obata, M. (1992) Migmatite and melt segregation at Cooma, New South Wales. *Trans. Roy. Soc. Edinburgh: Earth Sciences*, **83**, 95–106.

REFERENCES

Fisher, G.J. and Patterson, M.S. (1989) Dilatancy during rock deformation at high temperatures and pressures. *J. Geophys. Res.*, **94**, 17607–17.

Gapais, D., van den Driessche, J., Brun, J.P. and Richer, C. (1995) Pervasive ductile spreading and boudinage in extending orogens: Evidence from the French Variscan Belt. *TERRA Nova*, **7**, 121.

Glazner, A.F. and Ussler III, W. (1988) Trapping of magma at midcrustal density discontinuities. *Geophys. Res. Lett.*, **15** (7), 673–5.

Greenfield, J.E., Clarke, G.L., Bland, M. and Clark, D.J. (1996) In-situ migmatite and hybrid diatexite at Mt. Stafford, central Australia. *J. Metamorphic Geol.*, **14**, 413–26.

Grocott, J. and Wilson (this volume) Ascent and emplacement of granitic plutonic complexes in subduction-related extensional environments.

Grocott, J., Brown, M., Dallmeyer, R.D. *et al.* (1994) Mechanisms of continental growth in extensional arcs: An example from the Andean Plate-Boundary Zone. *Geology*, **22**, 391–4.

Hand, M. and Dirks, P.H.G.M. (1992) The influence of deformation on the formation of axial-planar leucosomes and the segregation of small melt bodies within the migmatitic Napperby Gneiss, Central Australia. *J. Struct. Geol.*, **14**, 591–604.

Harris, N., Ayres, M. and Massey, J. (1995) Geochemistry of granitic melts produced during the incongruent melting of muscovite: Implications for the extraction of Himalayan leucogranite magmas. *J. Geophys. Res.*, **100**, 15767–78.

Hartel, T.H.D. and Pattison, D.R.M. (1996) Genesis of the Kapuskasing (Ontario) migmatitic mafic granulites by dehydration melting of amphibole: The importance of quartz to reaction progress. *J. Metamorphic Geol.*, **14**, 591–611.

Hobbs, B.E., Mühlhaus, H.-B. and Ord, A. 1990. Instability, softening and localization of deformation, in *Deformation mechanisms, rheology and tectonics* (eds R.J. Knipe and E.H. Rutter), *Geol. Soc. Spec. Pub., London*, **54**, 143–65.

Holness, M.B. (this volume) The permeability of non-deforming rock.

Holtz, F. and Johannes, W. (1994) Maximum and minimum water contents of granitic melts implications for chemical and physical properties of ascending magmas. *Lithos*, **32**, 149–59.

Holtz, F., Scaillet, B., Behrens, H. *et al.* (1996) Water contents of felsic melts: Application to the rheological properties of granitic magmas. *Trans. Roy. Soc. Edinburgh: Earth Science*, **87**, 57–64.

Inger, S. and Harris, N.B.W. (1993) Geochemical constraints on leucogranite magmatism in the Langtang Valley, Nepal Himalaya. *J. Petrol.*, **34**, 345–68.

Johannes, W. (1988) What controls partial melting in migmatites? *J. Metamorphic Geol.*, **6**, 451–65.

Johannes, W., Holtz, F. and Möller, P. (1995) REE distribution in some layered migmatites: constraints on their petrogenesis. *Lithos*, **35**, 139–52.

Johnson, S.E., Vernon, R.H. and Hobbs, B.E. (1994) Deformation and metamorphism of the Cooma Complex, southeastern Australia. Geological Society of Australia, Specialist Group in Tectonics and Structural Geology, Field Guide No. 4, 1–89.

Jurewicz, S.R. and Watson, E.B. (1984) Distribution of partial melt in a felsic system: the importance of surface energy. *Contrib. Mineral. Petrol.*, **85**, 25–9.

Jurewicz, S.R. and Watson, E.B. (1985) The distribution of partial melt in a granitic system: The application of liquid phase sintering theory. *Geochim. Cosmochim. Acta*, **49**, 1109–22.

Laporte, D. (1994) Wetting behavior of partial melts during crustal anatexis: the distribution of hydrous silicic melts in polycrystalline aggregates of quartz. *Contrib. Mineral. Petrol.*, **116**, 489–99.

Laporte, D. and Watson, E.B. (1995) Experimental and theoretical constraints on melt distribution in crustal sources: the effect of crystalline anisotropy on melt interconnectivity. *Chem. Geol.*, **124**, 161–84.

Lerman, A. (1979) *Geochemical Processes: Water and Sediment Environments*, Wiley, New York.

141

Lindh, A. and Wahlgren, C. (1985) Migmatite formation at subsolidus conditions – an alternative to anatexis. *J. Metamorphic Geol.*, **3**, 1–12.

Lupulescu, A. and Watson, E.B. (1995) Tonalitic melt connectivity at low-melt fraction in a mafic crustal protolith. *EOS, Trans.*, AGU, **76**, 299–300.

McKenzie, D.P. (1984). The generation and compaction of partially molten rock. *J. Petrol.*, **25**, 713–65.

McLellan, E.L. (1988) Migmatite structures in the Central Gneiss Complex, Boca de Quadra, Alaska. *J. Metamorphic Geol.*, **6**, 517–42.

McLellan, E.L. (1989) Sequential formation of sub-solidus and anatectic migmatites in response to thermal evolution, eastern Scotland. *J. Geol.*, **97**, 165–82.

Maaløe, S. (1992) Melting and diffusion processes in closed-system migmatization. *J. Metamorphic Geol.*, **10**, 503–16.

Mancktelow, N. (1995) Deviatoric stress and the interplay between deformation and metamorphism. *Geological Society of Australia Specialist Group in Tectonics and Structural Geology, Abstracts*, **40**, 95–6.

Mehnert, K.R., Büsch, W. and Schneider, G. (1973) Initial melting at grain boundaries of quartz and feldspar in gneisses and granulites. *Neues Jahrb. Mineral. Abh.*, **4**, 165–83.

Miller, C.F., Watson, E.B. and Harrison, T.M. (1988) Perspectives on source, segregation and transport of granitic magmas. *Trans. Roy. Soc. Edinburgh: Earth Sciences*, **79**, 135–56.

Nyman, M.W., Pattison, D.R.M. and Ghent, E.D (1995) Melt extraction during formation of K-feldspar + sillimanite migmatites, west of Revelstoke, British Columbia. *J. Petrol.*, **36**, 351–72.

Oliver, N.H.S. and Barr, T.D. (1997) The geometry and evolution of magma pathways through migmatites of the Halls Creek Orogen, western Australia. *Min. Mag.* (in press).

Oliver, N.H.S., Valenta, R.K. and Wall, V.J. (1990) The effect of heterogeneous stress and strain on metamorphic fluid flow, Mary Kathleen, Australia, and a model for large-scale fluid circulation. *J. Metamorphic Geol.*, **8**, 311–31.

Ord, A. (1990) Mechanical controls on dilatant shear zones, in *Deformation Mechanisms, Rheology and Tectonics* (eds R.J. Knipe and E.H. Rutter), Geol. Soc. Spec. Publ., London, **54**, 183–92.

Paquet, J. and François, P. (1980) Experimental deformation of partially melted granitic rocks at 600–900 °C and 250 MPa confining pressure. *Tectonophysics*, **68**, 131–46.

Paquet, J., François, P. and Nedelec, A. (1981) Effect of partial melting on rock deformation: Experimental and natural evidences on rocks of granitic compositions. *Tectonophysics*, **78**, 545–65.

Paterson, S.R., Fowler, T.K., Jr. and Miller, R.B. (1996) Pluton emplacement in arcs: A crustal-scale exchange process. *Trans. Roy. Soc. Edinburgh: Earth Sciences*, **87**, 115–23.

Pattison, D.R.M. and Harte, B. (1988) Evolution of structurally contrasting anatectic migmatites in the 3-kbar Ballachulish aureole, Scotland. *J. Metamorphic Geol.*, **6**, 475–94.

Petford, N. (1996) Dykes or diapirs? *Trans. Roy. Soc. Edinburgh: Earth Sciences*, **87**, 105–14.

Pitcher, W. (1979) The nature of ascent and emplacement of granitic magmas. *J. Geol. Soc., London*, **136**, 627–62.

Powell, R. and Downes, J. (1990) Gamet porphyroblast-bearing leucosomes in metapelites: Mechanisms, phase diagrams, and an example from Broken Hill, Australia, in *High-temperature Metamorphism and Crustal Anatexis* (eds J.R. Ashworth and M. Brown), Unwin Hyman, London, pp. 105–23.

Ree, J.-H. (1994) Grain boundary sliding and development of grain boundary openings in experimentally deformed octachloropropane. *J. Struct. Geol.*, **16**, 403–18.

Robin, P.-Y.F. (1979) Theory of metamorphic segregation and related processes. *Geochim. Cosmochim. Acta*, **43**, 1587–600.

Rudnick, R.L. (1995) Making continental crust. *Nature*, **378**, 571–8.

REFERENCES

Rushmer, T. (1995) An experimental deformation study of partially molten amphibolite: Application to low-fraction melt segregation. *J. Geophys. Res.*, **100**, 15681–96.

Rushmer, T. (1996) Melt segregation in the lower crust: How have experiments helped us? *Trans. Roy. Soc. Edinburgh: Earth Sciences*, **87**, 73–83.

Rushmer, T., Rubie, D.C. and Connolly, J.A.D. (1995) Melt-induced fracturing as a function of rate and time: Implications for melt migration at the onset of reaction. *GSA Annual Meeting, Abst. with Programs*, **27**, 431.

Rutter, E.H. (1995) The role of deformation in the extraction of granitic melts from crustal protoliths, in *Deformation-enhanced Melt Segregation and Metamorphic Fluid Transport*, Abst. with Program.

Rutter, E.H (this volume) The influence of deformation on the extraction of crustal melts: a consideration of the role of melt-assisted granular flow.

Rutter, E. and Neumann, D. (1995) Experimental deformation of partially molten Westerly granite under fluid-absent conditions, with implications for the extraction of granitic magmas. *J. Geophys. Res.*, **100**, 15697–715.

Sawyer, E.W. (1991) Disequilibrium melting and the rate of melt-residuum separation during migmatization of mafic rocks from the Grenville Front, Quebec. *J. Petrol.*, **32**, 701–38.

Sawyer, E.W. (1994) Melt segregation in the continental crust. *Geology*, **22**, 1019–22.

Sawyer, E.W. (1996) Melt-segregation and magma flow in migmatites – Implications for the generation of granite magmas. *Trans. Roy. Soc. Edinburgh: Earth Sciences*, **87**, 85–94.

Scaillet, B., Holtz, F., Pichavant, M. and Schmidt, M. 1996. Viscosity of Himalayan leucogranites: Implications for mechanisms of granitic magma ascent. *J. Geophys. Res.*, **101**, 27691–9.

Schmid, S. (1992) Emplacement of granitoids in shear zones, in *Petrological and Structural Analysis of Plutonic Complexes: Reports of the Theoretical Lectures and Case Histories. IV Summer School*, Università degli Studi di Siena, Siena, Italy, pp. 33–7.

Shaw, H.R. (1980) Fracture mechanisms of magma transport from the mantle to the surface, in *Physics of Magmatic Processes*, (ed. R.B. Hargraves), Princeton University Press, Princeton, New Jersey, 201–64.

Sleep, N.H. (1988) Tapping of melt by veins and dikes. *J. Geophys. Res.*, **93**, 10255–72.

Srogi, L., Wagner, M.E. and Lutz, T.M. (1993) Dehydration partial melting and disequilibrium in the granulite-facies Wilmington Complex, Pennsylvania-Delaware Piedmont. *Am. J. Sci.*, **293**, 405–62.

Stevenson, D.J. (1989) Spontaneous small-scale melt segregation in partial melts undergoing deformation. *Geophys. Res. Lett.*, **16**, 1067–70.

Stromgard, K.E. (1973) Stress distribution during formation of boudinage and pressure shadows. *Tectonophysics*, **16**, 215–48.

Symmes, G.H. and Ferry, J.M. (1995) Metamorphism, fluid flow and partial melting in pelitic rocks from the Onawa contact aureole, central Maine, USA. *J. Petrol.*, **36**, 587–612.

Thompson, A.B. and Connolly, J.A.D. (1995) Melting of the continental crust: Some thermal and petrological constraints on anatexis in continental collision zones and other tectonic settings. *J. Geophys. Res.*, **100**, 15565–79.

Van der Molen, I. (1985) Interlayer material transport during layer-normal shortening, II, boudinage, pinch-and-swell and migmatite at Søndre Strømfjord Airport, west Greenland. *Tectonophysics*, **115**, 275–95.

Van der Molen, I. and Paterson, M.S. (1979) Experimental deformation of partially-melted granite. *Contrib. Mineral. Petrol.*, **98**, 7–22.

Vernon, R.H. and Collins, W.J. (1988) Igneous microstructures in migmatites. *Geology*, **16**, 1126–9.

Watson, E.B., Vicenzi, E.P. and Rapp, R.P. (1989) Inclusion/host relations involving accessory minerals in high-grade metamorphic and anatectic rocks. *Contrib. Mineral. Petrol.* **101**, 220–31.

143

Watt, G.R. and Harley, S.L. (1993) Accessory phase controls on the geochemistry of crustal melts and restites produced by dehydration melting. *Contrib. Mineral. Petrol.*, **114**, 550–66.

Watt, G.R., Burns, I.M. and Graham, G.A. (1996) Chemical characteristics of migmatites: Accessory phase distribution and evidence for fast melt segregation rates. *Contrib. Mineral. Petrol.*, **125**, 100–11.

Weber, C., Barbey, P., Cuney, M. and Martin, H. (1985) Trace element behavior during migmatization. Evidence for a complex melt-residuum-fluid interaction in the St. Malo migmatitic dome (France). *Contrib. Mineral. Petrol.*, **90**, 52–62.

Weinberg, R.F. (1996) The ascent mechanism of felsic magmas: News and views. *Trans. Roy. Soc. of Edinburgh: Earth Sciences*, **87**, 95–103.

Wernicke, B., Clayton, R., Ducca, M. *et al.* (1996) Origin of high mountains in the Continents: The southern Sierra Nevada. *Science*, **271**, 190–3.

White, A.J.R. and Chappell, B.W. (1990) Per migma ad magma downunder. *Geol. Jour.*, **25**, 221–5.

Wickham, S.M. (1987) The segregation and emplacement of granitic magmas. *J. Geol. Soc., London*, **144**, 281–97.

Williams, M.L., Hanmer, S., Kopf, C. and Darrach, M. (1995) Syntectonic generation and segregation of tonalitic melts from amphibolite dikes in the lower crust, Striding-Athabasca mylonite zone, northern Saskatchewan. *J. Geophys. Res.*, **100**, 15717–34.

Wilson, J.T. (1963) Hypothesis of earth's behavior. *Nature*, **198**, 925–9.

Wolf, M. and Wyllie, P.J. (1995) Liquid segregation parameters from amphibolite dehydration-melting experiments. *J. Geophys. Res.*, **100**, 15611–21.

CHAPTER SIX

Ascent mechanisms of granitic magmas: causes and consequences

J.D. Clemens, N. Petford and C.K. Mawer

A theory has only the alternative of being right or wrong. A model has a third possibility – it might be right, but irrelevant.

M. Eigen (1973, p. 618)

6.1 Introduction

The generation of most tonalite/dacite to alkali-feldspar granite/rhyolite (granitic) magmas involves major material input from the continental crust. This usually occurs through partial melting of a range of common crustal rock types, at pressures and temperatures characteristic of granulite-facies metamorphism (Read, 1948; Fyfe 1973a, 1973b; Wyllie *et al.*, 1976; White and Chappell, 1977; Conrad *et al.*, 1988; Thompson, 1990; Vielzeuf *et al.*, 1990). Phase equilibrium, petrological and geochemical evidence suggests that crustal melting normally occurs under fluid-absent conditions (Burnham, 1967; Brown, 1970; Clemens, 1990). Experiments and modelling have demonstrated that, under these circumstances, source rocks like metapelites, metagreywackes, meta-andesites and some amphibolites can yield up to 50 vol.% of H_2O-undersaturated granitic melt (Clemens and Wall, 1981; Clemens, 1984; Clemens and Vielzeuf, 1987; Conrad *et al.*, 1988; Rutter and Wyllie, 1988; Beard and Lofgren, 1991; Rapp *et al.*, 1991; Rushmer, 1991; Wolf and Wyllie, 1991). This melt is regarded as the primary material component of large-volume granitic magmas in a variety of tectonic environments.

Field studies, deformation experiments and modelling are being applied to understand how this partial melt may segregate from its restite or solid residue (e.g. Robin, 1979; van der Molen and Paterson, 1979; Paquet and François, 1980; Jurewicz and Watson, 1985; Nicolas, 1986; Lowell and Bergantz, 1987; McKenzie, 1987; Ribe, 1987; Turcotte, 1987; Dell'Angelo and Tullis, 1988;

Deformation-enhanced Fluid Transport in the Earth's Crust and Mantle. Edited by M.B. Holness. Published in 1997 by Chapman & Hall, London. ISBN 0 412 75290 5.

Sleep, 1988; Fountain *et al.*, 1989; Mawer, 1989; Stevenson, 1989; Cooper, 1990; Sawyer, 1991; Hand and Dirks, 1992; Maaløe, 1992; Harte *et al.*, 1993; Brown, 1994; Brown *et al.*, 1995; Laporte and Watson, 1995; Petford, 1995; Rushmer, 1995; Rutter and Neumann, 1995; Brown and Rushmer, this volume; Rutter, this volume). It seems that a wide variety of mechanisms may be involved, prominent among which are fracturing and compaction, but the exact controls are not yet clear. What *is* clear is that segregation does occur, that it can occur very rapidly and efficiently (Sawyer, 1991, 1994), and that this represents the birth of a granite. All of these processes operate mostly at deep crustal levels (20 to 40 km), yet we observe a great number of granitic batholiths and silicic volcanics that have been emplaced in the shallow crust (0 to 10 km). Thus, granitic magmas will commonly have traversed 10 to 40 km of crust to arrive at their emplacement sites.

6.2 Definitions and philosophy

At this point, it is critical to clearly distinguish between the processes of magma ascent and magma emplacement. Though the distinction between these processes is appreciated by some (e.g. Speer *et al.*, 1994; Vigneresse, 1995), recent exchanges on the Internet have very clearly demonstrated that the geological community labours under a considerable degree of confusion over the distinction. In proposing fracture-controlled ascent of granitic magmas, Clemens and Mawer (1992) and Petford *et al.* (1993) did not advocate emplacement as either dyke swarms or as sheeted complexes. The fact that such an interpretation has been read into these works shows how the mind may tenaciously cling to a misconception. We consider this point important enough to warrant separate discussion.

Perhaps the key to appreciating the difference between ascent and emplacement is the realization that overall magma transport distances are tens of kilometres, while large plutons are commonly nowhere near this in vertical extent (e.g. Hamilton and Myers, 1967; Neilson *et al.*, 1976; Sweeney, 1975, 1976; Lynn *et al.*, 1981; Hodge *et al.*, 1982; Vigneresse, 1988, 1990; Jébrak *et al.*, 1991). Emplacement involves formation of a pluton (or eruption) and is the mechanism (or mechanisms) by which the accumulated volume of magma is accommodated at its final resting place (in or on the crust). Ascent is the mechanism (or mechanisms) by which the magma traverses the tens of kilometres between its generation level and its emplacement level. With the possible exception of diapirism, there is no reason to suppose that ascent and emplacement must occur either by the same mechanism or at the same volumetric rate.

One of the most powerful ways of expressing a relationship in an unambiguous way is to use the formal language of the predicate calculus. The predicates of interest here (defined by capital letters) are ascent (A) and emplacement (E), with granitic magma defined as G_X. By introducing quantifiers we obtain the following axiomatic formulae:

$$(\exists_X)(G_X)$$
$$(\exists_X)(A_X \,\&\, E_X)$$
$$(\forall_X)[A_X \,\&\, E_X \,\&\, \neg(A_X = E_X)]$$

that state in an unambiguous way our contention that the ascent of granitic magmas is a discrete process independent of the emplacement of granitic magmas.

In thinking about granitic magmas, we might reflect on how we think about the migration of basaltic magmas and supercritical fluids, and ask ourselves: 'Why should we not apply similar thinking to all of these materials?'

6.3 Historical development of competing models

At present there are two competing and largely incompatible models for the ascent mechanisms of granitic magmas. Before the existence of high-resolution seismic surveys, granitic batholiths (from the Greek *bathos* = deep and *lithos* = rock) were usually pictured as vertically extensive bodies, continuing from outcrop to deep crustal levels. According to Cloos (1923) the batholith concept 'binds the largest portion of the Earth's crust directly to its great depths'. Geological cross-sections usually showed granitic bodies continuing to the lowest projected subsurface levels, commonly broadening as they did so. This view may have been partly inherited from the early Neptunian conception of granites as the oldest and deepest rocks of the Earth, precipitated first from the primordial ocean. This mindset possibly led later magmatists to imagine granitic magmas ascending through the crust as huge blobs of partially molten rock. Transformist petrology may well have assisted here, by introducing the notion of *in-situ* formation of batholiths, and suggesting that the commonly rounded pluton shape is, and always was, the shape that the granite took.

Early this century, prior to 1915, geometrical analogies were drawn with salt domes, and the concept of granitic diapirism was born (Mrazec, 1915; Nicolesco, 1929). Here it is important to realize that the early geometric analogies amounted to no more than the observation that intrusions (commonly dykes) of basalt or granite produced anticlinal structures in the clastic and evaporitic rocks that they had pierced. Such was the very loose definition of diapirism in those times. There was no detailed analysis of the structures either within the so-called 'diapir' or in its host rocks, or of patterns of schistosity development. Up-warping of the surrounding rocks was sufficient cause to apply the term diapirism. Indeed, the early descriptions of 'diapirs', formed by the intrusion of mafic dykes in Madagascar (Nicolesco, 1929) do not record the formation of any kind of foliation in the penetrated Mesozoic sediments. Marginal faulting, regional compression and the emplacement of the igneous bodies roughly along anticlinal axes were considered essential in the process of forming these 'diapiric' structures.

In contrast to the early French and Romanian references, Cloos (1923) pointed out that many granitic batholiths are actually laccolith-like and 'are connected to their magma chambers only by narrow, mostly vein-like, conduits, from which they broaden laterally into foreign and older formations'. Thus, Cloos had sound observational reasons for his doubts about the 'batholith nature' of most areally extensive granitic bodies. His views did not prevail, however, and the geological community still applies the batholith tag to granitic masses, perhaps without realizing the full implications.

Some recent work also postulates that batholiths extend from the surface to depths of 25 km or so (e.g. Hyndman, 1983). These inferences are mostly based on observations of plutons that are believed to have been tilted, exposing their root zones as well as their upper sections. Depth estimates are largely gained from geobarometry in the wall rocks and there is room for debate on the structural levels exposed, whether tilting has occurred, the temporal relations between wall-rock metamorphism and intrusion, and on the appropriateness and accuracy of the geobarometric techniques. In some crustal segments, an additional question surrounding this hypothesis is that of the origin of the magma. If the major input was crustal, there may be insufficient thickness of appropriate crust present to provide the inferred magma volume.

Despite only vague notions of what diapiric ascent really entails and the geological observations of dyke-fed, roughly tabular granitic massifs, the diapir concept was taken up with enthusiasm by magmatists. Eventually, with the application of physics to the problem, the early concepts evolved into the sort of buoyancy-driven diapirism with which we are now familiar. In some early models, the granite was not thought of as necessarily partially molten. It would rise, in any case, by virtue of its lower density than the surrounding strata, or be emplaced by some ill-defined process involving tectonic 'squeezing'. Modern magmatic diapir models mostly involve hot-Stokes flow of partially molten magma blobs through heated and softened wall rocks (e.g. Grout, 1932; Whitehead and Luther, 1975; Marsh, 1978, 1984; Spera, 1980; Bateman, 1984, 1985; Cruden, 1988; Mahon et al., 1988; Miller et al., 1988). The magma source region is commonly viewed as a regionally extensive partially molten layer, with periodic cuspate instabilities that spawn diapiric upwellings (e.g. Rickard and Ward, 1981).

In contrast to the common geometrical perception, increasing numbers of batholiths are proving to be shallow bodies (Leake, 1990) with vertical to inward-dipping contacts and floored at relatively shallow depths (e.g. Hamilton and Myers, 1967; Neilson et al., 1976; Sweeney, 1975, 1976: Hodge et al., 1982; Vigneresse, 1988, 1990; Jébrak et al., 1991). Apart from direct observation, the firmest evidence for this is the occurrence of distinct flat-lying or shallow-dipping reflectors beneath batholiths in seismic cross-sections (e.g. Lynn et al., 1981; Goleby et al., 1994; Evans et al., 1994). Highly evolved diapiric systems, sheets and laccoliths could all have tabular shapes, but there would be great contrasts in the mechanisms and structures involved in emplacement. Thus, it has become clear that emplacement mechanism may not equate with ascent mechanism. This, in turn, has led to a more active consideration of possible ascent mechanisms.

Since the 1980s, some workers have been postulating granitic magma ascent in dykes (e.g. Clemens, 1984, 1988; Wall et al., 1987; Clemens and Vielzeuf, 1987). The early 1990s saw a clutch of publications elaborating variants of this model, in which granitic magmas rise either by self-propagating fractures or by exploiting pre-existing faults and joints (Clemens and Mawer, 1992; Petford et al., 1993, 1994). Here too, the magma is thought to ascend by virtue of its density contrast with its wall rocks. The fundamental thermomechanical principles of self-propagating, buoyancy-driven magma fracture have been developed, in detail, by Weertman (1971, 1980), Knapp and Knight (1977), Pollard (1977), Delaney and Pollard (1981), Wilson and Head (1981), Spence et al. (1987), Turcotte (1987), Sleep (1988), Takada (1989, 1990), Spence and Turcotte (1990), and Lister and Kerr (1991). This mode of magma ascent is in accord with a growing number of geological observations of narrow feeder conduits for granitic plutons.

Diapirism and dyking then represent the two end-member processes competing for primacy in our thinking on granitic magma ascent. The differences in processes and products between the two are fundamental and their implications are profound for our understanding of the thermal, mechanical and chemical evolution of the continental crust. On the simplest level, diapir models, for the long-range transport of granitic magma, imply that there is something fundamentally different about the transport properties of partial melts formed from the mantle (mostly gabbroic and basaltic magmas) and from the crust (mostly granitic and rhyolitic magmas). Dyke models imply that all magmas, once separated from their sources, can penetrate the lithosphere in essentially the same manner, albeit at different rates, and that all magmas behave in a more-or-less mechanically similar fashion, irrespective of their composition.

We shall now examine field evidence concerning the occurrence of the two mechanisms and then briefly discuss certain theoretical considerations, illustrating what parameters govern each. The petrological and geodynamic implications of each mechanism will also be examined. In the final section, an attempt will be made to track the variations in magma crystallinity and physical properties during magma ascent, illustrating how these properties are largely governed by conditions in the magma source region (protolith). We will argue that most granitic magma ascent must be through fractures.

6.4 Field structural evidence

A granitic diapir must ascend through the crust by softening and causing plastic deformation of its wall rocks. In the wake of a diapir ascending by thermal softening of the crust (hot-Stokes flow), the wall rocks will show high-T metamorphism. The wall rocks will show radial flattening fabrics (concentric schistosity) and strong vertical shearing and stretching fabrics (asymmetric structures and lineations). A diapir tail will be a cylindrical, high-grade shear zone with core-up kinematic indicators throughout (Clemens and Mawer, 1992). This is distinctive and contrasts sharply with the forms and structures in ordinary shear zones. Also, a more-or-less pronounced rim syncline will accompany the shear zone and be most pronounced near the diapir root. Ideally, the diapir head will be accompanied by radial stretching lineations and extreme strain in the cap rock. Theoretical predictions and experimental demonstrations of these features are given by Ramberg (1967), Cruden (1988, 1990), Schmeling *et al.* (1988), England (1990), Schwerdtner (1990), Paterson and Fowler (1993) and Bateman (1995).

In order to correctly identify a structure as a granitic diapir, it is important to understand the structural consequences of diapiric ascent. Figure 6.1 shows a cartoon of a granitic diapir with its most important structural features indicated by numbers. Descriptions of these follow.

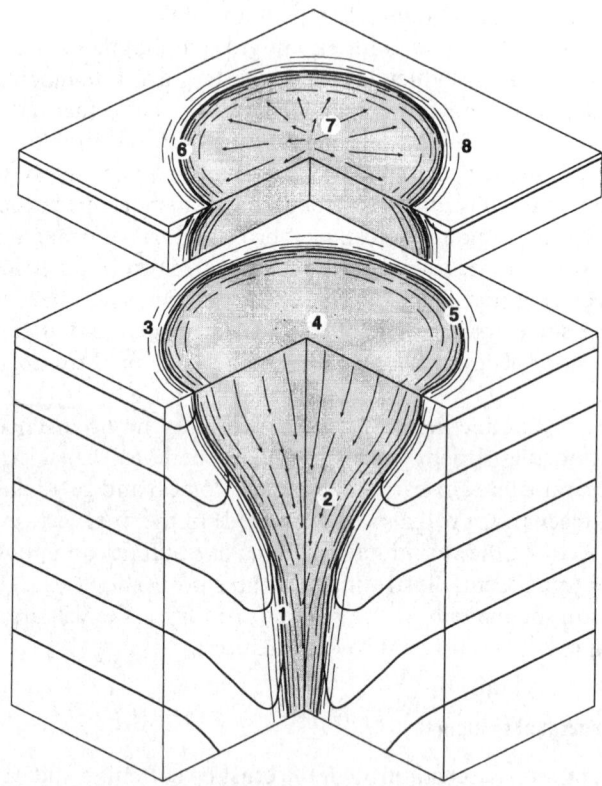

Figure 6.1 Cartoon of a granitic diapir showing the major structural features that should be developed in the granite and the surrounding country rock. Numbers refer to features mentioned in the text. The arrowheads on the lineations indicate plunge directions, rather than flow senses.

1. Around a diapir tail there should be a zone of intense high-temperature shearing, in both the wall rock and the granite. The contact between the two should dip steeply inward, within a relatively narrow zone of strong foliation and down-dip stretching lineation. In the strongly foliated zone, kinematic indicators (such as rotated porphyroclasts, foliation deflections, etc.) should invariably indicate a diapir-up sense of shear. Any layered or foliated host strata should have developed a rim syncline which will gradually decrease in tightness towards the diapir head.

2. Within the central zone of the diapir tail, a strong high-temperature stretching lineation should occur, plunging steeply toward the deeper part of the tail. Note that this lineation could be defined solely by a zone axis of mafic minerals if temperatures were high enough to permit recrystallization of quartz and feldspars. There should be little or no foliation within this zone.

3. At mid-levels, around and close to the body of the diapir, within its host rocks, marginal synthetic shear zones could develop. These could be vertical or could dip steeply either toward or away from the diapir. The diapir–host contact should have essentially the same dip as these shear zones. Kinematic indicators (such as S-C textures, foliation fish, etc.) would indicate diapir-up shear sense, and stretching lineations would plunge down-dip.

4. In the core of the diapir body, at mid-levels on horizontal sections, the granite may appear essentially undeformed. On vertical sections, however, a stretching lineation (but little or no foliation) should occur, though less intense than toward the diapir tail. As for the deeper levels, this lineation could be defined solely by a zone axis of mafic minerals if temperatures were sufficiently high, although one might expect stretched quartz grains, trails of fractured feldspar crystals, etc.

5. Around the contact with the host rocks at mid-levels, a zone of strong shearing would develop. The foliation and contact would be vertical or dip steeply inward or outward. A down-dip stretching lineation should be developed, though this might be suppressed if radial expansion at this level were important. If this were so, an essentially horizontal stretching lineation could develop, possibly accompanied by kinematic indicators with variable and inconsistent senses of shear (if deformation was partitioned), or with no kinematic indicators (if deformation was more or less homogeneous).

6. At high levels around a diapir head, in a zone around the diapir–host contact, shearing should again be concentrated. The shear zone should be narrower than around the tail, indicating lower temperatures of emplacement and stronger localization of deformation associated with ascent. The contact and shear zone foliation should dip outward at a shallow angle. The shear zone would have a meso- and micro-structural textural progression that indicates falling temperature during shearing (e.g. recrystallization followed by fracturing of feldspar, stretching of quartz grains, etc.). A down-dip stretching lineation would be developed.

7. Within the diapir head at such high levels, a strongly developed radial stretching lineation would occur. This would dip shallowly outward from and become intense toward the margin of the body.
8. At high levels, close to the body of the diapir and in the host rocks, marginal low-temperature synthetic shear and fault zones could develop. These could dip shallowly or steeply and either toward or away from the diapir core, depending on the degree of homogeneity of the host rocks in the carapace. However, kinematic indicators (such as shear bands, Riedel fracture assemblages, etc.) would still indicate diapir-up shear sense, and stretching lineations, fibre elongations, slickenside striae, etc., would plunge down-dip.

Field and literature searches for the above combinations of features have drawn complete blanks (e.g. Bateman, 1984; England, 1990; Clemens and Mawer, 1992), though confusion has sometimes been introduced when the differences between ballooning and diapirism have not been appreciated (e.g. Ramsay, 1989). In fact, this confusion is a long-standing tradition, since the earliest workers to posit granitic diapirism regarded any intrusive dome-like structure in sedimentary, hypabyssal or plutonic rocks as prima facie evidence of diapirism (e.g. Nicolesco, 1929). Diapir tracks (pipe-like, high-grade shear zones in which the cores have moved up) apparently do not exist. Sophisticated structural tests have been applied to putative granitic diapirs, which have universally failed (so far) to pass (e.g. Schwerdtner, 1990). Diapir enthusiasm engenders considerable ingenuity. The presence of horizontal or shallow-dipping structures, rather than a vertical lineation, in and around one pluton has been ascribed to helicoidal flow within the mass, with the 'diapir' screwing its way upward (Hippertt, 1994). Granitic bodies in the upper crust commonly have forms and structures indicating that diapirism cannot have been the emplacement mechanism, no matter how ascent may have occurred.

In contrast to diapirs, granitic dykes are commonly observed and mapped among rocks that come from many depth regions of the crust. The sizes of major granitic dykes vary from metres to hundreds of metres (Corry, 1988; Wada, 1994; Kerr and Lister, 1995). In migmatitic terranes, granitic veins occur on scales down to millimetres. Late differentiates of granitic plutons are commonly intruded as small dykes that range from a few centimetres to a metre wide. Thus, it is clear that granitic magmas can and do penetrate crustal rocks by brittle fracture mechanisms operating on a variety of scales and in a great range of mechanical properties of the wall rocks.

The bases of large granitic plutons are not commonly exposed. However, where they are, increasing numbers are proving to have one or more root zones fed by large dykes, e.g. the High Himalayan leucogranites (Scaillet et al., 1995) and the Chemehuevi Mountains plutons (John, 1988). The observations of Cloos (1923) have already been noted. Geophysical surveys are detecting examples of non-diapiric smaller plutons apparently fed by small, stalk-like structures (e.g. Vigneresse, 1990), and the ways in which the transition is made between dyke ascent and bulbous pluton growth have been investigated by Fowler (1994).

It would appear that the weight of field evidence fails to support diapiric ascent of granitic magmas. Is this because the evidence for diapirism has been erased by subsequent complex structural and metamorphic histories? Though this could be so in the middle to lower crust, we contend that the common preservation of thin, early dykes and dyke segments in these terranes argues against this interpretation of the geology. If small, thin, early dykes and dyke segments survive subsequent deformation, it is unreasonable to propose that major cylindrical shear zones and funnel-like diapir tails would be obliterated.

6.5 Theoretical considerations

6.5.1 Diapirs

Diapiric ascent of magma in the lithosphere can be modelled assuming both Newtonian and non-Newtonian rheology of wall rocks. In its simplest form, the buoyant ascent velocity of a spherical, isothermal diapir through isoviscous country rock can be approximated using Stokes law:

$$V = \frac{2\Delta\rho g r^2}{9\eta} \tag{6.1}$$

where V is the terminal velocity, r is the sphere radius, g is the acceleration due to gravity, $\Delta\rho$ (assumed to be $300 \, kg/m^3$) is the wall-rock density minus the magma density, and η is the wall-rock viscosity (Mahon et al., 1988). The critical parameter is the wall-rock viscosity. The high values expected for unmelted crustal rocks ($\sim 10^{24}$ Pa s; Cathles, 1975), combined with steady loss of thermal energy from the diapir to its surroundings result in rather slow ascent rates ($\sim 10^{-8}$ to 10^{-10} m/s, or at least 3×10^4 years for 1 km), and a velocity which is time-dependent. In a migmatitic middle crust, wall-rock viscosity will be lower and the ascent rate of a diapir correspondingly higher. However, the majority of great granitic batholiths intruded cool, brittle upper crust. Another noteworthy point is that the magma column in the tail of a diapir would provide an extra impetus to ascent, through the increased magmastatic head. This factor is not accounted for in equation 6.1, although, as pointed out by Marsh (1982), for low Reynolds number flow, the shape of the rising body has little effect on drag.

If crustal rheology is Newtonian (e.g. equation 6.1), a rising granitic diapir must expend considerable thermal energy on the surrounding rocks in order to soften them, lower their viscosity and allow continued ascent. Heat loss, crystallization and viscosity increase (in both magma and wall rock) all result in diapirs grinding to a halt at depths of at least 14 km. Born in the lower crust, a granitic diapir would suffer 'thermal death' in the middle crust. Furthermore, during hot-Stokes flow, deformation in the surrounding country rock should propagate

153

outwards over many body radii (Marsh, 1982). As pointed out in section 6.4, there appears to be little geological evidence for predicted ascent-induced deformation around granitic plutons. One argument for the lack of observed deformation in the vicinity of proposed diapirs is that ascent models based on hot-Stokes flow are unrealistic and oversimplified. Indeed, given that all rocks have temperature-dependent viscosities, the assumption of isoviscous wall rock (implicit in the application of Stoke's law) around a rising diapir that is transferring heat to its surroundings clearly will not hold, and a more sophisticated approach is required to model diapiric ascent based on non-Newtonian crust. In this case diapirs may ascend more rapidly than the above analysis suggests. In country rock exhibiting power-law behaviour, viscosity will decrease towards the diapir interface due to either thermal or strain-rate softening (Mahon et al., 1988; Weinberg and Podladchikov, 1994). Calculations by Mahon et al. (1988) suggest that thermal softening of temperature-dependent country rock allows a diapir to rise a maximum of 15 km in 10^4 yr ($V = 5 \times 10^{-6}$ cm/s) before freezing, with ascent velocity strongly dependent on the rheology and ambient temperature of surrounding country rock.

More recently, strain-rate softening has been suggested as an alternative to purely thermal weakening of wall rocks around a rising diapir (Weinberg and Podladchikov, 1994), as a means of transporting granitic magma through the crust, over time-scales comparable with cooling of large magma bodies ($\sim 10^5$ yr). This suggestion certainly merits further investigation, but is presently based on long extrapolations from laboratory experiments at unrealistically high T, and on materials (Westerly granite) that are probably poor models for the continental crust. It is interesting to note that a study of heated silicate glasses and melts (Webb and Dingwell, 1990) showed that, even with these relatively plastic materials, the onset of non-Newtonian behaviour occurred only at strain rates that brought the materials close to, and in some cases caused, brittle tensile failure. It may be important to note also that, where these experiments attained tensile failure, they did so at geologically small total strains, far smaller than those that would occur in and around the margins of rising diapirs. On the other hand, the lack of a confining pressure in these tests means that the materials would have fractured more easily than partially molten rocks at depth in the crust, though the pressure dependence of fracture toughness is not pronounced at crustal depths. It seems possible that a non-Newtonian diapir–crust system could evolve rapidly into a fracture transport regime.

Even assuming that the crust would continue to deform plastically, as a substance with a power-law exponent of 1.5 to 2.5, the modelling of Weinberg and Podladchikov (1994) predicts emplacement depths comparable to those obtained using models that involve thermal softening of the wall rocks. Though the calculated rates are faster than those based on hot-Stokes ascent ($c.\ 10^{-7}$–10^{-6} m/s), the diapirs would still ascend on a time-scale that would allow considerable conductive cooling. This is not particularly helpful if the aim is to produce a large-volume, shallow-level, silicic magma chamber that can undergo chamber-wide magma evolution processes.

The effect of pressure represents a further complication to this model, at depths typical for granite magma generation (*c.* 30 km or so). Given the temperatures and strain rates that must obtain, the main deformation mechanism, in and around non-Newtonian diapiric masses, is certain to be diffusion-accommodated dislocation creep. The effect of pressure on dislocation creep strain rates can be seen by considering the full form of the equation for the diffusion coefficient, D (after Lazarus and Nachtrieb, 1963):

$$D = D_o \exp(-Q/kT) \exp(-P\Delta V^*/kT) \qquad (6.2)$$

where Q is the activation energy, k is Boltzman's constant, T is the temperature in kelvins, P is pressure and ΔV^* is the activation volume. An increase in pressure will therefore lead to a decrease in the diffusion coefficient. Thus, any diffusion-controlled process (such as diffusion-accommodated dislocation creep) will proceed more slowly at higher pressure. For crustal pressures up to about 400 MPa, this effect will not be important, but will start to become so in the deep crust near the magma sources.

Thus, our current understanding of diapiric ascent implies that shallow-level felsic magma chambers should not exist. Against this prediction we have to weigh the known presence of active rhyolite magma chambers at just a few kilometres depth (Smith, 1979). We must also account for the huge plutons (>103 km^3) emplaced at similar depths, as inferred from geobarometric, structural and stratigraphical evidence (e.g. Phillips *et al.*, 1981). Pluton-spacing arguments, coupled with estimates of relative volumes of crust and mobile layer, have also been used to show that diapirism is unlikely to account for the observed volumes of granitic rock in the upper crust (Vigneresse, 1995). Interestingly, diapiric thermal death, by increasing crystallization (even with the latent heat that this provides), carries with it the important implication that diapiric ascent should give rise mainly to diapiric emplacement, since the magma would have little or no potential to migrate further once the diapiric process had finished. We have already mentioned the lack of field evidence for diapiric emplacement. In short, though physically possible within the modelling framework, diapirism does not appear to be the mechanism by which granitic magmas have ascended through the crust. What is required is a model ascent mechanism that does not depend on thermal transfer from magma to wall rock, that does not depend on special and unverified crustal behaviour, and that allows ascent at a rate sufficiently rapid to prevent significant conductive or convective heat loss from the magma to the walls.

6.5.2 Dykes

Buoyant ascent of magma in, or propagation of, a fracture (dyke) of width w is powered by the same $\Delta\rho$ as in diapirism. However, the factor limiting the vertical ascent velocity (V) is not wall-rock viscosity but the melt (or magma) viscosity (η_m):

$$V = \frac{gw^2\Delta\rho}{12\eta_m}. \qquad (6.3)$$

Varying $\Delta\rho$ within the likely range (200 to 400 kg/m^3) will only cause velocity to vary by a factor of 2. Thus, to assess the viability of this model, it is necessary to have an accurate idea of the viscosities of granitic magmas. Over many years, numerous incorrect and misleading statements and assumptions have been made about this parameter, resulting in a common perception that granitic magmas are extremely viscous. The two most important influences on η_m are the magma temperature (850 to 950 °C for most granitic magmas, prior to cooling; Clemens, 1990; Stevens and Clemens, 1993) and melt H_2O content (initially 2 to 4 wt% for the vast majority of granitic magmas; Clemens, 1984). With these realistic ranges of parameters, we can use the data in Dingwell *et al.* (1996) to predict that natural granitic melts will have viscosities in the range 10^4 to 10^6 Pa s (see also Thomas, 1994; Baker and Vaillancourt, 1995). Increasing pressure will lower the viscosity slightly, but the effect is very small at crustal pressures (Scarfe *et al.*, 1987). Crystal contents up to 30 vol.% should produce only a relatively small increase ($\leq$ an order of magnitude) in effective viscosity (Kerr and Lister, 1991; Koenders *et al.*, 1996).

Using the above (crystal-free) estimates of magma viscosity, equation 7.3 predicts dyke ascent at $\sim10^{-3}$ to 8 m/s (or, at most, about two weeks for 1 km) for dykes of 3 to 16 m width. A typical ascent rate would be 0.25 m/s. These values are between 10^{10} and 10^3 times faster than any diapir, ascending by either thermal or strain-rate softening of its wall rocks. The kinematic analysis of Spence and Turcotte (1990) has shown that dyke (fracture) propagation is commonly catastrophic. This holds true for fracture propagation with granitic magmas (Clemens and Mawer, 1992). That is, the effective strength of the crust is negligible and the fracture will be self-propagating, as long as it contains buoyant magma with a viscosity in the range of 10^3 to 5×10^5 Pa s.

As in the case of a diapir, magma freezing will limit the amount of ascent in a dyke (Bruce and Huppert, 1989). A granitic dyke must therefore attain a critical minimum width (w_c) in order that the flowing magma can advect heat faster than conduction through the walls, and thus avoid freezing. Petford *et al.* (1993, 1994) showed that the critical minimum dyke width varies with the fourth root of the magma viscosity (η_m) according to:

$$w_c = 1.5\left(\sqrt[4]{\frac{S_m}{S_\infty^2}}\right)^3 \cdot \left(\sqrt[4]{\frac{\eta_m\kappa H}{\Delta\rho g}}\right). \qquad (6.4)$$

Here $S_m = L/C(T_m-T_w)$ and $S_m = L/C(T_w-T_\infty)$, where L is the latent heat of solidification (3×10^5 J/kg), C is specific heat (1.2×10^3 J/kg/°C), T_∞ is the far-field temperature of the crust (assumed to be 200 °C), T_m is the initial magmatic temperature (850 to 950 °C), T_w is the temperature (~750 °C) at which the magma is effectively immobile (frozen), κ is the thermal diffusivity (8×10^{-7} m²/s) and H is the assumed dyke length of 20 km (2×10^4 m). Using the above values for the parameters in equation 6.4, and the other parameters as for earlier examples, we calculate that critical dyke widths will vary from 2.6 to 16.4 m for granitic magmas. Although some concern has been expressed as to whether propagating granitic fractures can grow to the required critical minimum width before freezing (Rubin, 1995), this is a common range actually observed for granitic dykes (e.g. Corry, 1988), and is good evidence that granitic dyke formation not only occurs, but the governing equations predict its occurrence on the scales actually observed. This is more than may be said of granitic diapirism, for which, though theoretically possible, there is no geological evidence.

The next question concerns the efficiency of fracture ascent of granitic melts. Can dykes supply granitic magma to a growing pluton at sufficiently high rates to allow for the formation of a molten mass faster than conductive cooling can cause solidification? Modelled as a horizontal sheet 3 km thick, emplaced into wall rocks at 100 MPa and 100 °C, the core of a granitic magma chamber would cool from 850 °C to the solidus (~700 °C) in 30 000 years. This modelling was performed using the program Contact (Spear et al., 1991). Other input parameters were: heat of crystallization = 1.7×10^5 J/kg, wall-rock heat capacity = 1.1×10^3 J/kg/K, thermal conductivity = 3 W/m/K, and rock density = 2750 kg/m³. The typical magma ascent rates (see above) can be used to calculate the time required for ascent through the crust – say 20 to 30 km. Thus, a continuously fed and extending dyke would penetrate this far in a matter of weeks. Once an ascent path was established, a dyke 1 km long (in plan) and only 3 m wide could supply a growing chamber (pluton), however emplaced, with 1000 km³ of magma in less than 300 years. Solidification of the chamber would occur on a time-scale around 100 times as long. Thus, there is little doubt that dyke transport of granitic magma can be both rapid and efficient. Indeed, the rate-limiting step in pluton formation will not be magma ascent, but either the rate of magma production in the source region or the rate at which space can be created to accommodate magma at the emplacement site, or magma supplied to the conduit at source (Petford, 1996). Given the various forces applied to potential host rocks by the magma, spaces should be able to open relatively rapidly – so magma supply is most likely to provide the overriding rate-limit.

6.6 The space problem

The previous passage evokes the time-honoured 'space problem' in granitic plutonism. Granitic magma ascent has traditionally been viewed as occurring through the upward movement of globular masses roughly equivalent in volume to the final plutons. Thus, the problem was how the crust could accommodate this volume displacement during ascent.

Diapirism offers the explanation of extreme flattening and folding of the wall rocks. However, as we have seen, this would produce structures that have, apparently, never been discovered in the rock record. It might be proposed that the displacement would be lessened if individual diapirs were small and that many ascended along a similar path to form an aggregate pluton. This is the 'nested diapir' hypothesis, recently resuscitated by Paterson and Vernon (1995). Unfortunately, small diapirs would cool very much faster than large ones and could not progress far, even if the crust were warmed by the passage of previous little diapirs. In any case, diapir tracks, large or small, seem to be non-existent.

If granitic magmas ascend through fractures then the only strain that needs to be accommodated is that necessary to open brittle fractures to widths of a few metres. Such widths can be accommodated by simple elastic strain release following fracture of stressed rocks. The abundance of dykes in the crust attests to the ease with which this is accomplished. So, the problem is reduced to the question of how emplacement occurs. This is not the subject of this paper, but we suggest that emplacement could occur by any one of a dozen or more mechanisms involving both brittle and plastic deformation. Our conclusion is that there is nothing particularly special about the ascent of granitic magma. It occurs in much the same way as does the ascent of basaltic magma. Batholiths and stocks of granite probably form on top of their feeder dykes in much the same way as large gabbro plutons grow on top of theirs.

6.7 Magma history during dyke ascent

At this point it is worthwhile considering what might happen to a granitic magma as it ascends in a dyke. This kind of analysis has previously been undertaken by Sykes and Holloway (1987), Johannes and Holtz (1990) and Holtz and Johannes (1994). We will trace the physical history of a magma formed by partial melting of a metagreywacke protolith in the lower crust, during granulite-facies metamorphism. Partial melting is assumed to take place at 1 GPa and 850 to 900 °C. The source contains 25 wt% biotite, which breaks down, by fluid-absent batch melting, to produce H_2O-undersaturated granitic melt plus a granulite-facies solid residue.

From experimentally verified modelling of these processes (Clemens and Vielzeuf, 1987; Johannes and Holtz, 1990), we can estimate that this will produce 25 to 32 vol.% melt. Figure 6.2 shows the variation of aH_2O with T for a pressure of 1 GPa (Johannes and Holtz, 1990). These H_2O activities were used to calculate melt H_2O contents using the solubility model of Burnham (1981). Mass balance relations were then used to calculate the melt proportions according to the method of Clemens and Vielzeuf (1987). As Figure 6.3 shows, the higher the T, the more melt, but the lower its H_2O content, since we are dealing with fluid-absent melting.

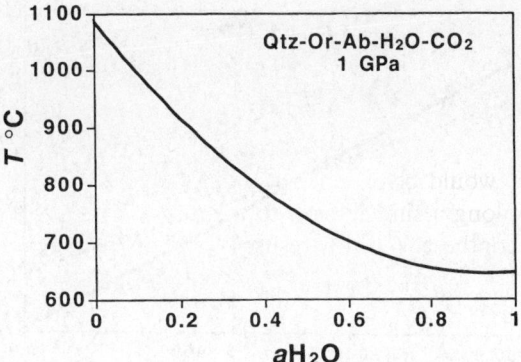

Figure 6.2 Plot of T versus the activity of H_2O for the haplogranite system. Data were derived from the experiments of Huang and Wyllie (1975), Johannes (1984) and Ebadi and Johannes (1991).

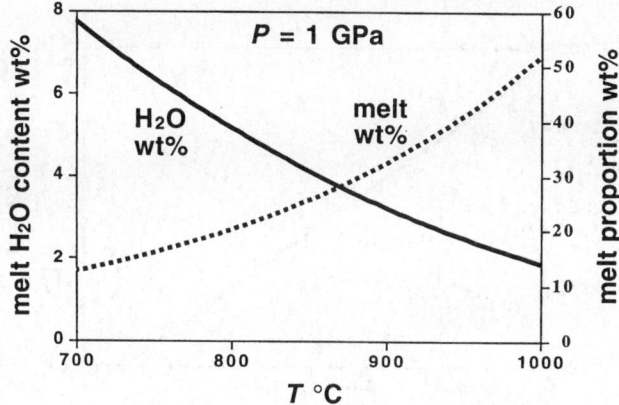

Figure 6.3 Plot showing potential melt proportion and melt H_2O content as a function of T, for fluid-absent partial fusion of a metagreywacke source rock (25 wt% biotite $\approx$ 1 wt% H_2O in the source rock) at 1 GPa.

6.7.1 Viscosity and density

Melt viscosity will decrease as T increases, despite the markedly lower H_2O content of the melt (Dingwell *et al.*, 1996). Melt density can be calculated using the partial molar volumes of silicate-melt oxide components, given by Bottinga *et al.* (1983), and the partial molar volume of H_2O in albite melt (Burnham and Davis, 1974). As shown in Figure 6.4, the melt density will decrease towards higher T and higher melt proportion. Thus, the density driving force for ascent will be greatest for the hotter, drier, less viscous magmas. At 870 °C and 1 GPa, the models predict that there will be 25 vol.% melt, with 4 wt% dissolved H_2O, a density of 2441 kg/m³, and a viscosity of $10^{4.9}$ Pa s. Figure 6.5 shows the varia-

159

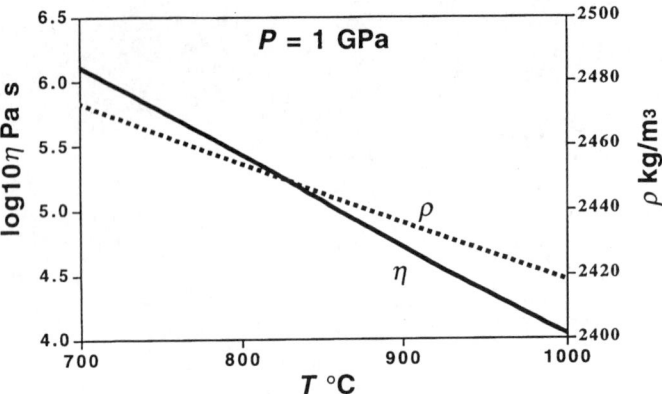

Figure 6.4 Plot showing the variations of granitic melt viscosity and density as a function of T during partial fusion.

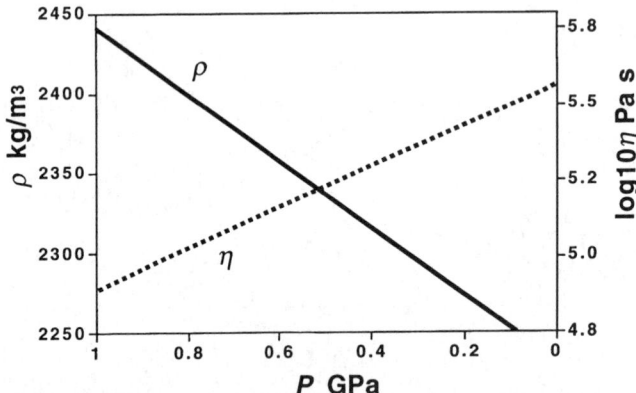

Figure 6.5 Variation in melt viscosity and density along an adiabatic magma ascent path (870 °C at 1 GPa to 807 °C at 0.1 GPa).

tions in viscosity and density as this melt ascends along an isentropic adiabatic cooling path with a slope of 0.07 °C/MPa (Sykes and Holloway, 1987).

6.7.2 Crystallinity and the presence of restite in granitic magmas

The restite unmixing model for the origin of chemical variation in granitic magmas (Chappell *et al.*, 1987) has been attacked on a variety of grounds (e.g. Wall *et al.*, 1987; Clemens, 1989). In fact, there is an implicit link between the restite unmixing and diapiric ascent models, at least for the restite model's origi-nators (B.W. Chappell, A.J.R. White and D. Wyborn, pers. comms., 1984–1994). In the restite-unmixing model, granitic magmas are seen as the products of convective homogenization of a partially molten protolith, emplaced by diapiric up-welling. As pointed out by Clemens and Vielzeuf (1987), Wall *et al.* (1987), Clemens (1989, 1990) and Clemens and Mawer (1992), this would mean that deep crustal source regions for granitic magmas would mostly be exhumed as diapirs and emplaced higher in the crust as restite-rich granitic plutons.

Let us assume that a granitic magma (melt plus any entrained crystals) ascends through a fracture at a rate of 10^{-2} m/s. Heat loss will be negligible for a dyke of 10 m width in the upper crust. At deeper levels, where T is higher, this critical width will be reduced. Thus, it is a reasonable proposition that the $P\text{-}T$ ascent path will approximate an isentropic adiabat. An ascent path with the same slope as the solidus (about 0.12 °C/MPa) will result in a constant crystal:melt ratio in the magma. The dP/dT slope of the magma's solidus is shallower than the adiabat so, by the time the magma arrives at its emplacement level (100 MPa, for example), it will possess a significant amount of constitutional superheat. Thus, during ascent, very little crystallization could occur and, indeed, it is possible that any entrained solids (restitic, early magmatic or xenocrystal) would be dissolved or substantially resorbed (Figure 6.6).

The degree of potential resorbtion of entrained crystals can be estimated if we know how magma crystallinity varies as a function of ΔT ($T\text{-}T$ solidus). From existing crystallization experiments on S-type granitic rocks (Clemens, 1981) it is

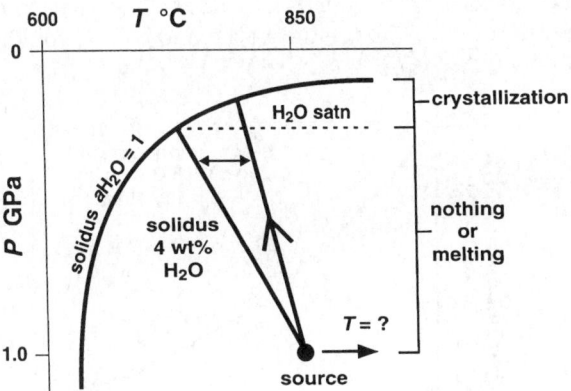

Figure 6.6 *P–T* plot showing the location of the H_2O-saturated granite solidus, and that for a melt H_2O content of 4 wt%. The H_2O-undersaturated liquidus will probably have a similar slope. The arrowed line represents a magma ascent path, with decompression from 1 GPa to 100 MPa and cooling of 63 °C. The actual location of the starting point will be at some higher T. The horizontal, dashed line marks the pressure at which H_2O saturation will occur, even in the absence of crystallization. Since the dP/dT slope of the ascent path is steeper than the solidus, the magma will contain constitutional superheat (double-ended arrow) at any P lower than that of the source region. Thus, most of the ascent path will be characterized by potential crystal resorbtion. Major crystallization will only set in if the degree of cooling is greater than shown.

possible to estimate the change in crystallinity for a given ΔT. The data are shown in Figure 6.7, and the following polynomial describes the variation:

$$\text{crystallinity (vol.\%)} = 0.005\Delta T^2 - 1.366\Delta T + 100 \qquad (6.5)$$

strictly valid over the temperature range of 740 to 875 °C, at 500 MPa.

The solidus temperatures of granitic magmas can be modelled from experiments in the H_2O-undersaturated haplogranite system. Johannes and Holtz (1990) place the haplogranite solidus (with 4 wt% H_2O) at 846 and 736 °C, at 1 GPa and 100 MPa respectively. Thus, if this model magma decompresses by 900 MPa while cooling from 870 to 807 °C, it will gain about 47 °C of superheat and could potentially resorb 47% restite. This calculation assumes that the crystallization rate remains approximately the same over the pressure range of interest, and is adequately modelled by the rate at 500 MPa. Naturally, a real magma ascending and resorbing crystals, in a closed system, will have a decreasing H_2O content and an increasing solidus T. Despite this, and the other uncertainties in the data and calculations, it is clear that there is considerable potential for crystal resorbtion in an adiabatically ascending granitic magma.

The particular mineral species that may be resorbed can be predicted if we know the shapes of the mineral saturation boundaries with respect to the ascent trajectory in P-T space. Figures 6.8 and 6.9 show the mineral saturation boundaries in an I-type granodiorite and an S-type monzogranite respectively. The model ascent path has a slope of 0.07 °C/MPa and begins at a pressure of 1 GPa, and a temperature of 900 °C for the I-type and 870 °C for the S-type. From

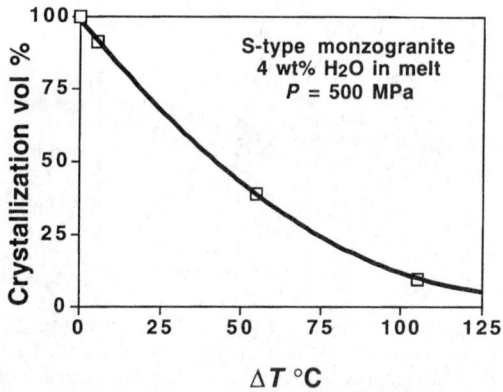

Figure 6.7 Plot of crystallinity (vol.%) as a function of temperature above the solidus (ΔT) for an S-type monzogranite magma containing 4 wt% H_2O, during crystallization at 500 MPa. The data points (open squares) are taken from the visual estimates of crystal contents in products of crystallization experiments on a monzogranite glass (Clemens, 1981). The reader is referred to Clemens and Wall (1981) for details of experimental methods.

162

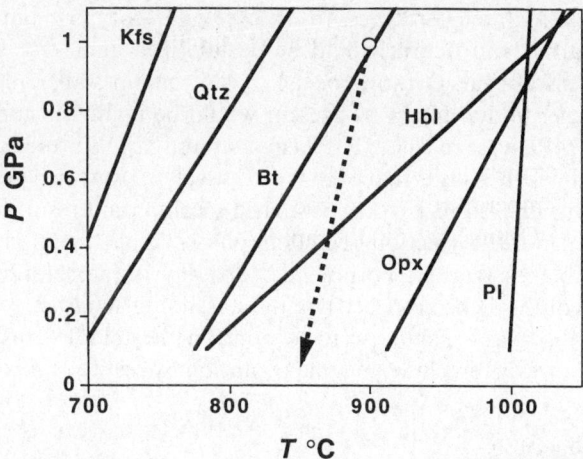

Figure 6.8 *P–T* plot showing the positions of mineral saturation boundaries for an I-type granodiorite magma with a melt H$_2$O content of 4 wt%. Original data, at 200 and 800 MPa, were taken from Naney (1977). For simplicity and lack of data at intermediate pressures, the boundaries have been assumed to be linear. The chosen ascent path is shown as the dashed line with an arrow, and the starting point indicated by a dot.

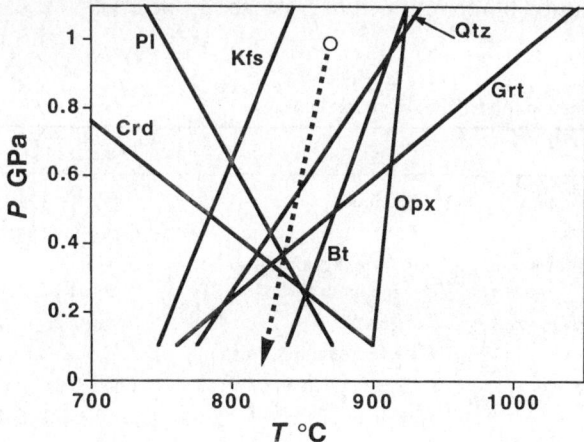

Figure 6.9 *P–T* plot showing the positions of mineral saturation boundaries for an S-type granite magma with a melt H$_2$O content of 4 wt%. Original data, at 100 and 500 MPa, were taken from Clemens (1981) and Clemens and Wall (1981). The simplifications in Figure 6.6 have also been applied here.

Figure 6.8 it can be seen that the minerals that become unstable or that will be partially resorbed during ascent include orthopyroxene, hornblende and biotite. Plagioclase should remain stable, while quartz and alkali feldspar should not be

163

present at these high temperatures. In the case of an S-type magma, Figure 6.9 shows that garnet and quartz would be destabilized and some resorbtion of biotite might also occur. Orthopyroxene would remain stable and any plagioclase and cordierite that might be present would be highly unstable until pressures of <400 MPa were reached. K-feldspar should not be present.

From the above, it is clear that even if a relatively crystal-rich granitic magma were initially mobilized in a dyke, adiabatic ascent could result in a relatively large increase in melt fraction. Mineral phases that typify granulite-facies restitic assemblages (garnet, hornblende, orthopyroxene, biotite, plagioclase and quartz) could all be resorbed by the ascending magma. This would be particularly effective if magma were to pond in the middle crust, and would severely limit, or even preclude, magma evolution by restite unmixing.

6.8   Conclusions

In this paper we distinguish between magma **ascent**, on one hand, and **emplacement**, on the other, since the mechanisms involved can differ markedly. Diapirism and dyking represent two competing end-member models for granitic magma ascent. The differences in processes and products between the two are fundamental, and their implications profound for the thermal, mechanical and chemical evolution of the continental crust (see Table 6.1).

Table 6.1   Implications of magma ascent mechanisms

Feature	Diapir	Dyke
Crustal differentiation	Precluded	Implicit
Structural state of protolith	Exhumed as magma	Quasi-undisturbed
Structural state of overlying crust	Regionally deformed/ metamorphosed	Regionally unaffected, locally contact metamorphosed
Shallow magma chambers	Precluded	Permitted
Depth of emplacement	Lower to middle crust only	Anywhere in crust
Cause of emplacement	Freezing/bouyancy	Local structure/rock type
Rate of pluton formation	Millions of years	Geologically instantaneous
Fabrics in pluton	Relate to ascent	Relate to emplacement
Syn-emplacement fabrics surrounding pluton	Very high strain, foliations and lineations	Variable, depending on emplacement mechanism
Magma viscosity at emplacement	Very high	Relatively low
Magma restite content	High	Low, variable
Differentiation site/time	During ascent	Post-emplacement (in magma chamber)
Differentiation mechanism	Mainly restite unmixing	Variable

N.B. Features in italics are those for which the characteristics of the two models most firmly favour dyking over diapirism as the main mechanism for grantoid magma ascent.

Field and literature searches have failed to reveal the structural and petrological features of crust through which granitic diapirs have passed. Indeed, if restite-rich diapirs were responsible for granite formation, then the observed differentiated crustal density structure could not have been produced. Granites in the upper crust commonly have forms and structures indicating that diapirism could not have been the emplacement mechanism. In contrast to diapirs, granitic dykes are common among rocks from a variety of crustal depths. Granitic magmas can and do penetrate the crust by brittle fracture mechanisms operating on a variety of scales and in wall rocks with a great range of mechanical properties. Many granitic batholiths are tabular in shape and, where their bases are exposed, they are apparently fed by large dykes. Thus the weight of evidence favours dyke ascent for the majority of granitic magmas. Calculations show that, while granitic diapirs could exist, their ascent rates are extremely slow and would result in small vertical transport distances and thus relatively deep emplacement. Modelling also shows that granitic magma ascent via fractures is geologically extremely rapid and efficient. The rate-limiting factor in granitic plutonism is not ascent rate but the rate of magma formation or the rate of deformation at the emplacement site. The 'space problem' associated with the ascent of a large blob of magma is removed if fracture ascent rates are considered, though the pluton must still be accommodated at the final emplacement site.

We have examined the variations in H_2O content, melt proportion, density, viscosity and crystal content of a granitic magma from its inception as partial melt in the lower crust to its emplacement in an upper crustal pluton. During partial melting, melt viscosity decreases as T increases. Melt density also decreases toward higher T and higher melt proportion. The dP/dT slope of a granitic magma's solidus is shallower than the isentropic adiabat that approximated the ascent path. Thus, by the time the magma arrives at emplacement level, it may possess significant constitutional superheat, and many tens of per cent of any entrained solids could be dissolved or partially resorbed.

Finally, we believe that more than sufficient evidence has now accumulated to enable us to abandon the restite and diapir paradigms of granitic magmatism, just as the transformist paradigm expired under the weight of contrary evidence and advances into new areas of knowledge. Now may also be the time to reconsider the utility of the batholith concept, that so channelled geological thinking through this century.

Acknowledgements

The authors gratefully acknowledge detailed and helpful reviews by Mike Brown, Simon Inger and Ernie Rutter, as well as gentle but firm editorial handling by Marian Holness.

References

Baker, D.R. and Vaillancourt, J. (1995) The low viscosities of F+H₂O-bearing granitic melts and implications for melt extraction and transport. *Earth Planet. Sci. Lett.*, **132**, 199–211.

Bateman, R. (1984) On the role of diapirism in the segregation, ascent and final emplacement of granitoid magmas. *Tectonophysics*, **110**, 211–31.

Bateman, R. (1985) Aureole deformation by flattening around a diapir during in-situ ballooning – the Cannibal Creek granite. *J. Geol.*, **93**, 293–310.

Bateman, R. (1995) Foliation development in the Cannibal Creek granite and its aureole heterogeneous shortening around a ballooning diapir. *J. Struct. Geol.*, **7**, 489.

Beard, J.S. and Lofgren, G.E. (1991) Dehydration melting and water-saturated melting of basaltic and andesitic greenstones and amphibolites at 1, 3, and 6.9 kb. *J. Petrol.*, **32**, 365–401.

Bottinga, Y., Richet, P. and Weill, D.F. (1983) Calculation of the density and thermal expansion coefficient of silicate liquids. *Bull. Minéral.*, **106**, 129–38.

Brown, G.C. (1970) A comment on the role of water in the partial fusion of crustal rocks. *Earth Planet. Sci. Lett.*, **9**, 355–8.

Brown, M. (1994) The generation, segregation, ascent and emplacement of granite magma, the migmatite-to-crustally-derived granite connection in thickened orogens. *Earth Sci. Rev.*, **36**, 83–130.

Brown, M. and Rushmer, T. (this volume) The role of deformation in the movement of granitic melts: a consideration of the role of melt-assisted granular flow.

Brown, M., Averkin Y.A., McLellan E.L. *et al.* (1995) Melt segregation in migmatites. *J. Geophys. Res.*, **B100**, 15655–79.

Bruce, P.M. and Huppert, H.E. (1989) Thermal controls of basaltic fissure eruptions. *Nature*, **342**, 665–7.

Burnham, C.W. (1967) Hydrothermal fluids at the magmatic stage, in *Geochemistry of Hydrothermal Ore Deposits*, Holt, Reinhart & Winston, New York, pp. 38–76.

Burnham, C.W. (1981) The nature of multicomponent aluminosilicate melts. *Phys. Chem. Earth*, **13**, 197–229.

Burnham, C.W. and Davis, N.F. (1974) The role of H₂O in silicate melts, II. Thermodynamic and phase relations in the system NaAlSi₃O₈–H₂O to 10 kilobars, 700° to 1100 °C. *Am. J. Sci.*, **274**, 902–40.

Cathles, L.M.I. (1975) *The Viscosity of the Earth's Mantle*, Princeton University Press, N.J.

Chappell, B.W., White, A.J.R. and Wyborn, D. (1987) The importance of residual source material (restite) in granite petrogenesis. *J. Petrol.*, **28**, 1111–38.

Clemens, J.D. (1981) *Origin and Evolution of Some Peraluminous Acid Magmas*, PhD Thesis, Monash University, Australia.

Clemens, J.D. (1984) Water contents of intermediate to silicic magmas. *Lithos*, **17**, 273–87.

Clemens, J.D. (1988) Volume and composition relationships between granites and their lower crustal source regions: an example. *Aust. J. Earth Sci.*, **35**, 445–8.

Clemens, J.D. (1989) The importance of residual source material (restite) in granite petrogenesis: a comment. *J. Petrol.*, **30**, 1313–16.

Clemens, J.D. (1990) The granulite–granite connexion, in *Granulites and Crustal Differentiation* (eds D. Vielzeuf and P. Vidal), Kluwer Academic Publishers, Dordrecht, pp. 25–36.

Clemens, J.D. and Mawer, C.K. (1992) Granitic magma transport by fracture propagation. *Technophysics*, **204**, 339–60.

Clemens, J.D. and Vielzeuf, D. (1987) Constraints on melting and magma production in the crust. *Earth Planet. Sci. Lett.*, **86**, 287–306.

Clemens, J.D. and Wall, V.J. (1981) Crystallization and origin of some peraluminous (S-type) granitic magmas. *Can. Mineral.*, **19**, 111–31.

Cloos, H. (1923) Das Batholithen problem. *Fortschr. Geol. Palaeon.*, **1**, 80 pp.

Conrad, W.K., Nicholls, I.A. and Wall, V.J. (1988) Water-saturated and -undersaturated melting of

REFERENCES

metaluminous and peraluminous crustal compositions at 10 kb: evidence for the origin of silicic magmas in the Taupo Volcanic Zone, New Zealand, and other occurrences. *J. Petrol.*, **29**, 765–803.

Cooper, R.F. (1990) Differential stress-induced melt migration: an experimental approach. *J. Geophys. Res.*, **B95**, 6979–92.

Corry, C.E. (1988) Laccoliths: mechanisms of emplacement and growth. *Geol. Soc. Am. Spec. Pap.*, **220**, 110 pp.

Cruden, A.R. (1988) Deformation around a rising diapir modeled by creeping flow past a sphere. *Tectonics*, **7**, 1091–1101.

Cruden, A.R. (1990) Flow and fabric development during the diapiric rise of magma. *J. Geol.*, **98**, 681–98.

Delaney, P.T. and Pollard, D.D. (1981) *Deformation of Host Rocks and Flow of Magma During Growth of Minette Dikes and Breccia-bearing Intrusions Near Ship Rock, New Mexico*, U.S. Geological Survey, Washington, D.C.

Dell'Angelo, L.N. and Tullis, J. (1988) Experimental deformation of partially melted granitic aggregates. *J. Metamorphic Geol.*, **6**, 495–515.

Dingwell, D.B., Romano, C. and Hess, K.-U. (1996) The effect of water on the viscosity of a haplogranitic melt under P-T-X conditions relevant to silicic volcanism. *Contrib. Mineral. Petrol.* (in press).

Ebadi, A. and Johannes, W. (1991) Beginning of melting and composition of first melts in the system Qz-Ab-Or-H_2O-CO_2. *Contrib. Mineral. Petrol.*, **106**, 286–95.

Eigen, M. (1973) in *The Physicists Conception of Nature* (ed. J. Mehra), D. Reidel, Dordrecht.

England, R.W. (1990) The identification of granitic diapirs. *J. Geol. Soc. London*, **147**, 931–3.

Evans, D.J., Rowley, W.J., Chadwick, R.A. *et al.* (1994) Seismic reflection data and the internal structure of the Lake District batholith, Cumbria, northern England. *Proc. Yorkshire Geol. Soc.*, **50**, 11–24.

Fountain, J.C., Hodge, D.S. and Shaw, R.P. (1989) Melt segregation in anatectic granites: a thermomechanical model. *J. Volc. Geotherm. Res.*, **39**, 279–96.

Fowler, T.J. (1994) Sheeted and bulbous pluton intrusion mechanisms of a small granitoid from southeastern Australia: implications for dyke-to-pluton transformation during emplacement. *Tectonophysics*, **234**, 197–215.

Fyfe, W.S. (1973a) The generations of batholiths. *Tectonophysics*, **17**, 273–83.

Fyfe, W.S. (1973b) The granulite facies, partial melting and the Archean crust. *Phil. Trans. R. Soc. Lond.*, **A273**, 457–61.

Goleby, B.R., Drummond, B.J., Korsch, R.J. *et al.* (1994) Review of recent results from continental deep seismic profiling in Australia. *Tectonophysics*, **232**, 1–12.

Grout, F.F. (1932) *Petrography and Petrology*, McGraw Hill, New York.

Hamilton, W. and Myers, W.B. (1967) The nature of batholiths. *U.S. Geol. Surv. Prof. Pap.*, **554-C**, 30 pp.

Hand, M. and Dirks, P.H.G.M. (1992) The influence of deformation on the formation of axial-planar leucosomes and the segregation of small melt bodies within the migmatitic Napperby Gneiss, Central Australia. *J. Struct. Geol.*, **14**, 591–604.

Harte, B., Hunter, R.H. and Kinny, P.D. (1993) Melt geometry, movement and crystallisation, in relation to mantle dykes, veins and metasomatism. *Phil. Trans. R. Soc. Lond.*, **A342**, 1–21.

Hippertt, J.F. (1994) Structures indicative of helicoidal flow in a migmatitic diapir (Bação Complex, southeastern Brazil). *Tectonophysics*, **234**, 169–96.

167

Hodge, D.S., Abbey, D.A., Harbin, M.A. *et al.* (1982) Gravity studies of subsurface mass distributions of granitic rocks in Maine and New Hampshire. *Am. J. Sci.*, **282**, 1289–324.

Holtz, F. and Johannes, W. (1994) Maximum and minimum water contents of granitic melts – implications for chemical and physical-properties of ascending magmas. *Lithos*, **32**, 149–59.

Huang, W-L. and Wyllie, P.J. (1975) Melting reactions in the system $NaAlSi_3O_8$-$KAlSi_3O_8$-SiO_2 to 35 kilobars, dry and with excess water. *J. Geol.*, **83**, 737–47.

Hyndman, D.W. (1983) The Idaho batholith and associated plutons, Idaho and Western Montana. *Geol. Soc. Am. Mem.*, **159**, 213–40.

Jébrak, M., Lequentrec, M.F., Maréschal, J.C. *et al.* (1991) A gravity survey across the Bourlamaque massif, southeastern Abitibi greenstone-belt, Quebec, Canada – the relationship between the geometry of tonalite plutons and associated gold mineralization. *Precambrian Res.*, **50**, 261–8.

Johannes, W. (1984) Beginning of melting in the granite system Qz-Or-Ab-An-H_2O. *Contrib. Mineral. Petrol.*, **86**, 264–73.

Johannes, W. and Holtz, F. (1990) Formation and composition of H_2O-undersaturated granitic melts, in *High-temperature Metamorphism and Crustal Anatexis* (eds J.R. Ashworth and M. Brown), Unwin Hyman, London, pp. 87–104.

John, B.E. (1988) Structural reconstruction and zonation of a tilted mid-crustal magma chamber: the felsic Chemehuevi Mountains plutonic suite. *Geology*, **16**, 613–17.

Jurewicz, S.R. and Watson, E.B. (1985) The distribution of partial melt in a granitic system: the application of liquid phase sintering theory. *Geochim. Cosmochim. Acta*, **49**, 1109–21.

Kerr, R.C. and Lister, J.R. (1991) The effects of shape on crystal settling and on the rheology of magmas. *J. Geol.*, **99**, 457–67.

Kerr, R.C. and Lister, J.R. (1995) Comment on 'On the relationship between dike width and magma viscosity' by Yutaka Wada. *J. Geophys. Res.*, **B100**, 15541.

Knapp, R.B. and Knight, J.E. (1977) Differential thermal expansion of pore fluids: fracture propagation and microearthquake production in hot pluton environments. *J. Geophys. Res.*, **B82**, 2515–22.

Koenders, M.A., Petford, N. and Clayton, D. (1996). Granular temperature and structural models for magmatic flows. Abstract, international meeting on Syntectonic Crystallization of Igneous Rocks: Structures, Mechanisms, Models. Waxenberg, Austria.

Laporte, D. and Watson, E.B. (1995) Experimental and theoretical constraints on melt distribution in crustal sources. *Chem. Geol.*, **124**, 161–84.

Lazarus, D. and Nachtrieb, N.H. (1963) Effects of high pressure on diffusion, in *Solids Under Pressure* (eds W. Paul and D.M. Warschauer), McGraw-Hill, New York, pp. 43–69.

Leake, B.E. (1990) Granite magmas: their sources, initiation and consequences of emplacement. *J. Geol. Soc. Lond.*, **147**, 579–89.

Lister, J.R. and Kerr, R.C. (1991) Fluid-mechanical models of crack propagation and their application to magma transport in dykes. *J. Geophys. Res.*, **B96**, 10049–77.

Lowell, R.P. and Bergantz, G. (1987) Melt stability and compaction in a partially molten silicate layer heated from below, in *Structure and Dynamics of Partially Solidified Systems* (ed D.E. Loper), Martinus Nijhoff Publishers, Dordrecht, pp. 383–400.

Lynn, H.B., Hale, L.D. and Thompson, G.A. (1981) Seismic reflections from the basal contacts of batholiths. *J. Geophys. Res.*, **B86**, 10633–8.

McKenzie, D. (1987) The compaction of igneous and sedimentary rocks. *J. Geol. Soc. Lond.*, **144**, 299–307.

Maaløe, S. (1992) Melting and diffusion processes in closed-system migmatization. *J. Metamorphic. Geol.*, **10**, 503–16.

REFERENCES

Mahon, K.I., Harrison, T.M. and Drew, D.A. (1988) Ascent of a granitoid diapir in a temperature varying medium. *J. Geophys. Res.*, **B93**, 1174–88.

Marsh, B.D. (1978) On the cooling of ascending andesitic magma. *Phil. Trans. R. Soc. Lond.*, **A288**, 611–25.

Marsh, B.D. (1982) On the mechanics of diapirism, stoping and zone melting. *Am. J. Sci.*, **282**, 808–53.

Marsh, B.D. (1984) Mechanics and energetics of magma formation and ascension, in *Explosive Volcanism: Inception, Evolution, and Hazards*, National Academy Press, Washington, D.C., pp. 67–83.

Mawer, C.K. (1989) Melt formation and migration in experimental and natural situations. *28th I.G.C. Abstracts*, **2**, 390.

Miller, C.F., Watson, E.B. and Harrison, T.M. (1988) Perspectives on the source, segregation and transport of granitoid magmas. *Trans. R. Soc. Edinburgh: Earth Sci.*, **79**, 135–56.

Mrazec, M.L. (1915) Les plis diapirs et le diapirisme en géneral. *Rumania Inst. Geol. Comptes Rendus*, **4**, 226–70.

Naney, M.T. (1977) *Phase Equilibria and Crystallization in Iron- and Magnesium-bearing Granitic Systems*, PhD Thesis, Stanford University, USA.

Neilson, D.L., Clark, R.G., Lyons, J.B. *et al.* (1976) Gravity models and modes of emplacement of the New Hampshire plutonic series. *Geol. Soc. Am. Mem.*, **146**, 301–18.

Nicolas, A. (1986) A melt extraction model based on structural studies in mantle peridotites. *J. Petrol.*, **27**, 999–1022.

Nicolesco, C.P. (1929) Anticlinaux diapirs sédimentaires, volcaniques et plutoniques. *Bull. Soc. Géol. Fr.*, **29**, 21–4.

Paquet, J. and François, P. (1980) Experimental deformation of partially melted granitic rocks at 600–900 °C and 250 MPa confining pressure. *Tectonophysics*, **68**, 131–46.

Paterson, S.R. and Fowler, T.K.J. (1993) Re-examining pluton emplacement processes. *J. Struct. Geol.*, **15**, 191–206.

Paterson, S.R. and Vernon, R.H. (1995) Bursting the bubble of ballooning plutons: a return to nested diapirs emplaced by multiple processes. *Geol. Soc. Am. Bull.*, **107**, 1356–80.

Petford, N. (1995) Segregation of tonalitic-trondhjemitic melts in the continental crust: the mantle connection. *J. Geophys. Res.*, **B100**, 15735–43.

Petford, N. (1996) Dykes or diapirs? *Trans. R. Soc. Edinburgh* (in press).

Petford, N., Kerr, R.C. and Lister, J.R. (1993) Dike transport of granitic magmas. *Geology*, **21**, 843–5.

Petford, N., Kerr, R.C. and Lister, J.R. (1994) Dike transport of granitic magmas: reply. *Geology*, **22**, 474–5.

Phillips, G.N., Wall, V.J. and Clemens, J.D. (1981) Petrology of the Strathbogie batholith: a cordierite-bearing granite. *Can. Mineral.*, **19**, 47–64.

Pollard, D.D. (1977) Derivation and evaluation of a mechanical model for sheet intrusions. *Tectonophysics*, **19**, 233–69.

Ramberg, H. (1967) *Gravity, Deformation and the Earth's Crust*, Academic Press, London.

Ramsay, J.G. (1989) Emplacement kinematics of a granite diapir: the Chindamora batholith, Zimbabwe. *J. Struct. Geol.*, **11**, 191–209.

Rapp, R.P., Watson, E.B. and Miller, C.F. (1991) Partial melting of amphibolite/eclogite and the origin of Archean trondhjemites and tholeiites. *Precambrian Res.*, **51**, 1–25.

Read, H.H. (1948) Granites and granites, *Origin of Granite*, Geol. Soc. Am. Mem., **28**, 1–19.

Ribe, N.M. (1987) Theory of melt segregation – a review. *J. Volc. Geotherm. Res.*, **33**, 241–53.

Rickard, M.J. and Ward, P. (1981) Paleozoic crustal thickness in the southern part of the Lachlan Orogen deduced from volcano and pluton-spacing geometry. *J. Geol. Soc. Aust.*, **28**, 19–32.

Robin, P.Y.F. (1979) Theory of metamorphic segregation and related processes. *Geochim. Cosmochim. Acta*, **43**, 1587–1600.

Rubin, A. M. (1995). Getting granitic dikes out of the source region. *J. Geophys. Res*, **B100**, 5911–29.

Rushmer, T. (1991) Partial melting of 2 amphibolites – contrasting experimental results under fluid-absent conditions. *Contrib. Mineral. Petrol.*, **107**, 41–59.

Rushmer, T. (1995) An experimental deformation study of partially molten amphibolite application to flow-melt fraction segregation. *J. Geophys. Res.*, **B100**, 15681–95.

Rutter, E.H. (this volume) The influence of deformation on the extraction of crustal melts: a consideration of melt-assisted granular flour.

Rutter, E.H. and Neumann, D.H.K. (1995) Experimental deformation of partially molten Westerly Granite under fluid-absent conditions, with implications for the extraction of granitic magmas. *J. Geophys. Res.*, **B100**, 15697–715.

Rutter, M. and Wyllie, P.J. (1988) Melting of vapour-absent tonalite at 10 kbar to simulate dehydration-melting in the deep crust. *Nature*, **331**, 159–60.

Sawyer, E.W. (1991) Disequilibrium melting and the rate of melt-residuum separation during migmatization of mafic rocks from the Grenville Front, Quebec. *J. Petrol.*, **32**, 701–38.

Sawyer, E.W. (1994) Melt segregation in the continental-crust. *Geology*, **22**, 1019–22.

Scaillet, B., Pêcher, A., Rochette, P. *et al.* (1995) The Gangotri granite (Garhwal Himalaya): laccolithic emplacement in an extending collisional belt. *J. Geophys. Res.*, **B100**, 585–607.

Scarfe, C.M., Mysen, B.O. and Virgo, D. (1987) Pressure dependence of the viscosity of silicate melts, in *Magmatic Processes: Physicochemical Principles* (ed. B.O. Mysen), The Geochemical Society, University Park, Penn. pp. 59–68.

Schmeling, H., Cruden, A.R. and Marquart, G. (1988) Finite deformation in and around a fluid sphere moving through a viscous medium: implications for diapiric ascent. *Tectonophysics*, **149**, 17–34.

Schwerdtner, W.M. (1990) Structural tests of diapir hypotheses in Archean crust of Ontario. *Can. J. Earth Sci.*, **27**, 387–402.

Sleep, N.H. (1988) Tapping of melt by veins and dykes. *J. Geophys. Res.*, **B93**, 10255–72.

Smith, R.L. (1979) Ash-flow magmatism, in *Ash Flow Tuffs* (eds C.E. Chapin and W.E. Elston), The Geological Society of America, Boulder, CO, pp. 5–27.

Spear F.S., Peacock S.M., Kohn M.J. *et al.* (1991) Computer programs for petrologic P-T-t path calculations. *Am. Mineral.*, **76**, 2009–12.

Speer, J.A., McSween, H.Y., Jr and Gates, A.E. (1994), Generation, segregation, ascent and emplacement of Alleghanian plutons in the southern Appalachians. *J. Geol.*, **102**, 249–67.

Spence, D.A. and Turcotte, D.L. (1990) Buoyancy-driven magma fracture: a mechanism for ascent through the lithosphere and the emplacement of diamonds. *J. Geophys. Res.*, **B95**, 5133–9.

Spence, D.A, Sharp, P.W. and Turcotte, D.L. (1987) Buoyancy-driven crack-propagation – a mechanism for magma migration. *J. Fluid Mech.*, **174**, 135–53.

Spera, F.J. (1980) Aspects of magma transport, in *Physics of Magmatic Processes* (ed. R.B. Hargraves), Princeton University Press, Princeton, NJ, pp. 265–323.

Stevens, G. and Clemens, J.D. (1993) Fluid-absent melting and the roles of fluids in the lithosphere: a slanted summary? *Chem. Geol.*, **108**, 1–17.

Stevenson, D. J. (1989) Spontaneous small-scale melt segregation in partial melts undergoing deformation. *Geophys. Res. Lett.*, **16**, 1067–70.

Sweeney, J.F. (1975) Diapiric granite batholiths in south-central Maine. *Am. J. Sci.*, **275**, 1183–91.

Sweeney, J.F. (1976) Subsurface distribution of granitic rocks, south-central Maine. *Geol. Soc. Am. Bull.*, **87**, 241–9.

REFERENCES

Sykes, M.J. and Holloway, J.R. (1987) Evolution of granitic magmas during ascent: A phase equilibrium model, in *Magmatic Processes: Physicochemical Principles* (ed. B.O. Mysen), The Geochemical Society, University Park, Penn., pp. 447–61.

Takada, A. (1989) Magma transport and reservoir formation by a system of propagating cracks. *Bull. Volcanol.*, **52**, 118–26.

Takada, A. (1990) Experimental study on propagation of liquid-filled crack in gelatin: shape and velocity in hydrostatic stress condition. *J. Geophys. Res.*, **B95**, 8471–81.

Thomas, R. (1994) Estimation of the viscosity and the water content of silicate melts from melt inclusion data. *Eur. J. Mineral.*, **6**, 511–35.

Thompson, A.B. (1990) Heat, fluids and melting in the granulite facies, *Granulites and Crustal Differentiation*, Kluwer Academic Publishers, Dordrecht, pp. 37–58.

Turcotte, D.L. (1987) Physics of magma segregation processes, in *Magmatic Processes: Physicochemical Principles* (ed. B.O. Mysen), The Geochemical Society, University Park, Penn., pp. 69–74.

Van der Molen, I. and Paterson, M.S. (1979) Experimental deformation of partially-melted granite. *Contrib. Mineral. Petrol.*, **70**, 229–38.

Vielzeuf, D., Clemens, J.D. and Pin, C. (1990) Granites, granulites and crustal differentiation, in *Granulites and Crustal Differentiation* (eds D. Vielzeuf and P. Vidal), Kluwer Academic Publishers, Dordrecht, pp. 59–86.

Vigneresse, J.-L. (1988) Forme et volume des plutons granitiques. *Bull. Soc. Géol. Fr.*, **8**, 897–906.

Vigneresse, J.-L. (1990) Thermal data and crustal structure, in *Granulites and Crustal Differentiation* (eds D. Vielzeuf and P. Vidal), Kluwer Academic Publishers, Dordrecht, pp. 551–68.

Vigneresse, J.-L. (1995) Crustal regime of deformation and ascent of granitic magma. *Tectonophysics*, **249**, 187–202.

Wada, Y. (1994) On the relationship between dike width and magma viscosity. *J. Geophys. Res.*, **B99**, 17743–55.

Wall, V.J., Clemens, J.D. and Clarke, D.B. (1987) Models for granitoid evolution and source compositions. *J. Geol.*, **95**, 731–50.

Webb, S.L. and Dingwell, D.B. (1990) The onset of non-Newtonian rheology of silicate melts. A fibre elongation study. *Phys. Chem. Minerals*, **17**, 125–32.

Weertman, J. (1971) Theory of water-filled crevasses in glaciers applied to vertical magma transport beneath ocean ridges. *J. Geophys. Res.*, **B76**, 1171–83.

Weertman, J. (1980) The stopping of a rising, liquid-filled crack in the Earth's crust by a freely slipping horizontal joint. *J. Geophys. Res.*, **B85**, 967–76.

Weinberg, R.F. and Podladchikov, Y. (1994) Diapiric ascent of magmas through power law crust and mantle. *J. Geophys. Res.*, **B99**, 9543–59.

White, A.J.R. and Chappell, B.W. (1977) Ultrametamorphism and granitoid genesis. *Tectonophysics*, **43**, 7–22.

Whitehead, J.A. and Luther, D.S. (1975) Dynamics of laboratory diapirs and plume models. *J. Geophys. Res.*, **B80**, 705–17.

Wilson, L. and Head, J.W. (III) (1981) Ascent and eruption of basaltic magma on the Earth and Moon. *J. Geophys. Res.*, **B86**, 2971–3001.

Wolf, M.B. and Wyllie, P.J. (1991) Dehydration-melting of solid amphibolite at 10 kbar: textural development, liquid interconnectivity and applications to the segregation of magmas. *Mineral. Petrol.*, **44**, 151–79.

Wyllie, P.J., Huang, W.-L., Stern, C.R. *et al.* (1976) Granitic magmas: possible and impossible sources, water contents, and crystallization sequences. *Can. J. Earth Sci.*, **13**, 1007–19.

171

Further reading

Shaw, H.R. (1965) Comments on viscosity, crystal settling, and convection in granitic magmas. *Am. J. Sci.*, **263**, 120–52.
Shaw, H.R. (1972) Viscosities of magmatic silicate liquids: an empirical method of prediction. *Am. J. Sci.*, **272**, 870–93.

CHAPTER SEVEN

Ascent and emplacement of granitic plutonic complexes in subduction-related extensional environments

John Grocott and Jeff Wilson

7.1 Introduction

In the Mesozoic magmatic arc of the Andean convergent plate boundary zone in northern Chile, granitic magmas (granite to tonalite, *sensu stricto*) were emplaced during regional extensional/transtensional deformation of the leading edge of the South American plate. These rocks provide an opportunity to study the relationship between local emplacement mechanisms and regional deformation in this type of tectonic setting. Our approach will be to review aspects of ascent and emplacement mechanisms of granitic magmas relevant to the emplacement of plutonic complexes in northern Chile together with the controls of overriding plate deformation at convergent plate boundary zones. This will provide a framework for the presentation and discussion of our data leading to a model linking granitic pluton emplacement to regional deformation in a subduction-related extensional environment.

7.1.1 Magma ascent and emplacement

Ascent and emplacement mechanisms of granitic magmas are controversial fields of geoscience research with the viability of diapiric and dyke-transport ascent mechanisms at the forefront of current debate. In research on magma emplacement, fault-jog dilation mechanisms have been emphasized recently and this has tended to overshadow the role played by magma over-pressure in pluton emplacement. There has also been renewed interest in pluton shape, particularly of high-level plutons and the nature of their continuation to depth. Clemens and Mawer (1992) proposed that granitic magma transport by dykes was an important ascent mechanism capable of supplying magma at rates suffi-

Deformation-enhanced Fluid Transport in the Earth's Crust and Mantle. Edited by M.B. Holness. Published in 1997 by Chapman & Hall, London. ISBN 0 412 75290 5.

cient to emplace upper crustal intrusions of batholithic proportions in less than 10^3 years. This conclusion was supported and extended by Petford *et al.* (1993) who recognized that the potential for very rapid ascent of granitic magma in dykes implies that pluton emplacement rates are controlled by the rate at which space for plutons can be created, for example by fault displacement, and the rate of magma generation in the source region. More recently, discussion of granitic magma ascent mechanisms has polarized between diapiric ascent and ascent by dyke transport (Paterson and Vernon, 1995; Petford, 1996). A consensus is emerging that diapiric ascent of granitic magmas is likely to have been restricted because granitic diapirs rise relatively slowly due to viscosity constraints imposed by crustal materials (Clemens *et al.*, this volume). In contrast, dyke-transport is able to convey large volumes of granitic magma rapidly from a source region to the upper crust by essentially the same mechanism long envisaged for the ascent of basic magmas (Vigneresse, 1995; Petford, 1996).

Many plutonic complexes emplaced at mid- to high-crustal levels cannot have been emplaced as ballooning plutons or as diapirs because they are not associated with the diagnostic structures and fabrics in the country rocks consistent with this hypothesis (Bateman, 1984; England, 1992; Schwerdtner, 1995; Clemens *et al.*, this volume). In recognition of this, re-evaluation of emplacement mechanisms has given prominence to the idea that space for pluton emplacement is created at dilational sites in fault systems (Hutton, 1988; Hutton *et al.*, 1990; Hutton, 1992; Hutton and Reavy, 1992). This notion has become popular and fault-dilation is widely cited as an emplacement mechanism for granitic magmas in many tectonic environments (Glazner, 1991; Tikoff and Teyssier, 1992; Karlstrom *et al.*, 1993). Emplacement of magmas passively at dilational sites at high-levels in fault systems has been linked to the dyke-transport magma ascent mechanism (Grocott *et al.*, 1994), with the implication that fault displacement is the rate-controlling step for emplacement and for magma ascent in dykes. However, fault-displacement rates are apparently too slow to allow a magma chamber to form in cold upper crust and instead each successively emplaced dyke would freeze to produce a sheeted dyke complex before a magma chamber could be established (Paterson and Fowler, 1993).

Hanson and Glazner (1995) examined in detail the thermal requirements for the emplacement of granitic plutons during extensional deformation and showed that shallow andesitic magma chambers can be maintained at extension rates typical of continental crust (*c.* 0.5 cm/yr), provided that a small magma chamber is created by a short initial burst of rapid extension facilitated by multiple dyking (*c.* 5 cm/yr). This would heat the upper crust and avoid the thermal problem identified by Paterson and Fowler (1993), prolonging the life of the magma chamber. This in turn removes the need for very small-scale sheeted construction in plutonic complexes formed by fault-dilation.

Plutonic complexes rarely have geometries that can be restored to reveal an original over-stepping or releasing-bend fault trajectory and this represents a weakness of fault-jog dilation emplacement models. For example, Ingram and

Hutton (1994) have shown that the 800 km-long Great Tonalite Sill, in south-eastern Alaska and British Columbia, was emplaced as a composite, sheeted pluton into a contractional shear zone when magma pressure exceeded the regional normal stress acting across the pluton margin. In these circumstances, magma emplacement exploits the anisotropy provided by pre-existing shear zones, but space is created by magma pressure, as predicted by theoretical models of dyke emplacement (Delaney et al., 1986; Lister and Kerr, 1991), rather than by fault-jog dilation. Consequently, magma pressure may have a more active role in pluton emplacement than is implied by fault-jog dilation models and magma supply and fault displacement rates need not necessarily be in step even in those cases where emplacement is mainly fault-controlled.

7.1.2 Overriding plate deformation

In continental magmatic arcs, crustal deformation at the leading edge of the over-riding plate is often thought of as primarily due to forces applied at plate bound-aries in response to plate kinematics and there have been several attempts to link overriding plate deformation at convergent plate boundary zones more-or-less directly to plate vectors (Dewey, 1980; Jarrard, 1986; Doglioni, 1991; Royden, 1993; Scheuber et al., 1994). Jarrard (1986) has proposed a seven-fold classifica-tion of modern arc systems which reveals that magmas are emplaced in structural environments that range from active spreading ridges in back-arc areas to very strong contractional arcs. Important controls on the structural setting of the over-riding plate, and the structural style of plutonic complexes emplaced within it, are the relative rates of slab roll-back velocity (V_r) and the trenchward velocity of the overriding plate (V_c) (Russo and Silver, 1996). When the overriding plate is advancing slowly, or is stationary, slab roll-back (Hamilton, 1995) leads to exten-sion in the overriding plate with the change in relative velocity $\Delta V = V_c - V_r < 0$ (Figure 7.1(b)). In essence, although subduction continues, roll-back leaves the overriding plate edge unsupported, allowing it to collapse towards the ocean

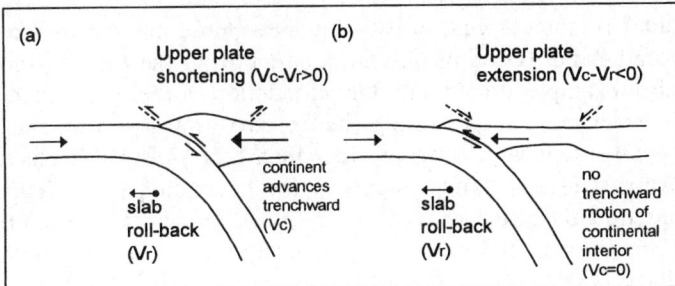

Figure 7.1 (a) Velocity of the continent toward the trench (V_c) is greater than the velocity of slab roll-back (V_r). The overriding plate is characterized by contractional deformation. (b) Slab roll-back (V_r) is greater than the velocity of the continent toward the trench (V_c). The overriding plate is char-acterized by extensional deformation. (After Russo and Silver, 1996.)

175

(Vissers *et al.*, 1995; Dokka and Ross, 1995). Extension may also be induced in the overriding plate by the viscous drag associated with flow in the mantle wedge above the down-going plate driven by its motion (Furukawa, 1993). In these circumstances plutonic complexes will be emplaced in an extensional environment in the overriding plate and extension may be sufficiently pronounced to generate core-complexes in the back-arc domain (Lister and Baldwin, 1993; Mpodozis and Allmendinger, 1993; Dokka and Ross, 1995). When the overriding plate advances rapidly trenchward, its velocity may exceed the roll-back velocity of the subducting plate with the change in relative velocity $\Delta V = V_c - V_r > 0$ (Figure 7.1(a)). In these circumstances marginal basins will be inverted and a cordillera will form by crustal thickening as the continent collides with the oceanic slab and subslab mantle (Russo and Silver, 1996).

7.1.3   Trench-parallel strike-slip faults

Trench-parallel strike-slip faults characterize the overriding plate at convergent plate boundary zones and they indicate that part of the oblique component of the plate motion vectors is routinely partitioned into overriding plate, strike-slip fault systems (Dewey, 1980; Woodcock, 1986; Sylvester, 1988). These strike-slip faults are commonly located along the axis of the arc, a position widely attributed to thermal weakening in the overriding plate. Glazner (1991) has proposed that oblique subduction and associated strike-slip faulting in the overriding plate are essential for plutonism in magmatic arcs, arguing that dilation of strike-slip faults is necessary to provide sites for magma to lodge in the crust. Conversely, according to Glazner (1991), during head-on convergence magmas ascend to the surface and the arc is dominated by volcanism. In fact, there seems to be no good reason to single out strike-slip environments as particularly favourable for plutonism since, even if dilation of fabric anisotropy by magma over-pressure is discounted as an emplacement mechanism, contractional fault systems, or extensional fault systems associated with retreating subduction boundaries, are equally likely to provide fault-dilation sites for granitic magma emplacement (Tikoff and Teyssier, 1992; Grocott *et al.*, 1994).

Unfortunately, the correlation between shear sense on trench-linked strike-slip faults and the direction of oblique subduction in the Andes is not always clear-cut. For example, dominantly NE subduction of the Nazca plate beneath the South American margin during the Tertiary (Pardo-Casas and Molnar, 1987), is inconsistent with observed increments of sinistral displacement on contemporary strike-slip fault systems in the overriding plate (Yáñez *et al.*, 1994). Consequently, although plate-vector modelling gives general predictions about deformation in the overriding plate, additional factors are involved. These include:

1. extensional collapse of the over-riding plate induced by gravitational instability caused by crustal thickening and/or retreat of the plate margin (Dokka and Ross, 1995);

2. helicoidal corner flow in the mantle wedge beneath the overriding plate which transmits stresses to the base of the plate by viscous drag (Furukawa, 1993; Yáñez et al., 1994);
3. strain partitioning in response to rheological changes caused, for example, by arc magmatism (Wilson, 1996);
4. heterogeneous response of the overriding plate due to the presence of 'buttresses' which may inhibit margin-parallel transport (Beck, 1988).

These factors are likely to hinder attempts to interpret overriding plate deformation and plutonism solely in terms of plate vectors. In this contribution we show in particular how factors (1) and (2) may have influenced magmatic arc evolution in the Andes between 25°S and 27°S.

7.2 Permian to Cretaceous magmatic arc 25°S to 27°S in northern Chile

Mechanisms of magma ascent and pluton emplacement clearly operate in a range of contractional to extensional structural settings in modern magmatic arcs (Jarrard, 1986). In this section we review pluton emplacement in Permian to Early Cretaceous magmatic arcs in the Andean margin of northern Chile in a well-constrained extensional to transtensional structural setting (Scheuber and Andriessen, 1990; Scheuber and Reutter, 1992; Brown et al., 1993; Scheuber et al., 1994, 1995). We argue that emplacement mechanisms in continental magmatic arcs are dominated more by the exploitation of fabric anisotropy by magma over-pressure than by fault-jog dilation and are not dependent on any particular state of stress in the overriding plate. From this we conclude that pluton shape and relation to fault systems owes more to emplacement level and crustal anisotropy than to the type of strain occurring regionally in the overriding plate.

7.2.1 Regional geology

Permian to Early Cretaceous magmatic arc rocks crop-out in a 50 km wide, trench-parallel belt between 25°S and 27°S in the Coastal Cordillera of northern Chile (Figure 7.2). Exposed basement comprises Devonian to Carboniferous low-grade, pelitic to psammitic metasedimentary rocks (Bahlburg and Breitkreuz, 1993) that are strongly deformed by a west-vergent fold-thrust belt (Bell, 1987). Permian plutonic complexes exposed close to the present-day coastline were emplaced into this basement following this contractional event and plutonic rocks with a similar age and geochemistry are exposed across a wide belt extending 120 km eastward to the Altiplano (Brown, 1991). In the Coastal Cordillera, eastward younging, Triassic to Early Cretaceous magmatic arc rocks are exposed to the east of the Permian plutonic complexes and form the main focus of this study.

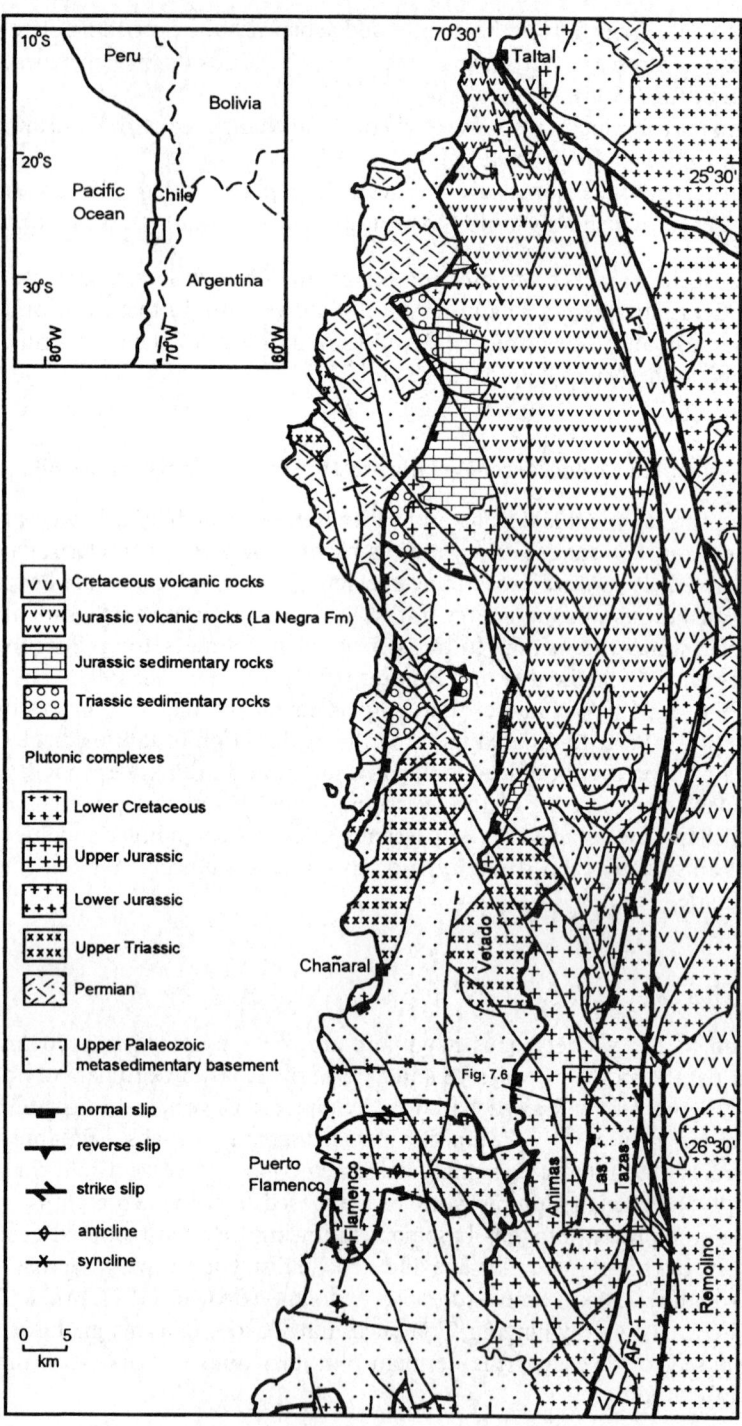

Cretaceous volcanic rocks

Jurassic volcanic rocks (La Negra Fm)

Jurassic sedimentary rocks

Triassic sedimentary rocks

Plutonic complexes

Lower Cretaceous

Upper Jurassic

Lower Jurassic

Upper Triassic

Permian

Upper Palaeozoic metasedimentary basement

normal slip

reverse slip

strike slip

anticline

syncline

7.2.2 Geochronology

The Permian and Triassic plutonic complexes have narrow metamorphic aureoles which are most evident where pelitic rocks of the metasedimentary basement form the country rock and andalusite has crystallized, but there is little field evidence for synplutonic ductile deformation in the aureole. In contrast, emplacement of Lower Jurassic to Lower Cretaceous plutonic complexes in the Coastal Cordillera of northern Chile was associated temporally and spatially with deformation in ductile shear zones and our recent geochronology and structural research has focused on these plutonic complexes. The high level of emplacement of the Jurassic and Lower Cretaceous plutonic complexes, and the consequent rapid cooling, has allowed precise $^{40}Ar/^{39}Ar$ geochronology to establish the relationship between deformation and emplacement of these magmas in the arc (Grocott et al., 1994; Dallmeyer et al., 1996). Undeformed plutonic rocks and mylonites in adjacent wall rocks yield $^{40}Ar/^{39}Ar$ hornblende and muscovite cooling ages and U–Pb zircon ages for the undeformed pluton that are, within error, identical (Dallmeyer et al., 1996). This leads to two conclusions that underpin our understanding of the relationship between deformation and magmatism in the Permian–Lower Cretaceous magmatic arc: (i) plutons were emplaced at high level and cooled rapidly; (ii) heat advected to high level during magmatism allowed transient ductile deformation to occur in pluton wall rocks.

7.2.3 Structural setting and emplacement of the plutonic complexes

Alongside the dilational-jog model for emplacement of plutonic complexes in continental magmatic published by Grocott et al. (1994), two further aspects of pluton emplacement are emphasized here. First, it is clearly not necessary to create space at a fault jog or overstep within a fault system in order to emplace plutonic complexes. Rather, magma pressure is capable of dilating any fracture, or pre-existing anisotropy, when it exceeds the normal effective stress acting across it and this allows intrusions to be emplaced as sheets by magma wedging (Delaney et al., 1986; Lister and Kerr, 1991; Ingram and Hutton, 1994). Secondly, where the crust at the level of emplacement has a low-angle fabric anisotropy, successively emplaced, concordant sheets may stack to give a 'cedar-tree'-type geometry and merge to form composite, apparently steep-sided intrusions of batholithic proportions (Evans et al., 1993; McCaffrey et al., 1996). In the following section, we re-evaluate dilational-jog pluton emplacement models for the Coastal Cordillera in the light of this perspective.

(a) Permian and Triassic plutonic complexes
A clear picture relating overriding plate deformation to tectonic setting for

Figure 7.2 Geological map of the area between c. 25° 30'S and 26° 30'S, Atacama region, northern Chile. (After Randall et al., 1996.)

179

Permian and Triassic time for the Andean margin between 25°S and 27°S has yet to emerge. Berg *et al.* (1983) proposed that the Devonian–Carboniferous sedimentary rocks were deformed by thick-skinned thrusting during Permian time, leading to melting in thickened crust and emplacement of plutons at high levels in a contractional setting. There is no evidence of terrane accretion at the Andean margin at this time (Mpodozis and Ramos, 1990), but contraction of the overrriding plate would still occur if westward velocity of the continent exceeded roll-back velocity of the subducting slab ($\Delta V > 0$). However, Late Triassic sedimentary and volcaniclastic basins are rift-related and the sedimentary sequences fine-up into a Lower Jurassic carbonate platform consistent with a phase of thermal subsidence following extension (Suarez and Bell, 1992). This mitigates against the idea that Late Triassic granitic rocks were emplaced in a contractional setting but this cannot be ruled out for the Permian plutonic complexes.

Permian and Triassic tonalite to leucogranite plutonic complexes exposed close to the present-day coast were emplaced into already strongly deformed Devonian–Carboniferous metasedimentary rocks (Figure 7.2). Structures within the metasediments include SW-vergent, major recumbent folds and low-angle mylonite belts (Bell, 1987) refolded by large-scale upright folds. Many contacts of the plutons with the wall rocks are steep brittle faults of the Atacama Fault System (Taylor *et al.*, 1997). When they are exposed, the roof and floor intrusive contacts are often concordant with pre-existing wall-rock foliation. Where composition of the country rocks is pelitic, andalusite and muscovite mark a narrow contact aureole. There is little or no evidence of synplutonic ductile deformation in the aureole and the granitic rocks are mostly unfoliated. Close to intrusive contacts at the western and eastern sides of the plutons, interleaving of thin, concordant granitic sheets with gently to moderately dipping metasedimentary rocks is common. We conclude that Permian and Triassic plutons were emplaced as sheets often parallel to the cleavage/bedding anisotropy of the Palaeozoic metasedimentary basement rocks. Sheet-like form, absence of direct association with over-stepping faults or dilational jogs and lack of internal and external synplutonic deformation are consistent with emplacement by dilation of anisotropy-controlled, mode 1 fractures induced by magma pressure at relatively high structural levels.

(b) Lower Jurassic plutonic complexes
The Lower Jurassic plutonic complexes of Flamenco and Caldera-Pajonales (Figures 7.2 and 7.3) were emplaced into Palaeozoic metasedimentary basement with a generally east-dipping, strong bedding/foliation anisotropy which predates pluton emplacement. The Flamenco complex is exposed close to its roof and all the exposed contacts are upper contacts (Figure 7.2). Mylonitic pelitic schists exposed in the metamorphic aureole overlying the complex contain stretched andalusite showing that the Palaeozoic shear zones were reactivated during or after pluton emplacement (Dallmeyer *et al.*, 1996). Alignment of

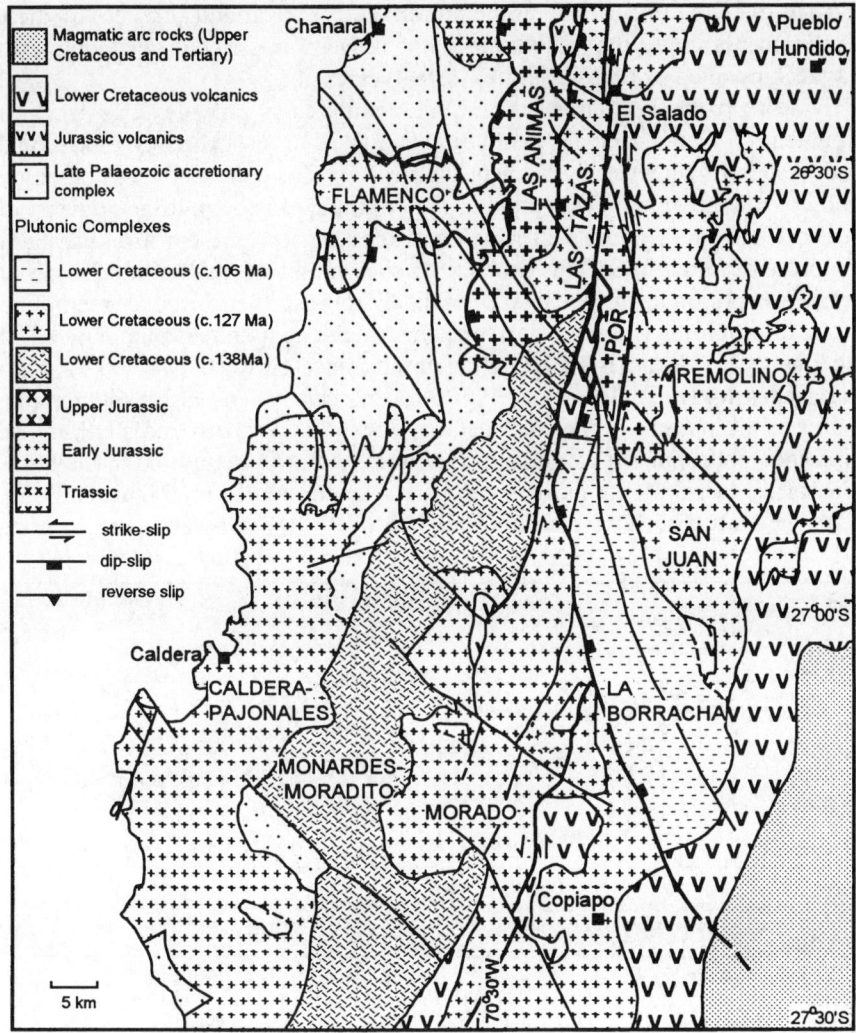

Figure 7.3 Geological map of the area between *c*. 26° 30′S and 27° 30′S, Atacama region, northern Chile. (After Dallmeyer *et al.*, 1996.)

stretched andalusite in the metamorphic aureole defines a NW–SE movement direction (Figure 7.5) and a top-east displacement direction is indicated by asymmetric boudins of thin psammite beds. Kinematic information for this reactivated shear zone is summarized in Figure 7.2. Granite sheets cut the deformed aureole rocks, but are themselves either weakly deformed or undeformed, and consequently we infer that deformation of the aureole was synplutonic. Allowing for later, mid-Cretaceous, rotation of Upper Jurassic basic

181

andesite dykes cutting the Flamenco pluton of *c.* 46° (Randall *et al.*, 1996), these displacements imply that emplacement of Lower Jurassic plutonic complexes was accompanied by east–west (arc-normal) extension.

Evidence from the Flamenco plutonic complex has provided the strongest arguments for emplacement at dilational jogs in an east-dipping extensional fault system (Dallmeyer *et al.*, 1996; Figure 7.4). This conclusion was based on study of the moderately east-dipping upper contact of the pluton southeast of Puerto Flamenco (Figure 7.2) which was thought to represent a dilated fault ramp. Re-examination of this contact has shown that the mylonitic schists that overlie the pluton at this locality are unlikely to be in their original orientation due to later folding (Figure 7.5(a)) and the roughly equidimensional outcrop of the Flamenco pluton is the result of interference between two sets of mid-Cretaceous folds (Taylor *et al.*, 1997). Large-scale, open, east–west trending synforms are present in metasedimentary rocks to the north and south of the Flamenco pluton and it can be inferred that the pluton is exposed in the hinge zone of an east–west trending, south-east-plunging antiform (Figures 7.2 and 7.5). The pluton has also been deformed by north to north-east-trending folds

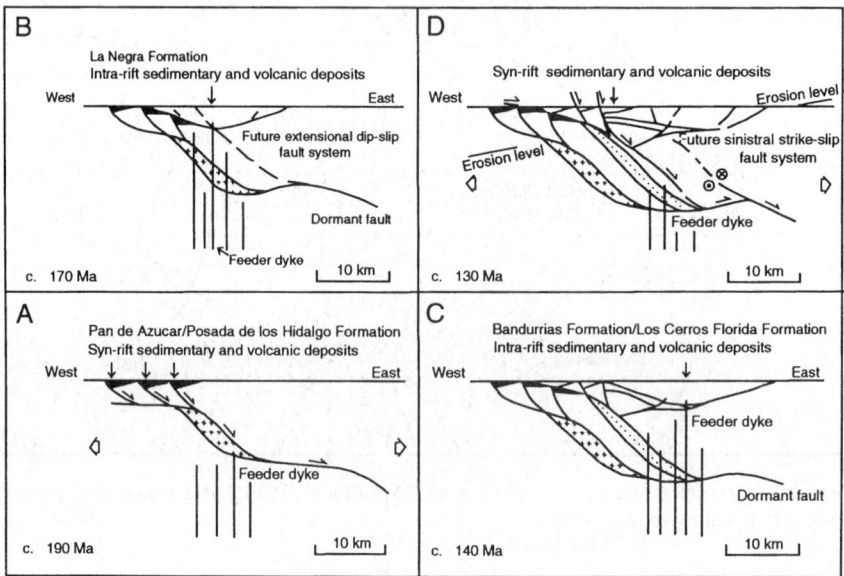

Figure 7.4 Crustal-scale cross-sections through a retreating convergent margin to illustrate the model of magma emplacement within a dilational jog in an extensional fault system and subsequent inboard volcanism as proposed by Grocott *et al.* (1994). Plutonic complexes are fed by dykes that transect the lower crust and they develop as magma accumulates at ramps in active extensional fault systems (A and D). Transient decrease in rate of roll-back relative to trenchward advance of the continent can lead to reduced extension in the upper plate and thereby lead to volcanism because insufficient space is created by fault displacement to accommodate magmas in the crust as plutons (B and C). (After Grocott *et al.*, 1994.)

182

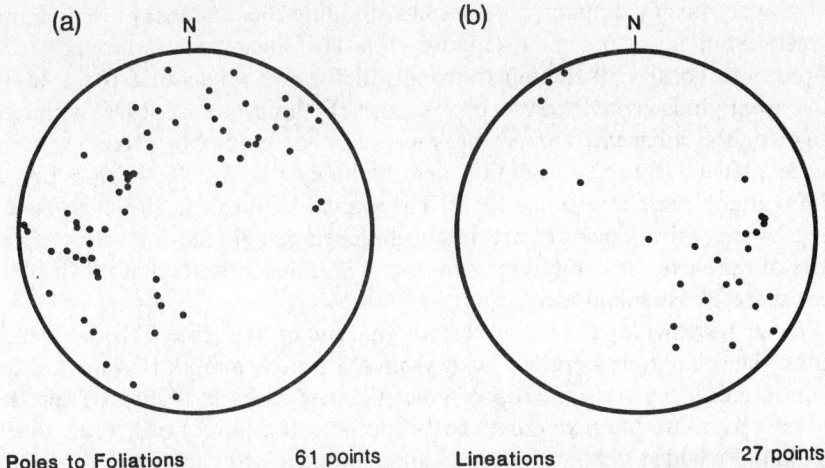

Figure 7.5 Structural data from mylonitic pelitic schists overlying the Flamenco plutonic complex. Lambert Equal Area Projections: (b) extension directions defined by stretched andalusite; (a) foliations; the evident non-cylindricity is due to interference between east–west and north-east–south-west trending folds, the vague π-girdle indicates that east–west trending folds plunge gently south-east.

and the dome-like structure defined by the pluton margins south-east of Puerto Flamenco (Figure 7.2) is interpreted as an interference pattern between these folds and the later, east–west antiform described above (Figure 7.5(a)).

The bases of the Lower Jurassic plutonic complexes are not exposed and we cannot be sure of their shape in three dimensions. Large-scale folds in the roof of the Flamenco complex, and folds of similar wavelength and amplitude in thinly layered metasedimentary rocks that overlie the complex over a wide area (Figure 7.2), are consistent with the control that a thick sheet of granitic rock would exercise on fold wavelength and amplitude in overlying metasedimentary rocks according to the wavelength–thickness relationships of buckle folding (Hudleston, 1986). On this basis we infer that the complex is a sheet-like intrusion emplaced parallel to a pre-existing sub-horizontal bedding/foliation anisotropy in Palaeozoic metasedimentary rocks. We conclude also that the foliation in the metasedimentary rocks was reactivated as a top-east shear zone during or just after emplacement during arc-normal extensional deformation. It remains feasible that the Flamenco complex was emplaced at a dilational jog during this deformation, as advocated by Grocott *et al.* (1994). However, this hypothesis has been weakened by the recognition of refolding of the pluton roof shear zone.

(c) Upper Jurassic-Las Animas complex

The north–south elongate, Upper Jurassic Las Animas quartz-diorite plutonic complex is exposed inboard of the Flamenco pluton. At its western margin, a

100 m wide, steeply dipping, north–south-trending shear zone is present mainly in metasedimentary basement (Figure 7.2). The shear zone is thought to be synplutonic because it contains strongly deformed migmatitic rocks in the aureole with muscovite $^{40}Ar/^{39}Ar$ plateau ages (Dallmeyer et al., 1996) within the range for the adjacent, Upper Jurassic pluton of c. 154 Ma ($^{40}Ar/^{39}Ar$ hornblende plateau isotope correlation age: Dallmeyer et al., 1996) to c. 160 Ma (U–Pb zircon age: Berg et al., 1983). The stretching fabric in the shear zone is steep. Kinematic indicators are not well-developed in the high-temperature rocks of the shear zone, but rare asymmetric boudins indicate that the displacement sense is east-side-down.

Traced northward along the eastern margin of the Late Triassic Vetado pluton the shear zone steps left away from the pluton margin (Figure 7.2) to a position along the western crop of volcaniclastic rocks of a Jurassic (pre-Las Animas) intra-arc basin where an east-dipping extensional brittle fault zone is exposed. Provided the over-step geometry is correctly interpreted and the ductile synplutonic fault and brittle extensional fault are correctly correlated, our interpretation implies that arc-normal extensional deformation continued into the Upper Jurassic. At the point where the shear zone at the western margin of the pluton steps left, the intrusive rocks pass northward into a folded sill complex that cuts arc volcanic rocks of the La Negra Formation (Figure 7.2).

Vertical, NE–SW-trending swarms of basaltic andesite dykes cut the Flamenco and Vetado plutonic complexes. The dykes have yielded Upper Jurassic whole rock $^{40}Ar/^{39}Ar$ ages indicating an emplacement age range similar to the Las Animas plutonic complex (Dallmeyer et al., 1996). Palaeomagnetic studies of the dykes in the Flamenco and Vetado plutons reveal mean clockwise rotations of c. 46° and c. 49° respectively, interpreted to be of mid-Cretaceous age (Randall et al., 1996). Restoration of this rotation implies that this dyke swarm was emplaced with a north–south orientation and therefore the dykes are also consistent with Upper Jurassic arc-normal extension.

We interpret the Las Animas plutonic complex to be emplaced as a concordant sheet within a steeply dipping synplutonic shear zone. Steep stretching lineations and east-down displacements in this shear zone together with the orientation of Upper Jurassic dyke swarms are all consistent with a component of arc-normal extension during the Upper Jurassic. The northward change in intrusive style, from a steeply dipping sheet emplaced in a shear zone within Palaeozoic basement rocks to a sill complex in Jurassic volcaniclastic rocks, implies that at a higher structural level the bedding anisotropy presented the weakest fabric element in these rocks and was exploited by magma over-pressure. There is no specific evidence that the pluton or intrusive elements of the sill complex occupy dilated fault-jogs.

(d) Lower Cretaceous Las Tazas complex
The Las Tazas plutonic complex was emplaced c. 130 Ma ago (Berg et al., 1983; Dallmeyer et al., 1996) just inboard of the Las Animas plutonic complex and is

bounded by the western and central branches of the Atacama Fault Zone (Figure 7.2). The complex contains two plutons, one of which is a composite N–S-trending vertical sheet with dimensions 60 km × 10 km (Wilson, 1996). A N–S-trending belt of upper amphibolite facies, mylonitic rocks up to 500 m wide characterized by dip-slip east-down displacements is exposed at the western sidewall of the Las Tazas plutonic complex (Brown *et al.*, 1993; this paper, Figure 7.2). Where the belt is widest, the mylonitic rocks are cut by syntectonic granitic sheets that vary from concordant and strongly foliated to strongly discordant with weak fabrics. At its western margin, the pluton cuts the mylonitic rocks and we interpret the annealed microstructure as a metamorphic effect of the adjacent pluton (Wilson, 1996). Within the western part of the pluton, weak magmatic state fabrics have linear and planar elements oriented parallel to those in the country rock mylonites (Wilson, 1996) and locally a zone of high-temperature crystal-plastic deformation is present in the pluton with similar fabric orientation (Brown *et al.*, 1993). We interpret the fabric in the mylonites at the western side of the pluton and the similarly orientated magmatic state/high-temperature crystal-plastic fabrics in the pluton to mean that dip-slip deformation characterized this part of the arc immediately before, during and immediately after emplacement of the western units of the Las Tazas complex. This is consistent with arc-normal extension of the margin continuing into Early Cretaceous time.

Late synplutonic strain in the Las Tazas complex is reflected by anisotropy of magnetic susceptibility (AMS) and magmatic state fabrics, characterized by a margin-parallel flattening with a weak, steeply south-plunging, linear element (Wilson, 1996; this paper, Figure 7.6). The AMS data show that late-synplutonic deformation partitioned into the pluton was an almost coaxial strain with a weak, oblique-slip non-coaxial component. The south-plunging linear element of the AMS fabric becomes shallower eastward due to an increasing strike-slip component in the non-coaxial portion of the strain. In addition, conjugate, melt-filled C′-type shear zones consistent with bulk horizontal shortening rework mylonites immediately adjacent to the complex to the west (Brown *et al.*, 1993).

In contrast to the annealed, high-temperature mylonites at the western margin of the pluton, the eastern margin is characterized by extremely fine grained ultramylonitic rocks with sub-horizontal stretching lineations. Deformation in these mylonites was polyphase, and dextral kinematic indicators (asymmetric quartz c-axis fabrics; oblique new-grain shape fabrics in recrystallized quartz) are overprinted by S-C fabrics consistent with sinistral strike-slip. The ultramylonites have greenschist facies mineralogy and, in contrast to the western mylonites they do not contain syntectonic granite sheets. Since they have not been annealed at high temperatures, but share the same *c.* 130 Ma ^{40}Ar/^{39}Ar hornblende cooling age as the western mylonites and the pluton itself (R.D. Dallmeyer, unpublished data), they are interpreted to have formed during cooling of the complex. The transition from high-temperature

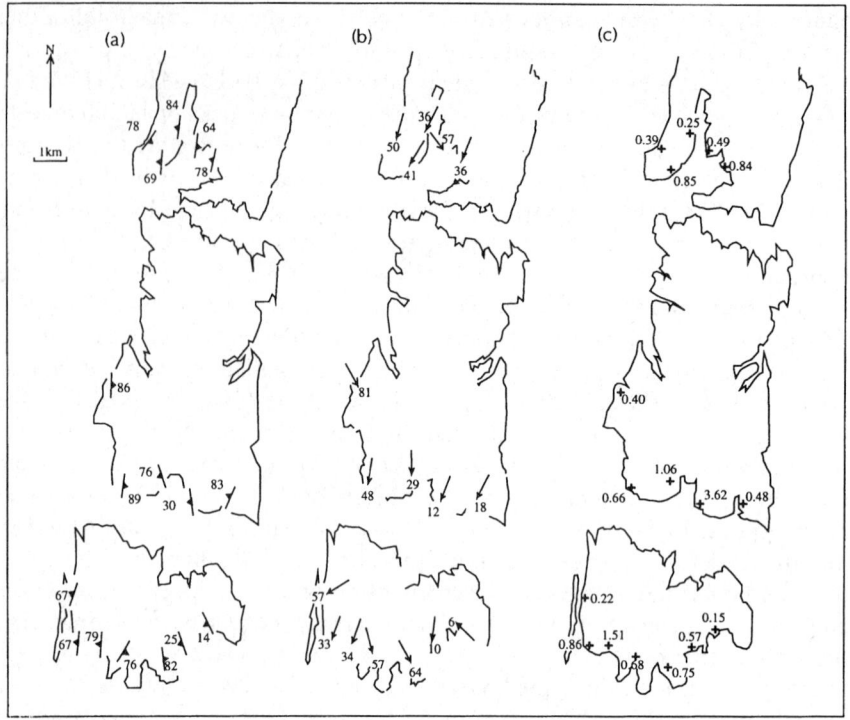

Figure 7.6 AMS data from the central part of the Las Tazas Complex. Maps of (a) AMS foliations; (b) AMS lineations; (c) AMS ellipsoid shape parameter Pflinn (analogous to the K shape parameter). Late synplutonic strain is characterized by a dominant margin-parallel flattening strain, as the majority of ellipsoids have a Pflinn 0<K<1, with a subordinate south-plunging linear element. The AMS data show that late-synplutonic deformation partitioned into the pluton was almost a coaxial strain with a weak, oblique-slip non-coaxial component.

dip-slip mylonites which (just) pre-date emplacement of the pluton at the western margin, through AMS and magmatic state fabrics with an increasing strike-slip component from west to east across the pluton, to low-temperature strike-slip mylonites in the eastern wall rocks of the pluton is consistent with growth of the plutonic complex from west to east and a transition in arc kinematics from dominantly dip-slip to dominantly strike-slip deformation during the emplacement of the Las Tazas complex.

Grocott *et al.* (1994) believed that the Las Tazas complex was emplaced at a dilational jog in an extensional fault system (Figure 7.4). However, the western mylonitic belt with dip-slip displacements is steeply dipping and there is insufficient vertical relief to demonstrate the required dilational-jog geometry (Wilson, 1996). Alternatively, the pluton may simply have dilated the vertical, N–S-trending belt of mylonitic rocks now exposed at its western margin when magma pressure overcame the normal stress acting across the fabric anisotropy in the ductile shear belt. Dilation of the anisotropy was facilitated by slip along a north-west-trending fault at the southern margin of the intrusion (Figure 7.7).

(e) Lower Cretaceous Remolino complex
The Remolino plutonic complex (Figure 7.3) was emplaced east of the Las Tazas complex at *c.* 127 Ma (Dallmeyer *et al.*, 1996). At its western margin a broad belt of greenschist facies, sinistral strike-slip mylonitic rocks has, within error, the same $^{40}Ar/^{39}Ar$ hornblende plateau isotope correlation age as the undeformed pluton. This mylonite belt can be traced south for at least 60 km on the western side of the Lower Cretaceous batholith (Arévalo, 1995; this paper, Figure 7.3). The eastern country rock of the pluton is characterized by a narrow belt of high-temperature mylonite with a vertical N–S-trending foliation and a down-dip stretching fabric. The steep stretching fabrics show that, although sinistral strike-slip is the dominant style of ductile deformation in the magmatic arc post-*c.* 130 Ma, dip-slip deformation continues to be associated with the early stages of the emplacement of arc plutonic complexes.

(f) Lower Cretaceous La Borracha complex
The La Borracha plutonic complex is the youngest plutonic complex in the Lower Cretaceous batholith between 25°S and 27°S and it has a $^{40}Ar/^{39}Ar$ hornblende plateau isotope correlation age of *c.* 106 Ma (Dallmeyer *et al.*, 1996). Like Las Tazas complex it appears to be a vertical sheet emplaced along the Atacama Fault Zone. However, the pluton trends NNW–SSE, rather than N–S, and it cuts the sinistral shear zone at the western margin of Remolino complex (Figure 7.3). A steep ductile shear zone is present in country rock at the western margin of the pluton with steeply north-plunging stretching lineations and east-

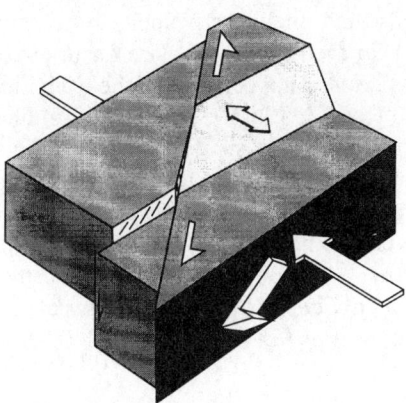

Figure 7.7 Emplacement model for the Las Tazas plutonic complex. The magma was emplaced under a dominant flattening strain by dilation of the Atacama Fault Zone accommodated by slip on a NW-trending fault. The Atacama Fault Zone was undergoing oblique dextral displacement at this time.

down kinematic indicators. Locally, sinistral strike-slip mylonitic deformation was superimposed on the dip-slip fabric. The trend and shape of the pluton, and the evidence for strike-slip, are consistent with emplacement at a releasing bend in the Atacama Fault Zone (Figure 7.3). La Borracha is the only plutonic complex between 25°S and 27°S to show clear map evidence for a fault-dilation emplacement mechanism.

7.3   Discussion

A transition from arc-normal extensional deformation to transtensional deformation occurred during Early Creatceous time (*c.* 130 Ma) in the Coastal Cordillera of northern Chile between 25°S and 27°S (Grocott *et al.*, 1994). In more detail, post-130 Ma transtensional deformation was characterized by an alternation of synplutonic ductile dip-slip deformation, associated with the early stages of the emplacement of plutonic complexes, and sinistral strike-slip ductile to brittle deformation associated with the cooling phase of the complexes. A further key feature of magmatism was the episodic emplacement of arc plutonic complexes during eastward propagation of deformation and plutonism from Triassic to Early Cretaceous time (Dallmeyer *et al.*, 1996). Each plutonic episode led to the construction of an arc batholith with distinctive field and geochemical characteristics (Mercado, 1978; Brown, 1991). We explain these features with a four-stage model for deformation and pluton emplacement in subduction-related extensional environments (Figure 7.8).

7.3.1   Dip-slip deformation in steep shear belts (Figure 7.8(a))

The overriding plate is fixed or advances only slowly towards the trench. Slab roll-back velocity exceeds the trenchward velocity of the continent ($V_c-V_r<0$). Extensional deformation in the overriding plate occurs due to: (i) upper plate topography which drives extensional collapse of the overriding plate during slab roll-back (cf. Dokka and Ross, 1995); (ii) viscous drag-induced trenchward flow of the asthenosphere (corner flow) in the mantle wedge (Furukawa, 1993). Slab roll-back is perpendicular to the trench line irrespective of the direction of convergence and, provided extension is driven primarily by gravitational collapse rather than by corner flow, strike-slip fault systems are dormant and arc-normal displacements accompany pluton emplacement. This situation corresponds to the Chilean Coastal Cordillera between 25°S and 27°S from Triassic to Early Cretaceous time (*c.* 130 Ma).

7.3.2   Magma ascent and emplacement during arc-normal extension (Figure 7.8(b))

Synkinematic intrusions exploit fabric anisotropy provided by extensional ductile shear zones formed in the overriding plate during stage 1. In the example

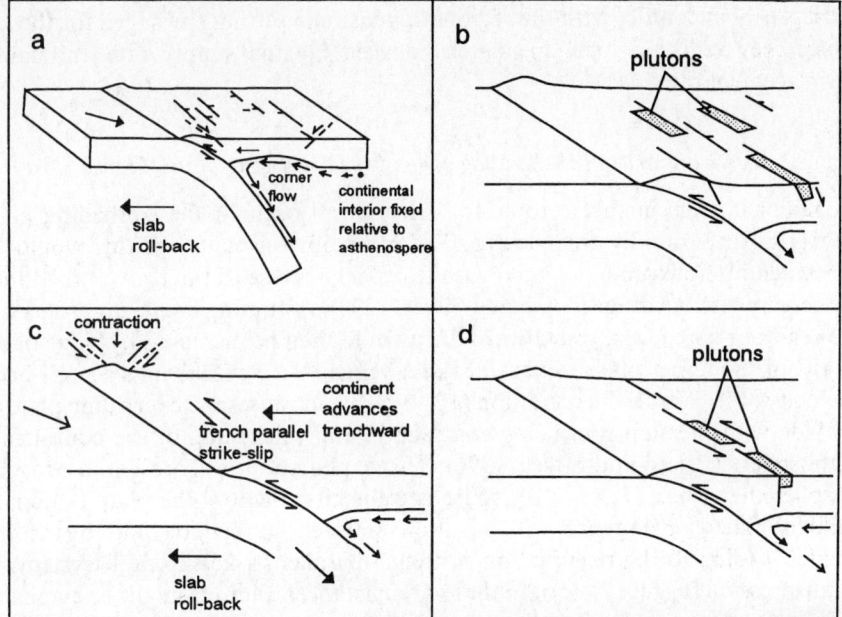

Figure 7.8 (a) Dip-slip deformation in steep shear belts formed during extensional collapse of the margin due to slab roll-back. (b) Magma ascent and emplacement by magma wedging into the fabric anisotropy provided by the dip-slip ductile shear zones. (c) Westward drift of the leading edge of the overriding plate, increased coupling between the overriding and the down-going plate leading to contraction and trench-parallel strike-slip in the upper plate. (d) Trench-linked strike-slip faulting, partitioning of the oblique component of convergence into the overriding plate expressed by increased activity on strike-slip, trench-parallel fault zones as each batholith was progressively emplaced.

shown (Figure 7.8(b)), the leading margin of the overriding plate drifts westward toward the trench during extensional deformation and plutonic complexes are emplaced in an east-younging sequence (cf. Figure 7.4). The emplacement of intrusions may be controlled by dilational jogs in the extensional fault system but, in the Andean margin, there is evidence that wall-rock fabric anisotropy has exercised an important, maybe dominant, control on pluton shape during emplacement.

Late-synplutonic flattening strains associated with the Las Tazas and other plutonic complexes in the Andean margin imply that the local fault displacement rate did not match the rate of magma supply. In these circumstances emplacement of batches of magma may induce a maximum principal stress acting at a high angle to the pluton walls, and/or on previously emplaced sheets. In the Las Tazas pluton, increments of flattening strain shown by magmatic state fabrics, AMS fabrics and conjugate, melt-filled C′ shear bands in the country rocks may reflect this contractional deformation. These strains are not necessarily in conflict with the regional extensional setting envisaged for the arc since they reflect a local mismatch between magma supply rate and fault-displacement rate.

7.3.3   Westward drift of the leading edge of the overriding plate (Figure 7.8(c))

A major change in the tectonic framework will occur if the overriding plate starts to drift rapidly trenchward. The most obvious cause of this would be continental break-up associated with transfer of stress to the base of the litho-sphere by viscous drag (Russo and Silver, 1996). Assuming break-up progresses to sea-floor spreading, trenchward drift would then be increased by ridge push. Drift of the upper plate trenchward at a velocity (V_c) exceeding slab roll-back velocity (V_r) leads to the condition ($V_c-V_r>0$) and causes the overriding plate to collide with the slab producing contraction and thickening of the continental lithosphere (Russo and Silver, 1996). This explanation of Cordilleran orogeny implies that when ($V_c-V_r>0$), stress is transmitted across the plate boundary and oblique convergence will partition between subducting and overriding plates leading to the reactivation of trench-parallel strike-slip fault systems in transpression (Figure 7.8(c)). In these circumstances, plutons might be emplaced at dilational jogs in strike-slip systems (Figure 7.8(d)) or by magma wedging which exploits the fabric anisotropy in these fault systems.

This discussion is relevant to the Andes at 25°S to 27°S because strike-slip displacements began at c. 130 Ma and coincide with the start of sea-floor spreading in the South Atlantic (Scotese, 1991). It is tempting to conclude that trenchward motion of the South American plate at this time (Brown et al., 1993) led to a situation where $V_c-V_r>0$ and to transpression in the upper plate. However, our evidence shows that the strain recorded in the Andean margin between c. 130 and c. 106 Ma was transtensional so that, although opening of the South Atlantic may well be implicated in important mid-Cretaceous trans-pression (Randall et al., 1996) which post-dates the emplacement of the La Borracha pluton at c. 106 Ma, it is difficult to invoke this as a direct cause for the development of a strike-slip component at c. 130 Ma.

An alternative explanation, consistent with transtension in the overriding plate post c. 130 Ma, is that during roll-back a strike-slip component of oblique convergence may still partition into the overriding plate due to corner-flow (Yáñez et al., 1994). Corner flow gives rise to an oblique mantle drag force at the base of the lithosphere when subduction is oblique and this may, in turn, acti-vate trench-parallel strike-slip faults. Provided the condition $V_c-V_r<0$ is main-tained and the margin also collapses gravitationally due to roll-back, the overriding plate will deform by some combination of arc-normal extension and strike slip (transtension). We infer that the change from arc-normal to transten-sional deformation reflects either a change from arc-normal to oblique conver-gence at c. 130 Ma, or an increase in the magnitude of an existing oblique convergence vector, and not a change from the condition $V_c-V_r<0$ to $V_c-V_r>0$.

7.3.4   Deformation partitioning during transtension

When $V_c-V_r < 0$, extensional deformation in the overriding plate is driven by a combination of corner flow-induced viscous drag at the base of the lithosphere and gravitational collapse as the subducting slab rolls back. During subduction-related arc magmatism, thermal erosion of the base of the lithosphere and addition of basaltic magmas to the lithosphere will tend to increase uplift and therefore accelerate gravitational collapse. Consequently, increase in arc-magmatism will favour extensional deformation in the overriding plate.

This idea can explain the observation that transtension in the post $c.$ 130 Ma arc is expressed by pre- and synplutonic dip-slip deformation while strike-slip deformation is characteristic of the cooling stage of each plutonic complex. We propose that an early high-temperature stage in the emplacement of a plutonic complex or group of complexes accelerates uplift of the arc, gives rise to increased topography and allows arc-normal, extensional collapse to dominate over arc-parallel strike-slip caused by viscous drag at the base of the lithosphere. The emplacement of the early units of each plutonic complex was therefore associated with dip-slip deformation in high-temperature mylonite belts. Towards the end of each emplacement cycle temperature declined and topography was reduced by extensional collapse and erosion. Consequently, the arc-parallel strike-slip component partitioned into the overriding plate due to corner flow increased relative to the arc-normal component and deformation became increasingly dominated by strike-slip. This cycle was then repeated by initiation of the next high-temperature episode of magma ascent and emplacement.

7.4   Conclusions

Re-evaluation of pluton emplacement mechanisms in the Coastal Cordillera between 25°S and 27°S indicates that Permian, Triassic and Lower Jurassic plutonic complexes were emplaced as subhorizontal sheets at high crustal levels. Subsequently, Upper Jurassic and earliest Lower Cretaceous plutonic complexes were emplaced mainly as vertical sheets by magma wedging, rather than at specific dilational-jog sites in fault systems. The plutons were emplaced in an extensional arc with an increasing sinistral strike-slip component in the Early Cretaceous from $c.$ 130 Ma onward. Transtensional deformation post-130 Ma is expressed as an alternation of dip-slip and strike-slip deformation through time. We interpret this as a consequence of episodic changes in slab roll-back velocity and/or the trenchward velocity of the overriding plate. Variations in trenchward velocity of the overriding plate may be determined indirectly by the rate of magma supply to the overriding plate.

Acknowledgements

J.W. acknowledges receipt of a Kingston University Research Studentship and J.G. a Royal Society Study Visit to Chile. We are grateful to the Geological

191

Survey of Chile (Servicio Nacional de Geología y Minería) for providing excellent logistical support. We have also benefited from discussions with E. Godoy, C. Mpodozis, C. Arévalo, C. Bonson, J.D. Clemens, K.J.W. McCaffrey, N. Petford and reviews by M.P. Atherton and R. D'Lemos.

References

Arévalo, C. (1995) Mapa Geológico de la Hoja Copiapó (1:100 000): Región de Atacama. Servicio Nacional Geología y Minería, Documentos de Trabajo No. 8.

Bahlburg, H. and Breitkreuz, C. (1993) Differential response of a Devonian–Carboniferous platform-deeper basin system to sea-level change and tectonics, N. Chilean Andes. *Basin Research*, **5**, 21–40.

Bateman, R. (1984) On the role of diapirism in the segregation, ascent and final emplacement of granitoids. *Tectonophysics*, **10**, 211–31.

Beck, M.E. (1988) Analysis of Late Jurassic–Recent paleomagnetic data from active plate margins of South America. *Journal of South American Earth Sciences*, **1**, 39–52.

Bell, C.M. (1987) The origin of the Upper Palaeozoic Chañaral melange of northern Chile. *Journal of the Geological Society, London*, **144**, 599–610.

Berg, K. and Baumann, A. (1985) Plutonic and metasedimentary rocks from the Coastal Range of northern Chile: Rb-Sr and U-Pb isotopic systematics. *Earth and Planetary Science Letters*, **75**, 101–15.

Berg, K., Breitkreuz, C., Damm, K.-W., Pichowiak, S. and Zeil, W. (1983) The North-Chilean Coast Range – an example for the development of an active continental margin. *Geologische Rundschau*, **72**, 715–31.

Brown, M. (1991) Comparative geochemical interpretation of Permian-Triassic plutonic complexes of the Coastal Range and Altiplano (25°30′–26°30′S), northern Chile. *Geological Society of America Special Paper*, **265**, 157–77.

Brown, M., Diáz, F. and Grocott, J. (1993) Displacement history of the Atacama Fault System 25°00–27°00 S, northern Chile. *Geology Society of America Bulletin*, **105**, 1165–74.

Clemens, J.D. and Mawer, C.K (1992) Granitic magma transport by fracture propagation. *Tectonophysics*, **204**, 339–60.

Clemens, J.D., Petford, N., and Mawer, C.K (this volume) Ascent mechanisms of granitic magmas: causes and consequences.

Dallmeyer, R.D., Brown, M., Grocott, J. *et al.* (1996) Mesozoic magmatic and tectonic events within the Andean plate boundary zone, 26–27°30′S, north Chile: constraints from $^{40}Ar/^{39}Ar$ mineral ages. *Journal of Geology*, **104**, 19–40.

Delaney, P.T., Pollard, D.P., Ziony, J.I. and McKee, E.H. (1986) Field relations between dikes and joints: emplacement processes and paleostress analysis. *Journal of Geophysical Research*, **91**, 4920–38.

Dewey, J.D. (1980) Episodicity, sequence and style at convergent plate boundaries, in *The continental crust and its mineral deposits* (ed. Strangeway, D.W.), Geological Society of Canada Special Paper **20**, 553–73.

Doglioni, C. (1991) A proposal for the kinematic modelling of W-dipping subduction – possible applications to the Tyrrhenian-Appenines system. *Terra Nova*, **3**, 423–34.

Dokka, R.K and Ross, T.M. (1995) Collapse of south-western North America and the evolution of early Miocene detachment faults, metamorphic core complexes, the Sierra Nevada orocline, and the San Andreas fault system. *Geology*, **23**, 1075–8.

REFERENCES

England, R.W. (1992) The genesis, ascent and emplacement of the Northern Arran Granite: implications for granitic diapirism. *Geological Society of America Bulletin*, **204**, 606–14.

Evans, D.J., Rowley, W.J., Chadwick, R.A. *et al.* (1993). Seismic reflections within the Lake District Batholith, Cumbria, northern England. *Journal of the Geological Society, London*, **150**, 1043–6.

Furukawa, Y. (1993) Magmatic processes under arcs and formation of the volcanic front. *Journal of Geophysical Research*, **98**, 8309–19.

Glazner, A.F. (1991) Plutonism, oblique subduction and continental growth: an example from the Mesozoic of California. *Geology*, **19**, 784–6.

Grocott, J., Brown, M., Dallmeyer, R.D. *et al.* (1994) Mechanisms of continental growth in extensional arcs: an example from the Andean plate-boundary zone. *Geology*, **22**, 391–4.

Hamilton, W.B. (1995) Subduction systems and magmatism, in *Volcanism associated with extension at consuming plate margins* (ed. Smellie, J.L.) Geological Society Special Publication, **81**, 3–28.

Hanson, R.B. and Glazner, A.F. (1995) Thermal requirements for extensional emplacement of granitoids. *Geology*, **23**, 213–16.

Hudleston, P.J. (1986) Extracting information from folds in rocks. *Journal of Geological Education*, **34**, 237–45.

Hutton, D.H.W. (1988) Granite emplacement mechanisms and tectonic controls; inferences from deformation studies. *Transactions of the Royal Society of Edinburgh: Earth Sciences*, **79**, 615–31.

Hutton, D.W.H. (1992) Granite sheeted complexes: evidence for the dyking ascent mechanism. *Transactions of the Royal Society of Edinburgh: Earth Sciences*, **83**, 377–82.

Hutton, D.H.W. and Reavy, R.S. (1992). Strike-slip tectonics and granite petrogenesis. *Tectonics*, **11**, 960–7.

Hutton, D.H.W., Dempster, T.J., Brown, P.E. and Becker S.D. (1990) A new mechanism of granite emplacement: intrusion in active extensional shear zones. *Nature*, **343**, 452–5.

Inger, S. and Harris, N.B.W. (1992). Geochemical constraints on leucogranitic magmatism in the Langtang Valley, Himalaya. *Journal of Petrology*, **34**, 345–68.

Ingram, G.M. and Hutton, D.H.W. (1994) The Great Tonalite Sill: Emplacement into a contractional shear zone and implications for Late Cretaceous to early Eocene tectonics in southeastern Alaska and British Columbia. *Geological Society of America Bulletin*, **106**, 715–28.

Jarrard, R.D. (1986) Relations among subduction parameters. *Reviews of Geophysics*, **24**, 217–84.

Karlstrom, K.E., Miller, C.F., Kingsbury, J.A. and Wooden, J.L. (1993) Pluton emplacement along an active ductile thrust zone, Piute Mountains, southeastern California: Interaction between deformational and solidification processes. *Geological Society of America Bulletin*, **105**, 213–30.

Lister, G.S. and Baldwin, S.L. (1993) Plutonism and the origin of metamorphic core complexes. *Geology*, **21**, 607–10.

Lister, J.R. and Kerr, R.C. (1991) Fluid-mechanical models of crack propagation and their application to magma transport in dykes. *Journal of Geophysical Research*, **96**, 10049–77.

McCaffrey, K.J.W., Evans, D.J. and Petford, N. (1996). Scale-invariant geometry of tabular sheet intrusions. *Journal of the Geological Society*, **154**, 1–4.

Mercado, M. (compiler) (1978) Instituto de Investigaciónes Geológicas: Geología de la Cordillera de la Costa entre Chañaral y Caldera, scale 1:100 000.

Mpodozis, C. and Allmendinger, R.W. (1993) Extensional tectonics, Cretaceous Andes, northern Chile (27°S). *Geological Society of America Bulletin*, **5**, 1462–77.

Mpodozis, C. and Ramos, V. (1990) The Andes of Chile and Argentina, in *Geology of the Andes and*

193

its relation to hydrocarbon and mineral resources (eds Ericksen, G.E., Cañas-Pinochet, M.T. and Reinemund, J.A.) *American Association of Petroleum Geologists Circum-Pacific Earth Science Series*, **11**, 59–91.

Pardo-Casas, F. and Molnar, P. (1987) Relative motion of the Nazca (Farallon) and South American Plates since late Cretaceous time. *Tectonics*, **6**, 233–48.

Paterson, S.R. and Fowler, K.T., Jr. (1993) Re-examining pluton emplacement processes. *Journal of Structural Geology*, **15**, 191–206.

Paterson, S. R. and Vernon, R.H. (1995) Bursting the bubble of the ballooning plutons: A return to nested diapirs emplaced by multiple processes. *Geology Society of America Bulletin*, **107**, 1356–1380.

Petford, N. (1996) Dykes or diapirs? *Transactions of the Royal Society of Edinburgh: Earth Sciences*, **87**.

Petford, N., Kerr, R.C., and Lister, J.R. (1993) Dike transport model for transport of granitoid magmas. *Geology*, **21**, 845–8.

Randall, D.E., Taylor, G.K. and Grocott, J. (1996) Major crustal rotations in the Andean margin: Paleomagnetic results from the Coastal Cordillera of northern Chile. *Journal of Geophysical Research*, **101 (B7)**, 783–98.

Royden, L.H. (1993) The tectonic expression of slab pull at convergent boundaries. *Tectonics*, **12**, 303–25.

Russo, R.M. and Silver, P.G. (1996). Cordillera formation, mantle dynamics and the Wilson Cycle. *Geology*, **24**, 511–14.

Scheuber, E. and Andriessen, P.A.M. (1990) The kinematic and geodynamic significance of the Atacama Fault Zone, northern Chile. *Journal of Structural Geology*, **12**, 243–57.

Scheuber, E. and Reutter, K.J. (1992) Magmatic arc tectonics in the Central Andes between 21° and 25°S. *Tectonophysics*, **205**, 127–40.

Scheuber, E., Hammerschmidt, K and Friedrichsen, H. (1995) $^{40}Ar/^{39}Ar$ and Rb-Sr analyses from ductile shear zones from the Atacama Fault Zone, northern Chile: the age of deformation. *Tectonophysics*, **250**, 61–87.

Scheuber, E., Bogdanic, T., Jensen, A. and Reutter, K.-J. (1994) Tectonic development of the north Chilean Andes in relation to plate convergence and magmatism since the Jurassic, in *Tectonics of the southern central Andes* (eds Reutter, K.-J., Scheuber, E. and Wigger, P.J.), Springer-Verlag, Berlin.

Schwerdtner, W.M. (1995) Local displacement of diapir contacts and its importance to pluton emplacement study. *Journal of Structural Geology*, **17**, 907–10.

Scotese, C.R. (1991) Jurassic and Cretaceous plate tectonic reconstructions. *Palaeogeography, Palaeoclimatology, Palaeoecology*, **87**, 493–501.

Suarez, M. and Bell, C.M. (1992) Triassic rift-related sedimentary basins in northern Chile (24°–29°S). *Journal of South American Earth Sciences*, **6**, 109–21.

Sylvester, A.G. (1988) Strike-slip faults. *Geology Society of America Bulletin*, **100**, 1666–1703.

Taylor, G.K., Grocott, J., Pope, A. and Randall, D.E. (1997) Mesozoic fault systems, deformation and fault block rotation in the Andean forearc 25°–27°S. *Tectonophysics* (in press).

Tikoff, B. and Teyssier, C. (1992) Crustal-scale, en échelon 'P-shear' tensional bridges: a possible solution to the batholithic space problem. *Geology*, **20**, 927–30.

Vigneresse, J.L. (1995) Crustal regime of deformation and ascent of granitic magma. *Tectonophysics*, **249**, 187–202.

Vissers, R.L.M., Platt, J.P. and Van der Waal, D. (1995) Late orogenic extension of the Betic Cordillera and the Alboran domain: a lithospheric view. *Tectonics*, **14**, 786–803.

REFERENCES

Wilson, J. (1996) The emplacement of the Las Tazas plutonic complex, northern Chile. Unpublished Ph.D. Thesis, Kingston University, 223 pp.

Woodcock, N.H. (1986) The role of strike-slip fault systems at plate boundaries. *Philosophical Transactions of the Royal Society of London* **A 317**, 13–29.

Yáñez, G., Mpodozis, C. and Tomlinson, A.J. (1994) Eocene dextral oblique convergence and sinistral shear along the Domeyko Fault System: a thin viscous sheet approach with asthenospheric drag at the base of the crust. *7th Congreso Geológico Chileno, Actas*, **2**, 1478–82.

CHAPTER EIGHT

Lithological, structural and deformation controls on fluid flow during regional metamorphism

Colin M. Graham, Alasdair D.L. Skelton, Mike Bickle and Coleen Cole

8.1 Introduction

In this chapter we review the state of knowledge concerning fluid flow during regional metamorphism. Prograde metamorphism is usually accompanied by extensive devolatization, and pelitic rocks typically produce ~ 10% by volume of a dominantly $H_2O–CO_2$ fluid phase during heating to amphibolite facies conditions (Walther and Orville, 1982). This fluid is generated continuously or episodically, and is lost rapidly by compaction, porous flow and/or hydrofracture in hot, ductile metamorphic rocks. The passage of this fluid may modify the petrological, geochemical, thermal and rheological state of rocks along the flow-path, and it is these modifications which may in turn be used to quantify the direction and magnitude of fluid flow. However the extent, geometry and significance of metamorphic fluid flow is controversial, and several key aspects of the flow regime, including fluid pathways, transport mechanisms (pervasive or channelled), and the relationships between fluid movement, deformation, metamorphism and rock structure, are not well understood. In this chapter we consider:

- theoretical and experimental constraints on rock permeability, compaction and flow;
- recent models which describe and quantify the geometry and magnitude of fluid flow, together with their assumptions, limitations, and geological applications;
- field evidence for the magnitude, transport mechanisms, timing, and lithological and deformational control of regional metamorphic fluid flow.

We use recent work on fluid flow in greenschist facies regional metamorphic rocks of the south-west Scottish Highlands as an illustrative case study.

Deformation-enhanced Fluid Transport in the Earth's Crust and Mantle. Edited by M.B. Holness. Published in 1997 by Chapman & Hall, London. ISBN 0 412 75290 5.

8.2 Expulsion of fluid from metamorphic rocks

Fluid production in metamorphic belts is governed by the rate of change of pressure and temperature and the continuous and discontinuous reactions involving volatile-bearing phases in metamorphic lithologies. These are determined largely by the externally imposed tectonics of the metamorphic terrain and the composition of the metamorphic rocks, respectively. The fundamental questions concerning metamorphic fluid flow are as follows:

- How fast and over what length scales will a porous metamorphic rock compact to expel internally generated pore fluid?
- What is the scale, geometry and nature of the flow paths of the escaping fluid (pervasive porous flow, channelled porous flow, flow along cracks or flow along shear zones moderated by deformation)?
- What is the role of deformation in controlling the geometry of fluid flow?
- Is fluid introduced into the metamorphic belt from external sources?
- Does the flow itself influence the petrology, reaction history and fluid production within the metamorphic belt?

In this section we address the compaction of porous media at the site of fluid generation, and its relation to permeability structure and the geometry of fluid flow.

8.2.1   Compaction of porous media and expulsion of pore fluid

The escape of fluid from any ductile porous medium such as a metamorphic rock is driven by the density difference between the solid and fluid phases, and the rate at which it escapes is governed by the relationship between permeability and porosity and the viscosities of the solid and fluid phases. McKenzie (1984, 1985) shows that porous media will initially tend to compact over a characteristic length scale, the compaction length (δ_c), with a characteristic time for fluid loss from the scale of the compaction length (τ_c) or from the whole compacting region (τ_h). Theoretical and numerical modelling of compacting porous media show that the porosity structure is inherently unstable and porosity waves develop in one, two or three dimensions (Scott and Stevenson, 1984; McKenzie, 1985; Barcilon and Richter, 1986; Wiggins and Spiegelman, 1995). Porosity waves are high porosity zones or channels which move through the medium. Fluid escape from metamorphic rocks is likely to be more complex than modelled by devolatilization-induced compaction of homogeneous porous media because:

- the natural permeability structure of metamorphic rocks is both inhomogeneous and anisotropic;
- fluid may move along brittle fractures (veins) in addition to interconnected pore networks;

197

- metamorphic belts are generally deformed by tectonically imposed strains in addition to compaction;
- the evolving petrological state of rocks may cause large changes to the relationship between permeability and porosity (e.g. reaction-enhanced porosity/permeability) during devolatilization.

It is the existence of these potential complexities that make it vital to combine field constraints with a theoretical understanding of the possible controls on fluid movement.

Despite these complexities and the poor knowledge of some of the key governing parameters, the calculation of the theoretical characteristics of compaction in metamorphic rocks provides important boundary conditions on the scales, driving forces and geometries of metamorphic fluid flow. The compaction length (McKenzie, 1985) is:

$$\delta_c = \left[\frac{(\xi + {}^4/_3\eta)K_\phi}{\mu} \right]^{1/2} \tag{8.1}$$

where ξ is the bulk viscosity and η the shear viscosity of the solid, μ is the fluid viscosity and K_ϕ is the permeability. The term $(\xi + {}^4/_3\eta)$ is not well known for crustal materials but is of the same order of magnitude as the bulk viscosity. This is probably between 10^{18} and 10^{20} Pa s from field constraints (Kruse et al., 1991), although some experiments yield higher values (10^{25} Pa s; Gleason and Tullis, 1995). The more problematic term in equation 8.1 is the permeability, K_ϕ, which is dependent on porosity, ϕ. In a static, texturally equilibrated, rock the permeability depends on grain size, porosity and the fluid–solid dihedral angle, θ (Holness, this volume). If $\theta < 60°$ then a connected porosity will exist, however small the porosity, and permeability may be calculated numerically (e.g. Cheadle, 1989). If $\theta > 60°$, the porosity will become interconnected only at a finite value of porosity which increases with increasing dihedral angle. Experimental determination of dihedral angles in monomineralic metamorphic rocks for various fluids at various conditions indicates that dihedral angles range from ~ 40° to 100° (Holness, 1996).

For many pore structures, including parallel cracks, permeability depends on porosity through a relationship of the form (Holness, this volume):

$$K_\phi = \frac{a^2\phi^n}{C} \tag{8.2}$$

where a is a length scale (crack spacing or grain size), ϕ is porosity, C is a constant, and the exponent, n, is generally in the range 2 to 3. Assuming Cheadle's (1989) values for low dihedral angles ($n = 2$, C $= 3000$) and a grain diameter of ~ 0.1 mm, a value of 10^{19} Pa s for the solid viscosity term $(\xi + {}^4/_3\eta)$ and a fluid viscosity of 10^{-4} Pa s, equations 8.1 and 8.2 imply a compaction length of ~ 600 m for a porosity of 10^{-3} or ~ 1 m for a porosity of 10^{-6}. The compaction time constant, τ_c, is ~ 30 years and is independent of porosity if $n = 2$. If $n > 2$ then τ_c is proportional to $\phi^{(n/2-1)}$ and compaction times and compaction length increase with increasing porosity. Since fluid is produced by devolatilization reactions, metamorphic rocks will initially start compacting over smaller length scales once the critical porosity is reached, and both the porosity structure and fluid loss processes will be inherently unstable, time varying and inhomogeneous.

Alternatively we may ask how large a porosity is required for metamorphic rocks to lose the fluid produced by devolatilization reactions. If it is assumed that metamorphic rock loses 10% fluid by volume uniformly over a fixed time period, then it is possible to use equations 8.1 and 8.2 with $n = 2$ and $C = 3000$ to calculate the porosity at the top of a layer of given thickness necessary to drain the fluid. Table 8.1 shows such calculations for layer thicknesses of 10 m, 1 km and 10 km, and time-scales of 10^5–10^7 a. The second column shows the fluid flux ($\omega\phi$) at the top of the layer, the third column the porosity at the top of the layer and the fourth and fifth columns the compaction length and time for this maximum porosity region. The short compaction lengths and times suggest that such uniform flow conditions will never become established. If fluid is lost by porous flow, the rock will develop short-lived channels or high porosity waves with transient porosities higher than the calculated steady state porosity.

8.2.2 Permeability and flow geometry

Layered metamorphic rocks have a heterogeneous, anisotropic and potentially highly dynamic permeability structure which may control the expulsion of fluid and preclude the development of the unstable porosity waves predicted for homogeneous porous media by the compaction theory discussed above. Since permeability is a power law function of porosity (equation 8.2), both the porosity and the dihedral angle are likely to vary widely in adjacent lithological units (on scales from centimetres to tens of metres), leading to order-of-magnitude variations in permeability between contrasted lithologies. Flow is likely to

Table 8.1 Example compaction calculations

Time (Ma)	Layer thickness	$\omega\phi \, (m^3 \, m^{-2} \, a^{-1})$	ϕ	δ_c (m)	τ_h (a)	$\delta^{18}O$ diffusion distance (m)
10	10 km	10^{-4}	7×10^{-5}	42	7300	38
	1 km	10^{-5}	2×10^{-5}	13	2300	21
	10 m	10^{-7}	2×10^{-6}	1.3	230	7
1	10 km	10^{-3}	2×10^{-4}	134	2320	21
	1 km	10^{-4}	7×10^{-5}	42	734	12
	10 m	10^{-6}	7×10^{-6}	4	73	4
0.1	10 km	10^{-2}	7×10^{-4}	424	734	12
	1 km	10^{-3}	2×10^{-4}	134	232	7
	10 m	10^{-5}	2×10^{-5}	13	23	2

be channelled along more permeable units, as many field observations confirm (Yardley *et al.*, 1991a, 1991b; Yardley and Lloyd, 1995). This can be illustrated by calculation of flow vectors in a simple layered porous medium with a factor of 10 difference in porosity between layers which dip at a given angle to the horizontal (Figure 8.1). Equation 8.2 predicts a permeability contrast of 100 between such layers (or more, if $n > 2$) and if flow is driven by buoyancy, the piezometric driving force along the layers is proportional to the sine of the angle of dip. In such a case lateral flow along the more permeable layer will exceed vertical flow through the less permeable layer if the dip is greater than 0.6° and is an order of magnitude greater if the dip is greater than 6°. Results for various permeability contrasts are illustrated in Figure 8.1. The geochemical boundary layer profiles in metamorphic rocks reviewed below are consistent with such porosity and permeability contrasts, and although fluid escape is probably driven by gravitational buoyancy forces, flow paths in rocks will be determined largely by the structure and relative permeabilities of rock layers in the meta-

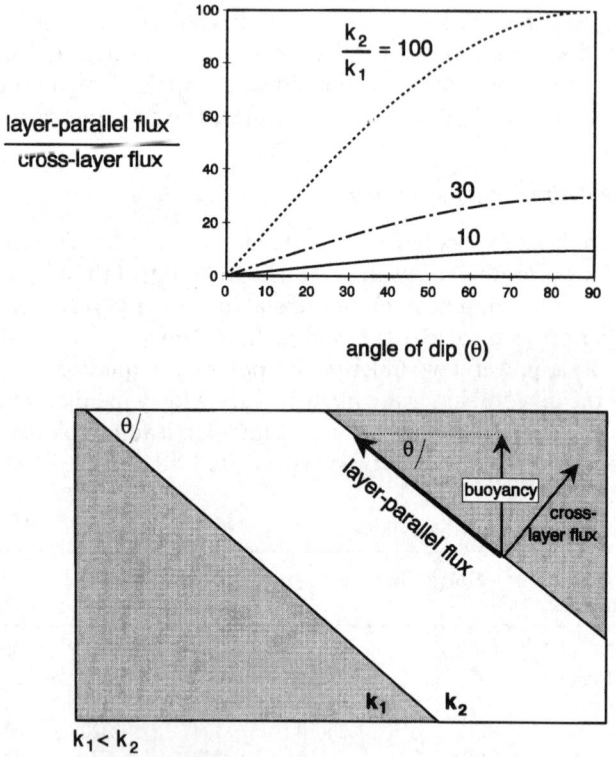

Figure 8.1 Plot and illustration showing the effect of angle of dip and permeability on buoyancy driven flow, contrasting flux along a permeable layer with flux upward through the host rock. Note that layer-parallel flow dominates for most real geometries.

200

morphic belt. The roles of lithology and structure in controlling the geometry of regional metamorphic fluid flow are illustrated and discussed below.

8.2.3 Thermal driving forces and flow geometry

Ferry (1994), among others, has suggested that thermal gradients across tens of kilometres in regional metamorphic terrains can cause significant lateral fluid flow. It is important to consider the extent to which thermally generated buoyancy forces are important driving mechanisms for fluid flow in regional metamorphic belts. It has long been realized that plutonic heat sources dominate fluid circulation around high-level cooling plutons (e.g. Taylor and Forester, 1971) and in the oceanic crust (e.g. Lister, 1972). However, these systems operate in brittle, hydrostatically pressurized upper crust. Convection of metamorphic fluid in the ductile middle and lower crust has also been proposed (Etheridge et al., 1983). However, the requirement for fluid to flow downwards against the lithostatic pressure gradient and for the permeability to be simultaneously sustained over the scale of convection cells in a heterogeneous, deforming matrix presents severe constraints on the scale, efficacy and viability of convection in the deep crust (e.g. Wood and Walther, 1986; Oliver, 1996). The calculations above show that ductile metamorphic rocks will compact rapidly over small length scales, and it follows that large-scale convection of fluid in the deep crust of the sort proposed by Etheridge et al. (1983) is implausible. In rocks with compaction lengths of metres to hundreds of metres, fluid flow will be determined by the pressure gradients arising on such scales, and the buoyancy forces across ~ 100 m (~ 2 MPa if dipping vertically, to 0.2 MPa if layering is dipping at 6°) are large compared with horizontal pressure difference of ~ 20 Pa across 100 m arising from a thermal gradient of ~ 30 °C/km.

8.2.4 The role of deformation in fluid flow

Deformation almost certainly plays an important role in the expulsion of fluid from metamorphosing rocks, but the mechanisms by which deformation drives fluid flow are controversial. The association of fluid flow with shear zones is well established (McCaig, this volume), but the quantitative relationship between flow and fold structures on various scales is largely unexplored (but see Ferry, 1994), due to the magnitude and detail of the data sets necessary to establish such relationships. Relationships between flowpaths and folds might merely reflect the role of varying lithology and hence varying permeabilities within a pre-existing and lithologically heterogeneous fold structure, rather than indicating any dynamic mechanistic relationship between compaction, flow and deformation. Therefore the timing of fluid production and flow in relation to deformation and metamorphism is crucial in establishing deformation/flow relationships. This question has seldom been satisfactorily addressed because of the difficulties in accurately establishing the timing of potentially short-lived flow events during protracted periods of deformation and regional metamorphism. A comprehensive review and classification of structural controls of fluid flow during regional metamorphism is presented by Oliver (1996).

8.3   The petrological and geochemical record of fluid flow

Because the controls on fluid movement in metamorphic belts are many and complex, it is essential to use direct observations from the geological record to determine which are the important processes. Important objectives are: (i) to track the passage of fluid; (ii) to measure the volume of fluid which has passed through a unit area of rock (time-integrated fluid flux); (iii) to measure the relative fluid fluxes through adjacent rock layers; (iv) to determine the timing and number of flow events and their relationship to metamorphic and deformational events. All methods of quantifying fluid flow in metamorphic rocks involve measuring the geochemical, isotopic or petrological changes which occur in response to fluid flow. In many studies, non-dimensional fluid/rock ratios have been calculated from observed changes in rock composition, but these calculations do not constrain the necessary parameter, the time-integrated fluid flux, which is dimensional (e.g. m^3/m^2) and crucial if flow patterns are to be mapped. If the flow direction can be inferred, calculation of time-integrated fluid fluxes is straightforward, and follows similar logic to that of fluid/rock ratios (e.g. Ferry, 1989, 1992, 1994). However, where flow geometry depends on compaction, buoyancy, thermal gradients, deformation and permeability contrasts, as outlined above, this approach becomes implausible.

The basic method for calculating time-integrated fluid fluxes, reviewed here, is to identify some initial chemical, isotopic or petrological heterogeneity and then to identify the advective displacement of this heterogeneity in response to flow of fluid through the rock (e.g. Ogata, 1964; Hofmann, 1972; Rye et al., 1976; Ferry, 1980; Bickle and McKenzie, 1987). This method is subject to a number of limitations and difficulties:

- the altered chemical or petrological state of the rock represents the time-integrated sum of all the fluid flow events which may have occurred during the history of the rock;
- heterogeneities in metamorphic rock are characteristically limited to the two-dimensional lithological layering and it is thus only possible to directly monitor transport perpendicular to layering;
- transport characteristics are often inferred by fitting simplified transport models to chemical, isotopic or petrological profiles. It is important to recognize that goodness of fit of geochemical or petrological data to a model may not exclude alternative models.

The first two limitations are not insuperable. It is possible to use the petrology, structure and mineral fabrics of rocks to constrain the infiltration history and the timing of fluid flow relative to metamorphism and deformation. Two- or three-dimensional structural complexities present in the rock may allow reconstruction of flux vectors in three dimensions (e.g. Skelton *et al.*, 1995) and study of boundary layer profiles preserves information about relative layer-parallel fluxes in adjacent rock units (e.g. Bickle and Baker, 1990a).

8.4 Chromatographic theory

Understanding the controls on advective tracer transport requires solutions to the one-dimensional differential equation which describes transport through porous media by advection and diffusion with some kinetic law relating fluid to solid compositions. The general equation for transport of a chemical component (concentration C_f in the fluid phase) in one-dimension (z) with time (t) may be written (cf. Lichtner, 1988):

$$\phi \frac{\partial C_f}{\partial t} = -\omega\phi \frac{\partial C_f}{\partial z} + D\phi \frac{\partial^2 C_f}{\partial z^2} - \frac{\partial C_s}{\partial t} \frac{(1-\phi)\rho_s}{\rho_f} \qquad (8.3)$$

for tortuosity ≈ 1 and where ω is the Darcy flow velocity, D is the diffusion coefficient, ϕ is porosity and ρ_f and ρ_s are the respective densities of the fluid and solid phases. Note that in equation 8.3, change in concentration with time (left-

Table 8.2 Glossary of parameters for chromatographic theory

Symbol	Quantity	Units
C_f	Concentration of chemical component in fluid	Various
C_f'	Dimensionless concentration of chemical component in fluid	None
C_s	Concentration of chemical component in solid	Various
C_s'	Dimensionless concentration of chemical component in solid	None
C_{f1}	Concentration of chemical component in fluid upstream of front	Various
C_{f2}	Concentration of chemical component in fluid downstream of front	Various
$C_{s.1}$	Concentration of chemical component in solid upstream of front	Various
$C_{s.2}$	Concentration of chemical component in solid downstream of front	Various
D	Diffusion coefficient	$m^2 s^{-1}$
h	Length scale	m
K_v	Fluid/solid partition coefficient	None
N_D	Damköhler number	None
Pe	Peclet number	None
V_f	Velocity of front	$m s^{-1}$
t	Time	s
t'	Dimensionless time	None
z	Distance	m
z'	Dimensionless distance	None
ϕ	Porosity	None
κ	Exchange rate	s^{-1}
ρ_f	Density of fluid	$g m^{-2}$
ρ_s	Density of solid	$g m^{-2}$
ω	Fluid velocity	$m s^{-1}$

hand side) is written as a function of advection, diffusion and reaction/exchange (first, second and third terms respectively on right-hand side).

The behaviour of chemical components (e.g. isotopes) in real systems, in which the porosity and mineralogy of the rock evolve during the infiltration event, may be very complex as fluid–solid exchange and thus transport of any species may be a function of the transport of several other components (e.g. Hofmann, 1972; Lichtner, 1988). However in many cases it has proved possible to obtain solutions to equation 8.3 if several simplifying assumptions are used. A crucial consideration is that the various parameters cannot all be determined independently from solutions for given boundary conditions. This is best illustrated by transformation of the variables to appropriate dimensionless constants. We illustrate two of the more important simplifications below.

In circumstances in which fluid–solid exchange is rapid compared to transport of the chemical component, for small ϕ, equation 8.3 may be simplified to:

$$\frac{1}{K_v}\frac{\partial C_f}{\partial t} = -\omega\phi\frac{\partial C_f}{\partial z} + D\phi\frac{\partial^2 C_f}{\partial z^2} \tag{8.4}$$

where K_v is the fluid/solid partition coefficient. If the dimensionless transformations given below:

$$z = hz'$$

$$t = \frac{h^2}{D\phi K_v}t' \tag{8.5}$$

$$C_f = C_{f,1} + (C_{f,2} - C_{f,1})C_f'$$

are applied to equation 8.4, we obtain:

$$\frac{\partial C_f'}{\partial t'} = Pe\frac{\partial C_f'}{\partial z'} + \frac{\partial^2 C_f'}{\partial z'^2} \tag{8.6}$$

where h is an appropriate length, and the subscripts 1 and 2 denote upstream and downstream. Solutions to equation 8.6 depend on one dimensionless constant, the Peclet Number (Pe) given by:

$$Pe = \frac{\omega\phi h}{D\phi} \tag{8.7}$$

as well as dimensionless time and distance and the boundary conditions. Thus geochemical profiles perturbed by advective and diffusive transport may be used to constrain only four variables (e.g. $C_{f,1}$, $C_{f,2}$, t' and Pe) for given boundary conditions including the shape of the initial geochemical profile. By rearranging equation 8.5 and knowing the solid/fluid partition coefficient, K_v, it is possible to

204

recover the time integrated fluid flux ($\omega\phi t$) and diffusion distance ($2\sqrt{D\phi t}$) (cf. Bickle and Baker, 1990a). Further, by considering a moving frame of reference, equation 8.6 becomes the diffusion equation with the velocity of a front, V_f, given by:

$$V_f = \omega\phi K_v \tag{8.8}$$

Integrating equation 8.8 over time gives an expression for the position of the front, $V_f t$:

$$V_f t = \omega\phi t K_v \tag{8.9}$$

Note that $\omega\phi t$ is the time-integrated fluid flux.

An alternative simplification is to consider the consequences of kinetically limited fluid–solid exchange in the absence of diffusive transport.

Although equation 8.3 may be solved numerically for any specified kinetic law relating fluid and solid exchange processes, available information is generally insufficient to justify more complex models. Therefore solutions to equation 8.3 have been sought for the simplest kinetic exchange models. For example the kinetic law for exchange of a chemical tracer may be expressed as a linear function of fluid–solid compositions by:

$$\frac{\partial C_s}{\partial t} = \kappa(C_s - C_{s,2}) \tag{8.10}$$

where κ is an exchange rate (dimensions of t^{-1}) and the term in brackets is a measure of how far the solid is out of equilibrium with the fluid. Equation 8.10 can be applied to more general situations than those with linear kinetics because the rates of many diffusively controlled exchange processes can be approximated by linear functions over a significant range of their parameter space (e.g. Bickle, 1992).

Equations 8.3 and 8.10 may be transformed to dimensionless parameters by:

$$z = hz'$$

$$t = \frac{h}{\omega\phi} t'$$

$$C_f = C_{f,1} + (C_{f,2} - C_{f,1})C_f'$$

$$C_s = C_{s,1} + (C_{s,2} - C_{s,1})C_s' \tag{8.11}$$

With these transformations, equations 8.3 and 8.11 reduce to:

$$\phi\frac{\partial C_f'}{\partial t'} = -\frac{\partial C_f'}{\partial z'} + N_D C_f' \tag{8.12}$$

205

In a fashion similar to equation 8.6, solutions to equation 8.12 depend on one dimensionless constant, the Damköhler Number (N_D), which is given by:

$$N_D = \frac{\kappa h}{\omega \phi} \qquad (8.13)$$

in addition to the dimensionless time and distance and the boundary conditions. Likewise, geochemical profiles perturbed by advective transport may be used to constrain only four variables (e.g. $C_{f,1}$, $C_{f,2}$, t' and N_D) for given boundary conditions including the shape of the initial geochemical profile (Bickle, 1992).

8.4.1 Model limitations

There are several limitations to these models. First, they are one-dimensional, and therefore yield only 1-D time-integrated fluid fluxes, Peclet and Damköhler numbers. To model fluid flow in 3-D, we must resolve at least three 1-D fluid flow vectors. This type of modelling is discussed below. Secondly, the input data (e.g. isotopic ratios, trace element concentrations, reaction progress), and therefore the output data (e.g. fluid fluxes, N_{Pe}, N_D) are time-integrated. Thirdly, these models assume pervasive infiltration of a medium with constant porosity and permeability, which is unlikely to hold during regional metamorphism (see sections 8.6.3 and 8.6.4 below). However, although a mineral/fluid reaction may be accompanied by a change in porosity, this will occur upstream of the reaction front, and is therefore unlikely to affect the downstream propagation rate.

8.4.2 Modelling fluid flow from geochemical gradients

Although observation of geochemical fronts provides unambiguous proof of fluid infiltration, absence of fronts does not preclude infiltration. In particular, infiltration in situations with more complex physics than that assumed above may result in alteration profiles with continuous gradients in chemistry, isotope composition or reaction progress. Fletcher and Hofmann (1974) discuss infiltration of tracers where fluid–rock partition coefficients are a function of tracer concentration and show how initial profiles may be sharpened or distended depending on the relationship between partition coefficient, tracer concentration and the shape of the initial profile. This is because the tracer transport velocity is a function of partition coefficient (see equation 8.8) and thus varies across the profile if the partition coefficient is a function of rock composition.

Infiltration-driven reactions may not develop identifiable fronts for reasons other than kinetically limited reaction rates. In general, divariant reactions will be stable across a range of fluid compositions at any given pressure and temperature and, depending on the relationship between the stoichometry of the reaction and fluid composition, may develop as broad zones. Univariant reactions may also be stabilized across significant distances if either temperature, pressure or any other intensive variable changes significantly with distance. Baumgartner and Ferry (1991) and Ferry and Dipple (1991) discuss univariant devolatilization reactions driven by infiltration of H_2O–CO_2 fluids in terrains with temperature and/or pressure gradients. If the temperature or pressure gradient along the flow path is such that the fluid in equilibrium with the univariant reaction becomes more enriched downstream by the component produced by the forward reaction then the univariant reaction may be stabilized across a significant distance. Baumgartner and Ferry (1991) show how the time-integrated fluid flux may be calculated from reaction progress if both the temperature and pressure remained fixed during the fluid flow event and fluid is buffered along the flow path by rocks of identical univariant phase assemblage. The time-integrated fluid flux along the flow path, z, is given by:

$$\omega\phi t = \frac{\xi \cdot (1 - X_{CO_2})}{\left[\left(\dfrac{\partial X_{CO_2}}{\partial T}\right)_P \cdot \dfrac{\partial T}{\partial z} + \left(\dfrac{\partial X_{CO_2}}{\partial P}\right)_T \cdot \dfrac{\partial P}{\partial z}\right]} \tag{8.14}$$

where ξ is reaction progress, X_{CO_2} is the volume fraction of CO_2 in the fluid phase, T is temperature and P is pressure. It should be noted that although equation 8.14 would provide an accurate measure of fluid flux for a given physical setting, it is not clear how closely conditions in a metamorphic belt would correspond to the necessary assumptions. Important among these assumptions are:

- the fluid is buffered by identical assemblages upstream so that the term $\partial X_{CO_2}/\partial z$ is determined;
- $\partial T/\partial z$ and $\partial P/\partial z$ along the flow path;
- an initial reactant mineral assemblage and/or reaction pathway, which is seldom if ever supported by petrological or textural information;
- local, grain-scale fluid–solid equilibrium (compare Wood and Walther, 1986 with Lasaga and Rye, 1993);
- temperatures and pressures are fixed in time along the flow path, whereas rocks evolve continuously in pressure and temperature during metamorphism (England and Richardson, 1977);
- reactions are modelled as univariant although rocks invariably contain additional components.

The significance of limited divariance has yet to be evaluated. Like all the methods of monitoring fluid flow in metamorphic rocks the assumptions need to be tested by independent methods of quantifying fluxes.

8.5 Application of chromatographic theory: evidence for controls of regional metamorphic fluid flow

8.5.1 Introduction

A comprehensive listing of studies which have used chromatographic theory to compute fluid fluxes and resolve time-integrated fluid fluxes is given in Table 8.3 together with their geological settings. These studies may variously provide empirical constraints on:

• 1-D time-integrated fluid fluxes;
• flow geometries;
• diffusion;
• reaction/exchange kinetics;
• mechanisms of fluid flow.

   Because equation 8.3 has no analytical solution, each of these studies adopts the geologically most appropriate approximation to that equation.

   For the common example of a 'semi-infinite' layer (e.g. marble horizon, metabasite dyke), the critical factor which constrains choice of boundary conditions is the flow geometry, and the alteration profile may contain information on the boundary conditions. Uniform flow conditions, in which fronts migrate down stream (Figure 8.2), are favoured by low permeability contrasts between layers, and indicate significant cross-layer flow. Pinned boundary conditions, in which concentrations are fixed at boundaries between low and high permeability layers by flow channelled through the permeable layers, are characterized by front migration inwards towards the interior of the low permeability layer (Figure 8.2), indicating dominantly layer-parallel flow. The limiting case, where fronts are symmetrical, is consistent with either pure layer parallel flow or no flow. Layers used to measure cross-layer flow are thus generally relatively impermeable.

   Front broadening relates to diffusion and/or finite exchange rates. Where reaction/exchange rates are rapid, front-broadening characteristics can either be used to compute the diffusion distance ($2\sqrt{D\phi t}$) or to compare diffusive vs. advective flux rates using the dimensionless Peclet number (Pe, equation 8.7). Where reaction/exchange rates are sluggish, front-broadening characteristics can either be used to characterize empirical $\kappa-t-\phi$ interdependencies or to compare advection vs. reaction/exchange time constants using the dimensionless Damköhler number (N_D, equation 8.13). Pe and N_D can either be resolved by comparison between upstream and downstream front-broadening characteristics (Skelton et al., 1997) or by variance mapping of $Pe-N_D$ space (Baker and Spiegelman, 1995).

8.5.2 Evidence for structural and lithological controls

In the examples listed in Table 8.3(a), application of chromatographic theory enables time-integrated fluid fluxes to be constrained for the 1-D cross-layer

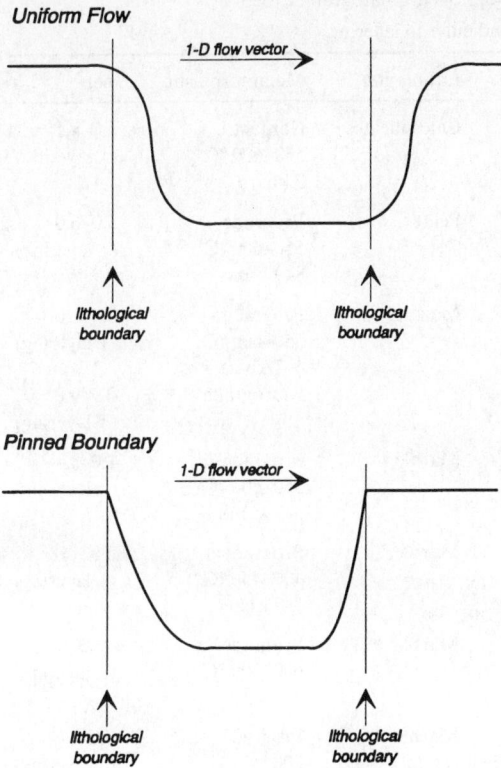

Figure 8.2 'Uniform flow' and 'pinned boundary' models. The solid curve is a geochemical profile which is displaced and broadened from the lithological boundaries by the 1-D cross-layer fluid flow vector, as shown.

component of fluid flow. This component is small (typically $\omega\phi t < 10 \text{ m}^3/\text{m}^2$) and is characterized by sharp fronts. Note that these fluxes are comparable in magnitude to those predicted for escape of pore fluid during compaction of porous media (Table 8.1). Because pinned boundary conditions prevail and these small fluxes tend to drain large source regions, we infer that these examples represent the small cross-layer component of flow associated with a strong or dominant component of layer-parallel flow.

In the examples listed in Table 8.3(b), application of chromatographic theory enables time-integrated fluid fluxes to be constrained parallel to (i) the geothermal gradient, (ii) lithological layering and (iii) regional structure. For example, Ferry (1994), Bowman *et al.* (1994), and Jamtveit *et al.* (1992b) measure fluxes parallel to the geothermal gradient, Fein *et al.* (1994) measure fluxes parallel to lithological layering, Oliver *et al.* (1994) measure fluxes parallel to a shear zone, McCaig *et al.* (1995) and Bowman *et al.* (1994) measure fluxes parallel to a thrust plane, and Skelton *et al.* (1995) resolve fluxes parallel to the

Table 8.3 Applications of chromatographic theory
(a) Fluid flow perpendicular to layering

Study area	Lithologies	Metamorphism	$\omega\phi t$	References
Mary Kathleen, Australia	Calc-silicates	Contact 550–600 °C 2 kbars	0.7–8.1 ($\perp$ layering)	Cartwright (1994)
Vermont, USA	Pelite	Prograde 550–600 °C 8–9 kbars	3.0–6.0 ($\perp$ layering)	Kohn and Valley (1994)
Naxos, Greece	Marble	Prograde 550–600 °C 5–7 kbars Retrograde	0.2–2.0 ($\perp$ layering) 0.9–9.0 ($\perp$ layering)	Baker et al. (1989)
Naxos, Greece	Marble	Prograde 550–600 °C 5–7 kbars	1.0 ($\perp$ layering)	Bickle and Baker (1990a)
Sifnos, Greece	Marble	Prograde 400–450 °C 5–7 kbars	0.0 ($\perp$ layering)	Ganor et al. (1989)
Lizzies Basin, Nevada	Marble	Prograde 600–700 °C 5–7 kbars	<0.2 ($\perp$ layering)	Bickle et al. (1995)
Seward Peninsula, Alaska	Marble	Prograde 800 °C 8 kbars	0.001 ($\perp$ layering)	Todd and Evans (1993)
Maine, USA	Marble	Prograde 400–550 °C 3.5 kbars	2.4±0.6 ($\perp$ layering)	Bickle et al. (1994)
Naxos, Greece	Marble	Prograde 550–600 °C 5–7 kbars	0.9 ($\perp$ layering)	Baker and Spiegelman (1995)
Adamello, Italian Alps	Marble	Contact 400–500 °C 1 kbars	0.5 ($\perp$ layering)	Gerdes et al. (1995)
Vermont, USA	Metabasite Marble	Prograde 475–550 °C 7–8 kbars	1.4–5.6 ($\perp$ layering)	Skelton (1997)
SW Highlands, Scotland	Metabasite	Prograde 440–500 °C 9–10 kbars	62.1±1.3 ($\perp$ layering)	Skelton et al. (1997)
Adirondacks, New York State, USA	Skarn	Contact 500–700 °C 2–7 kbars	0.0 ($\perp$ layering) >1.5×10^4 ($\parallel$ layering)	Gerdes and Valley (1994)

Table 8.3 Applications of chromatographic theory (*continued*)

(b) Fluid flow parallel to layering

(i) Parallel to lithological layering

Study area	Lithologies	Metamorphism	$\omega\phi t$	References
SW Highlands, Scotland	Marble	Retrograde ~400 °C	2.4–28.6 (∥ layering)	Fein *et al.* (1994)

(ii) Parallel to geothermal gradient

Study area	Lithologies	Metamorphism	$\omega\phi t$	References
Maine and Vermont, USA	Pelite Psammite Marble	Prograde 400–550, 475–550 °C 3.5, 7–8 kbars	$0.02–42 \times 10^3$ (∥ layering)	Ferry (1994)
Oslo Rift, Norway	Marble Pelite	Contact 370–410 °C 0.5 kbars	$>2.0 \times 10^3$ (∥ layering)	Jamtveit *et al.* (1992b)
Adirondacks, New York State, USA	Skarn	Contact 500–700 °C 2–7 kbars	0.0 (⊥ layering) $>1.5 \times 10^4$ (∥ layering)	Gerdes and Valley (1994)

(iii) Parallel to regional structures

Study area	Structure	Metamorphism	$\omega\phi t$	References
SW Highlands, Scotland	Antiform	Prograde ~475 °C ~10 kbars	>400 (max.) 100 (av.)	Skelton *et al.* (1995)
Mary Kathleen, Australia	Shear zone	Contact 500–700 °C 2–3 kbars	$0.02–4.0 \times 10^4$	Oliver *et al.* (1994)
Gavarnie Thrust, Central Pyrenees	Thrust		$>1.8–3.0 \times 10^3$	McCaig *et al.* (1995)
Glarus shear zone, Switzerland	Shear zone	300–400 °C	$>3.0 \times 10^3$	Bowman *et al.* (1994)

axial surface of an antiform from several 1-D component flux vectors in different orientations. All of these studies record high time-integrated fluid fluxes ($10^2–10^4$ m^3/m^2) which are argued to be layer-parallel. However, the significance of these evaluations is controversial, particularly where the boundary conditions (e.g. initial position of the geochemical front, relative timing of front displacement) are disputed, arbitrary or unknown. Relating to this, only Jamtveit *et al.* (1992b) demonstrate the correlation of geochemical fronts (i.e. front displacement $\propto K_v$; Figure 8.3), and therefore their simultaneous displacement from a common initial position.

211

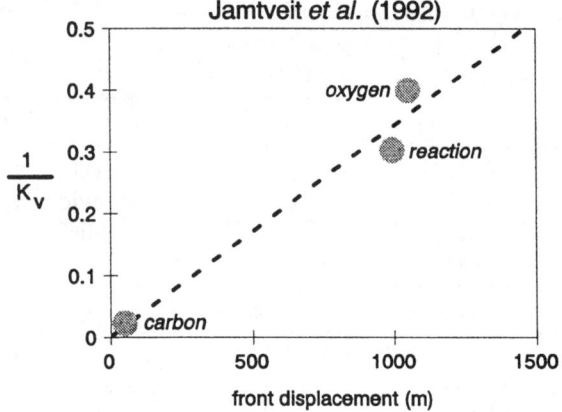

Figure 8.3 A plot of K_v vs. front displacement for oxygen and carbon isotope fronts and an infiltration driven reaction front about a pluton in the Oslo rift, from the data of Jamtveit et al. (1992b).

Channelling or focusing of flow of regional metamorphic fluids through permeable rock layers, zones of deformation (e.g. fold axes, thrusts), or during contact metamorphism, is implicit in all these studies. Metamorphic fluid flow is thus largely parallel to lithological layering or regional structure.

8.5.3 Lithological vs. deformation controls

Many studies listed in Table 8.3 involve isotopic (O, C, Sr) profiles across regionally metamorphosed marble layers. With the exception of those studies in which the marbles are associated with shearing and thrusting (Bowman et al., 1994; McCaig et al., 1995), 1-D cross-layer components of fluid flow are small ($< 10 \, m^3/m^2$) or zero. These results are consistent with low permeability or impermeability respectively of unsheared marbles, and with the importance of deformation in enhancing the permeability of marbles. Experimental measurement of calcite-fluid dihedral angles in texturally equilibrated calcite matrices (Holness and Graham, 1995; Holness, this volume) showed that grain-edge flow ($\theta < 60°$) is only possible for strong brines and intermediate H_2O–CO_2 fluids at low pressures, showing that regional metamorphic marbles under hydrostatic stress should generally be impermeable to pervasive fluid flow. However, only the studies of Ganor et al. (1989) and Gerdes and Valley (1994) in Table 8.3 (see also Bucher-Nurminen, 1981, 1989; Taylor and Bucher-Nurminen, 1986) may be interpreted to support such a conclusion. In contrast, other studies (e.g. Baker et al., 1989; Bickle and Baker, 1990a; Baker and Spiegelman, 1995 in Table 8.3; see also Rye et al., 1976; Burkhard and Kerrich, 1988; Burkhard et al., 1992; Wickham and Peters, 1992) argue for non-zero permeability of regionally metamorphosed marbles. Holness and Graham (1995) argued that examples of apparently conflicting field (Table 8.3) and experimental evidence reflect

212

the importance of deformation in enhancing marble permeability. Proposed permeability enhancement mechanisms include microcracking, fluid inclusion drag during grain boundary migration, or dynamically maintained grain boundary fluid films (Holness and Graham, 1995). Similar deformation-induced mechanisms of permeability enhancement have been experimentally reproduced in feldspar aggregates (Tullis *et al.*, 1996) and rock salt (Urai *et al.*, 1986) and partially molten mantle (Bai *et al.*, this volume). Direct evidence regarding the mechanisms of fluid transport in rocks may best be established by textural and microchemical studies (McCaig and Knipe, 1990).

8.5.4 *Evidence for textural and micro-structural controls*

In some studies (Table 8.3), geochemical and petrological data relating to fluid flow have been integrated with textural studies in order to provide direct evidence of mechanisms of permeability enhancement and fluid flow. A good example of textural control of permeability, fluid flow and front advection in non-deforming rock is provided by infiltration of vein-controlled retrograde brines into marble wall rocks in the SW Scottish Highlands (Fein *et al.*, 1994; see Holness, this volume). Infiltration was accompanied by replacement of calcite by dolomite, first along grain-edges and then along grain boundaries, and by advection of an ^{18}O front. For the P–T–X (fluid) conditions of infiltration, $\theta > 60°$, and pervasive fluid flow occurred by calcite dissolution. In each marble band, isotopic and reaction fronts are coincident because oxygen isotope exchange between carbonate and fluid was kinetically inhibited in the absence of mineral reaction and/or recrystallization at the low temperatures of infiltration.

A good example of textural control of fluid flow in deforming rock is provided by carbonate mylonites along the Gavarnie Thrust in the central Pyrenees, which provided the conduit for large fluxes of fluid expelled during Tertiary thrusting (Table 8.3; McCaig *et al.*, 1995; McCaig, this volume). Textural studies (McCaig *et al.*, 1993) have identified several transient, syndeformational features and mechanisms occurring within a regime of dominantly plastic deformation which enhanced permeability, including:

- fractures, microcracks and grain boundary crack networks;
- dynamically-maintained grain boundary and grain-edge porosity on various scales;
- grain boundary migration.

8.6 Discussion and SW Scottish Highlands case study

8.6.1 *Introduction*

Numerous geological examples of the application of chromatographic theory (Table 8.3) have been used to demonstrate the dominance of layer-parallel fluid flow in metamorphic rocks, reflecting the potential roles of lithology, structure and texture in controlling fluxes and pathways. The application of chromato-graphic theory requires geochemical and petrological data at the hand-specimen to outcrop scale, and may be used to constrain and identify the roles of lithology and structure in channelling fluid flow at scales from metric to kilometric. However, in the absence of textural and microstructural information these models do not constrain the fluid transport mechanisms operating at the sub-grain to grain scale. In the following discussion we consider in more detail the case study of greenschist facies metamorphic rocks of the SW Scottish Highlands, in which chromatographic theory has been used to constrain the macro-scale controls of lithology and structure, and textural and structural observations used to deduce not only micro-scale processes and mechanisms of metamorphic fluid transport, but also the timing of fluid flow in relation to metamorphism and deformation.

8.6.2   Macro-scale controls: lithology and structure

In the SW Scottish Highlands, greenschist facies metabasite sills enclosed within phyllites and psammites have been used as quantitative sensors of fluid fluxes and flow paths (Skelton et al., 1995). Infiltration of H_2O–CO_2 fluid caused carbonation of the metabasite sills according to the reaction (Graham et al., 1983):

$$3 \text{ amphibole} + 2 \text{ epidote} + 10 \text{ CO}_2 + 8 \text{ H}_2\text{O} = 3 \text{ chlorite} + 10 \text{ calcite} + 21 \text{ quartz}$$
$$(8.15)$$

The asymmetrical arrangement of carbonated margins and unreacted, carbonate-free interiors was used to constrain flow directions, and front advec-tion distances used to constrain one-dimensional fluxes. The use of field-based geochemical techniques to determine progress of carbonation reactions, and thus the location of reaction fronts, made it possible for the first time to map fluxes and flow patterns on a regional scale, in a 25 km-long across-strike traverse.

The resulting 1-D and 3-D time-integrated flux profile across the SW Highland terrain (Figure 8.4) demonstrates a clear correlation between fluxes, as sensed by the metabasite sills, and host lithology. Low and variable fluxes in psammite-hosted metabasites are consistent with layer-parallel flow through deformed metabasite margins within undeformed and impermeable psammites (see quartz-fluid dihedral angle data of Holness, 1992, 1993). Fluid flow is domi-nantly channelled through permeable phyllites.

The role of local and regional structure is also well demonstrated in this study. Interlayered phyllites, psammites and metabasite sills are folded about regional (kilometre)-scale and local (100 m)-scale isoclinal and semi-recumbent antiforms whose axes lie within thick phyllitic units. Asymmetric carbonation of metaba-site sill margins and associated reaction front development record infiltration of an H_2O–CO_2 fluid, from which the direction and magnitude of the time-inte-

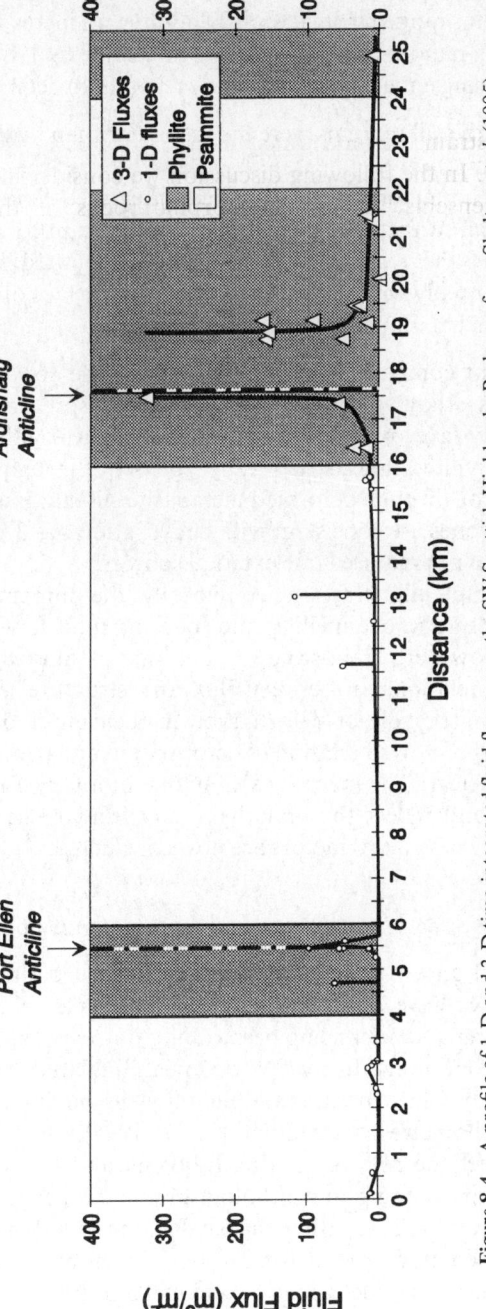

Figure 8.4 A profile of 1-D and 3-D time-integrated fluid flux across the SW Scottish Highlands. Redrawn from Skelton *et al.* (1995).

grated fluid flux can be constrained. 1-D time-integrated fluid fluxes perpendicular to the sill margins ranged from 0 to >100 m³/m² with increasing proximity to the axial region of a major antiform. By resolving some 69 1-D time-integrated fluid flux vectors, using equation 8.9, Skelton *et al.* (1995) determined that:

- fluxes parallel to the sill margins were much greater than fluxes perpendicular to the sill margins;
- the direction of the regional-scale 3-D time-integrated fluid flux vector was parallel to the axial surface of the regional recumbent antiform;
- the magnitude of the regional-scale 3-D time-integrated fluid flux vector increased exponentially with proximity to the axial surface of the semi-recumbent antiform to a maximum value of >400 m³/m².

Several important conclusions regarding macro-scale controls of fluid flow emerged from this study. Fluid flow was upwards (presumably buoyancy-driven), dominantly layer-parallel (except at fold hinges), lithologically channelled through phyllites, and structurally focused through axial zones of regional antiforms or through deformed metabasite sill margins. Average fluxes through high-flux zones are consistent with dehydration of ~3 km of underlying pelitic rocks, and do not require large external sources.

The above example illustrates unequivocally the important controls of lithology and structure in channelling and focusing fluid flow during regional metamorphism. However, grain-scale mechanisms of fluid transport are not resolved. While a relationship between flux and structure is clearly demonstrated, the mechanistic role of deformation in focusing and enhancing fluid flow, and the timing of flow in relation to deformation and metamorphic crystallization, remain unclear. The macro-scale factors, lithology and structure, may be circumstantial and reflect the underlying and fundamental importance of fabric and texture. These latter factors are now considered.

8.6.3 *Micro-scale processes: deformational control of fluid flow*

In reconciling field and experimental evidence for the permeability of meta-carbonate rocks, we have argued that deformation may play a vital role in dynamically creating and enhancing permeability in rocks which are impermeable under static conditions. In the SW Scottish Highlands we have noted the relationship between fold structures and fluid flow in metamorphic rocks at the macro (metres to kilometres) scale (Skelton *et al.*, 1995). The latter relationships might merely reflect the role of varying lithology and hence varying permeabilities within a pre-existing and lithologically heterogeneous fold structure (see sections 8.2.1 and 8.2.2), rather than indicating any dynamic mechanistic relationship between flow and deformation. Therefore a knowledge of the timing of flow in relation to deformation and metamorphism is crucial in establishing deformation/flow relationships. Micro-textural evidence regarding infiltration-driven reaction may provide such evidence.

In the SW Scottish Highlands, the carbonation reaction (equation 8.15) in metabasic sills, used to recognize reaction fronts, provides a means of constraining the timing of infiltration relative to deformation, metamorphism and fabric development, and therefore the role of deformation in fluid advection. Brodie and Rutter (1985) review mechanisms by which the resistance of a rock to deformation may be modified by metamorphic change. These include the formation of fine-grained reaction products, transformation-enhanced crystalline plasticity, and lowering of effective confining pressures as fluid pressure is increased. These mechanisms are all relevant to greenschist facies infiltration-driven reaction in the metabasic sills. Reaction progress in sill margins is observed to be associated with penetrative deformation and formation of a schistose fabric, whereas uncarbonated sill interiors are massive, implying a direct relationship between infiltration, reaction and penetrative deformation. One possible explanation of these observations is that the sill margins are deformed prior to fluid infiltration, and that fluid flowed solely within the sill margins. This creates an important dilemma, in that asymmetric carbonation would reflect asymmetric deformation and could not be used to constrain the direction of fluid flow.

This dilemma has been addressed in the SW Scottish Highlands through petrological investigation of partially carbonated metabasites. Chlorite, and in some instances calcite, is observed to be aligned parallel to the regional fabric, supported by massive interlocking amphiboles, epidotes and feldspars, consistent with the conclusion that infiltration-driven reaction preceded deformation. However, it is also possible to interpret the alignment of chlorite as a mimicking of pre-existing schistose fabrics (e.g. Brodie and Rutter, 1985), and therefore to conclude that fluid flow and resulting reaction post-dated deformation. Independent textural evidence indicates that reaction consequent to the infiltration of H_2O–CO_2 fluid also affects the undeformed metabasic rocks which abound in the SW Scottish Highlands. An example of undeformed metadolerites and metagabbros with relict igneous textures containing large amphiboles partially or wholly replaced by decussate calcite and chlorite as a result of the infiltration-driven carbonation reaction is illustrated in Figure 8.5. These textures demonstrate unequivocally that infiltration-driven reaction occurred in the absence of deformation, perhaps as a result of increasing fluid pressure with a consequent lowering of effective confining pressure which opened up grain boundaries. This process may have initiated deformation following mechanisms identified by Brodie and Rutter (1985), with a positive feedback between deformation, fabric formation and further infiltration (see below). We conclude that while fluid flow, infiltration and deformation proceeded synchronously and are intimately linked in a general sense, infiltration, front advection and chlorite-carbonate forming reaction proceeded initially through undeformed, low permeability metabasite, 'softening' the rocks for subsequent deformation and layer-parallel infiltration. We argue that infiltration proceeds by more than one mechanism, and this has important implications for the validity of assumptions

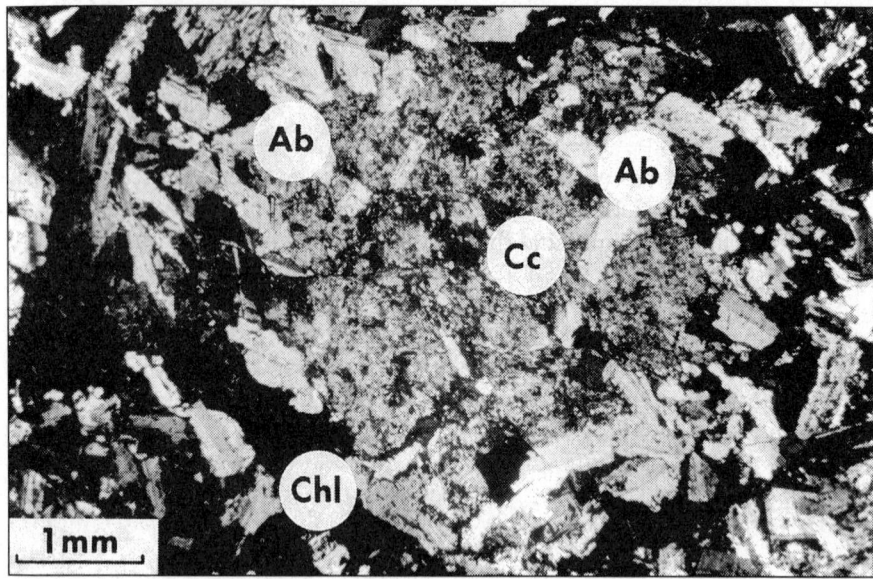

Figure 8.5 A photomicrograph of undeformed Type III (reacted) metabasite from the SW Scottish Highlands showing original igneous pyroxene, pseudomorphed by amphibole, then replaced by calcite and chlorite.

of constant porosity and permeability implicit in the application of chromato-graphic theory to constrain fluid fluxes.

8.6.4 *Fabric-parallel fluid flow and its relation to microvein fluid transport*

That both lithology and structure control regional-scale fluid flow geometries suggests that fluid flow is controlled at the grain scale by some more funda-mental process. Furthermore, if fluid flow were strictly layer- or lithology-parallel, the direction of flow would have to reverse at fold closures, implying an implausible component of downward flow. From field evidence of regional-scale focused fluid flow through shear zones, along fold axial planes, and through mica-rich lithologies we can predict that fluid flow may be controlled by grain-scale mineral alignment (i.e. rock fabric). We can make the same prediction from field evidence of regional-scale lithology-parallel fluid inasmuch as rock fabric is commonly layer-parallel in regional metamorphic terrains. The ability of rock fabric and lithology to control the direction of fluid flow at the micro- and macro-scales provides a powerful mechanism for the focusing of regional metamorphic fluid flow, with a natural unifying link to the identified roles of structure and deformation.

Of obvious interest is the less common situation in which rock fabric is oblique to rock layering. In such a situation, the roles of grain-scale rock fabric and regional-scale rock layering in controlling fluid flow can be distinguished and compared. In the SW Scottish Highlands, Skelton *et al.* (1997) calculate $\omega\phi t = 62.1 \pm 1.3$ m³/m² for flow across a greenschist facies metabasite sill in the SW Scottish Highlands (Table 8.3), about an order of magnitude greater than typical cross-layer fluxes (Table 8.3). This particular sill is on the inverted limb of the antiform, which Skelton *et al.* (1995) argue forms a regional-scale conduit for upward fluid flow. Because the axial-planar fabric is generally inclined at 5–10° to anticlinal fold limbs and thus to sill margins, flow parallel to the axis of the antiform will also be oblique to the sill margin, resulting in a significant component of cross-layer flow (Figure 8.6). The principal conclusion from this is that rock fabric is more important than mineralogy in controlling and directing fluid flow at the grain scale. The general predominance of layer-parallel flow (Table 8.3) may thus be a reflection of the predominance of layer-parallel rock fabric, in addition to the role of permeability contrast.

This interpretation is of particular relevance to resolving the important question of whether fluid flow is along grain-boundaries or close-spaced fractures at the grain-scale (cf. McCaig *et al.*, 1993; Hippertt, 1994; Cartwright and Buick, 1994). In a foliated metamorphic rock, closely spaced microfractures are likely to develop parallel to the rock fabric. Depending on the intensity of this fabric,

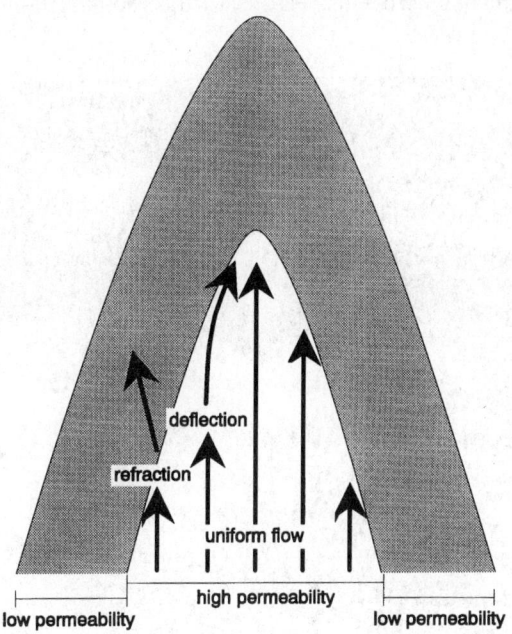

Figure 8.6 Schematic illustration of fluid flow across a folded low permeability layer. Note that regional-scale 'uniform flow' may reflect combined 'deflection' and 'refraction' on the outcrop-scale. Redrawn from Skelton *et al.* (1995).

219

grain-boundaries and closely spaced micro-fractures will converge and become indistinguishable, and the distinction between fluid flow along grain-boundaries and close-spaced fractures becomes blurred.

Field evidence of fluid flow through micro-fractures in zones of high fluid flux is evident in the SW Scottish Highlands. Discrete, fabric-parallel, millimetre-scale calcite–quartz microveins occur in all deformed and infiltrated lithologies (Figure 8.7) (Cole and Graham, 1994). Two principal controls on microvein development are evident. The first is lithological. Microveins occur in phyllites and deformed, carbonated metabasites but occur to a much lesser extent in non-schistose psammites. The observation that microveins occur preferentially in lithologies subject to the fabric-producing deformation demonstrates a relationship between fluid flow geometry, deformation and rock fabric.

Microvein development is also governed by a structural control related to the recumbent antiform, since the density of microveins increases towards the axial region where the largest time-integrated $H_2O–CO_2$ fluid fluxes have also been measured. The microveins become closely spaced on a millimetre- to centimetre-scale in the vicinity of the anticlinal hinge. Regardless of the host lithology and the frequency of microvein distribution, vein thickness and mineralogy are consistent across the antiformal structure. We can infer a direct correlation between fluid flux and microvein density within the regional-scale isoclinal fold. The deformation of microveins relative to the regional fabrics also provides

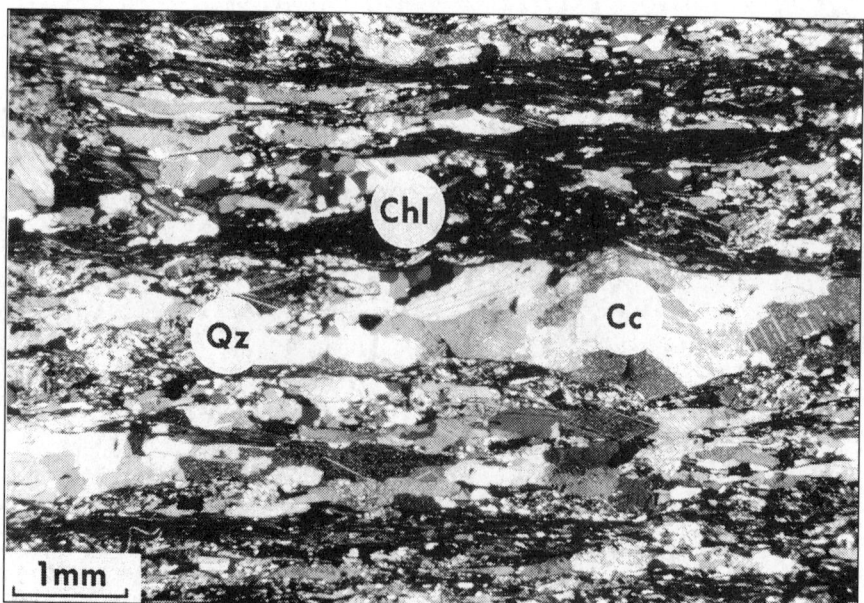

Figure 8.7 A photomicrograph of deformed Type III (reacted) metabasite from the SW Scottish Highlands showing millimetre-scale quartz–calcite microveins.

220

useful information regarding the timing of fluid flow in the SW Scottish Highlands and elsewhere.

In Vermont, Skelton (1996) mapped regions of transient high fluid fluxes through early Acadian fold closures from observations of thermal-acceleration of garnet crystal growth. His interpretation is consistent with regional-scale layer-parallel fluid flow on the inclined fold limbs predicted from symmetrical carbonation of metabasite dykes by reaction 8.15. Abundant fabric-parallel 2–3 mm microveins predominate throughout these high flux regions.

The formation of apparently dilational axial planar microveins in fold hinges is surprising, inasmuch as the axial plane is normally considered to be perpendicular to the direction of maximum compressive stress. Microvein formation in this orientation would therefore seem to require transiently high fluid pressures exceeding the maximum compressive stress, coupled with low tensile strength of the axial planar fabric.

The association of anticlinal fold closures, rock fabric, microvein development and zones of high fluid flux may be important in the dewatering of regional metamorphic terrains. It is in this association that the unifying and complementary controls of structure, lithology, fabric and deformation in focusing the flow of buoyant metamorphic fluid are most clearly demonstrated. The introduction of new material in microveins in such zones of high fluid flux will result in major synmetamorphic modification of rock chemistry.

8.7 Conclusions

The application of chromatographic theory has successfully constrained the magnitude and direction of metamorphic fluid flow in several recent studies. In many of these, fluid flow is buoyancy-driven, layer parallel, lithologically channelled and structurally focused at the macroscale, and flow geometry is determined by porosity and permeability contrast between different lithologies. Fluxes are usually consistent with fluid sourced from compaction of nearby or underlying devolatilizing rocks.

When such studies are integrated with textural and fabric studies in order to constrain underlying transport and infiltration-driven reaction mechanisms, and the important question of the timing of fluid flow is addressed, it emerges that the likely underlying grain-scale control of fluid flow is the rock fabric. The predominance of layer-parallel flow is consistent with the predominance in deformed regional metamorphic rocks of layer-parallel rock fabric. The ability of rock fabric and lithology to control the direction of fluid flow at the micro- and macro-scales provides a powerful mechanism for the focusing of regional metamorphic fluid flow.

Experimental and field/textural evidence indicates that deformation enhances grain-scale fluid flow in a dynamic sense by several possible mechanisms. However, it is also clear that infiltration and reaction can enhance permeability in undeformed rocks, and modify the resistance of rocks to deformation.

Infiltration, metamorphic reaction and deformation are therefore intimately interlinked. Deformation can enhance fluid flow both dynamically, by opening up the rock, and also in a more static sense by creating a rock fabric along which subsequent flow is focused.

Fabric-parallel microveins are important conduits of metamorphic fluid flow in zones of high fluid flux, which may cause major synmetamorphic modification of rock chemistry. Microveins are axial planar and concentrated in fold hinges, implying the transient development of high fluid pressures. The association of anticlinal fold closures, rock fabric, microvein development and zones of high fluid flux may be important in the dewatering of regional metamorphic terrains.

Acknowledgements

We thank Rainer Abart, Marian Holness, Ian Main, Andy McCaig and Nick Oliver for thorough and helpful reviews at short notice, and Marian Holness for editorial forbearance. One of the authors, A.D.L.S., acknowledges support of a NERC Research Fellowship. Research at Cambridge on fluid flow in metamorphic rocks is supported by NERC.

References

Bai, Q., Jin, Z.-M. and Green, II, H.W. (this volume) Experimental investigation of the rheology of partially molten peridotite at upper mantle pressures and temperatures.

Baker, J. and Spiegelman, M. (1995) Modelling an infiltration-driven geochemical front. *Earth Planet. Sci. Lett.*, **136**, 87–96.

Baker, J., Bickle, M.J., Buick, I.S. *et al.* (1989) Isotopic and petrological evidence for the infiltration of water-rich fluids during the Miocene M2 metamorphism on Naxos, Greece. *Geochim. Cosmochim. Acta*, **53**, 2037–50.

Barcilon, V. and Richter, F.M. (1986) Non-linear waves in compacting media. *J. Fluid Mechanics*, **164**, 429–48.

Baumgartner, L.P. and Ferry J.M. (1991) A model for coupled fluid-flow and mixed-volatile mineral reactions with applications to regional metamorphism. *Contrib. Mineral. Petrol.*, **106**, 273–85.

Bickle, M.J. (1992) Transport mechanisms by fluid-flow in metamorphic rocks: oxygen and strontium decoupling in the Trois Seigneurs Massif - a consequence of kinetic dispersion? *Am. J. Sci.*, **292**, 289-316.

Bickle, M.J. and Baker, J. (1990a) Advective-diffusive transport of isotope fronts: an example from Naxos, Greece. *Earth Planet. Sci. Lett.*, **97**, 78–93.

Bickle, M.J. and Baker, J. (1990b) Migration of reaction and isotopic fronts in infiltration zones: assessments of fluid flux in metamorphic terrains. *Earth Planet. Sci. Lett.*, **98**, 1–13.

Bickle, M.J. and McKenzie, D.P. (1987) The transport of heat and matter by fluids during metamorphism. *Contrib. Mineral. Petrol.*, **95**, 384–92.

Bickle, M.J., Chapman, H.J., Fallick, A.E. and Ferry, J.M. (1994) Oxygen, carbon and strontium isotopic constraints on fluid movement during the Acadian metamorphism in South-central Maine. *Mineral. Mag.*, **58A**, 90–1.

Bickle, M.J., Chapman, H.J., Wickham S.M. and Peters, M.T. (1995) Strontium and oxygen isotope profiles across marble-silicate contacts, Lizzies Basin, East Humboldt Range, Nevada: constraints on metamorphic permeability contrasts and fluid flow. *Contrib. Mineral. Petrol.*, **121**, 400–13.

REFERENCES

Bowman, J.R., Willett, S.D. and Cook, S.J. (1994) Oxygen isotopic transport and exchange during fluid flow: one-dimensional models and applications. *Am. J. Sci.*, **294**, 1–55.

Brodie, K.H. and Rutter, E.H. (1985) On the relationship between deformation and metamorphism, with special reference to the behaviour of basic rocks, in *Metamorphic Reactions* (eds A.B. Thompson and D.C. Rubie), Advances in Physical Geochemistry, Springer-Verlag, New York, **4**, 138–79.

Bucher-Nurminen, K. (1981) The formation of metasomatic reaction veins in dolomitic marble roof-pendants in the Bergell intrusion (Province Sondrio, Northern Italy). *Am. J. Sci.*, **281**, 1197–222.

Bucher-Nurminen, K. (1989) Reaction veins in marbles formed by a fracture-reaction-seal mechanism. *Europ. J. Mineral.*, **1**, 701–14.

Burkhard, M. and Kerrich, R. (1988) Fluid regimes in the deformation of the Helvetic nappes, Switzerland, as inferred from stable isotope data. *Contrib. Mineral. Petrol.*, **99**, 416–29.

Burkhard, M., Kerrich, R., Maas, R. and Fyfe, W.S. (1992) Stable and Sr-isotope evidence for fluid advection during thrusting of the Glarus nappe (Swiss Alps). *Contrib. Mineral. Petrol.*, **112**, 293–311.

Cartwright, I. (1994) The two-dimensional pattern of metamorphic fluid flow at Mary Kathleen, Australia: fluid focusing, transverse dispersion, and implications for modelling fluid flow. *Amer. Mineral.*, **79**, 526–35.

Cartwright, I. and Buick, I.S. (1994) Channelled fluid infiltration and variation in permeability in Reynolds Range marbles, Australia. *J. Geol. Soc. London*, **151**, 583–6.

Cheadle, M.J. (1989) Properties of texturally equilibrated two-phase aggregates. Unpublished PhD thesis, University of Cambridge.

Cole, C. and Graham, C.M. (1994) Stable isotope and textural evidence on the mechanisms of metamorphic fluid infiltration within a zone of structurally-focused high fluid flux. *Mineral. Mag.*, **58A**, 187–8.

England, P.C. and Richardson, S.W. (1977) The influence of erosion upon the mineral facies of rocks from different metamorphic environments. *J. Geol. Soc. London*, **134**, 201–13.

Etheridge, M.A., Wall, V.J. and Vernon, R.H. (1983) The role of the fluid phase during regional metamorphism and deformation. *J. Metam. Geol.*, **1**, 205–26.

Fein, J.B., Graham, C.M., Holness, M.B. *et al.* (1994) Controls on the mechanisms of fluid infiltration and front advection during regional metamorphism: a stable isotope and textural study of retrograde Dalradian rocks of the SW Scottish Highlands. *J. Metam. Geol.*, **2**, 249–60.

Ferry, J.M. (1980) A case study of the amount and distribution of heat and fluid during metamorphism. *Contrib. Mineral. Petrol.*, **71**, 373–85.

Ferry, J.M. (1989) Contact metamorphism of roof pendants at Hope Valley, Alpine County, California, USA. *Contrib. Mineral. Petrol.*, **101**, 402–17.

Ferry, J.M. (1992) Regional metamorphism of the Waits River Formation, eastern Vermont: delineation of a new type of giant hydrothermal system. *J. Petrol.*, **33**, 45–94.

Ferry, J.M. (1994) Overview of the petrological record of fluid flow during regional metamorphism in northern New England. *Am. J. Sci.*, **294**, 905–88.

Ferry, J.M. and Dipple, G.M. (1991) Fluid flow, mineral reactions, and metasomatism. *Geology*, **19**, 211–14.

Fletcher, R.C. and Hofmann, A.W. (1974) Simple models of diffusion and combined diffusion infiltration metasomatism, in *Geochemical Transport and Kinetics* (eds A.W. Hofmann, B.J. Giletti, H.S. Yoder and R.A. Yund), Carnegie Inst. Washington Publ. **634**, 243–59.

Ganor, J., Matthews, A. and Paldor, N. (1989) Constraints on effective diffusivity during oxygen isotope exchange at a marble-schist contact, Sifnos, Greece. *Earth Planet. Sci. Lett.*, **94**, 208–16.

Gerdes, M.L. and Valley, J.W. (1994) Fluid flow and mass transport at the Valentine wollastonite deposit, Adirondack Mountains, New York State. *J. Metam. Geol.*, **12**, 589–608.

Gerdes, M.L., Baumgartner, L.P., Person, M. and Rumble, D. (1995) One and two dimensional models of fluid flow and stable isotope exchange in the Adamello contact aureole, southern Alps, Italy. *Amer. Mineral.*, **80**, 1004–19.

Gleason, G.C. and Tullis, J. (1995) A flow law for dislocation creep of quartz aggregates determined with the molten salt cell. *Tectonophysics*, **247**, 1–23.

Graham, C.M., Greig, K.M., Sheppard, S.M.F. and Turi, B. (1983) Genesis and mobility of the H_2O–CO_2 fluid phase during regional greenschist and epidote amphibolite facies metamorphism: a petrological and chemical isotope study in the Scottish Dalradian. *J. Geol. Soc. London*, **140**, 577–99.

Hippertt, J.F.M. (1994) Grain boundary microstructures in micaceous quartzite: significance for fluid movement and deformation processes in low metamorphic grade shear zones. *J. Geol.*, **102**, 331–48.

Hofmann, A.W. (1972) Chromatographic theory of infiltration metasomatism and its application to feldspars. *Am. J. Sci.*, **272**, 69–90.

Holness, M.B. (1992) Equilibrium dihedral angles in the system quartz-CO_2-H_2O-NaCl at 800 °C and 1-15 kbar: the effects of pressure and fluid composition on the permeability of quartzites. *Earth Planet. Sci. Lett.*, **114**, 171–84.

Holness, M.B. (1993) Temperature and pressure dependence of quartz-aqueous fluid dihedral angles: the control of adsorbed H_2O on the permeability of quartzite. *Earth Planet. Sci. Lett.*, **117**, 363–77.

Holness M.B. (1996) The controls of surface chemistry on pore topologies in texturally equilibrated materials, in *Fluid flow and transport in rocks: mechanisms and effects* (eds B. Jamtveit and B.W.D. Yardley), Chapman & Hall (in press).

Holness, M.B. (this volume) The permeability of non-deforming rocks.

Holness, M.B. and Graham, C.M. (1995) P-T-X effects on equilibrium carbonate-H_2O-CO_2-NaCl dihedral angles: constraints on carbonate permeability and the role of deformation during fluid infiltration. *Contrib. Mineral. Petrol.*, **119**, 301–13.

Jamtveit, B., Bucher-Nurminen, K. and Stijfhoorn, D.E. (1992a) Contact metamorphism of layered shale-carbonate sequences in the Oslo Rift: I. Buffering, infiltration, and the mechanisms of mass transport. *J. Petrol.*, **33**, 377–422.

Jamtveit, B., Grorud, H.F. and Bucher-Nurminen, K. (1992b) Contact metamorphism of layered shale-carbonate sequences in the Oslo Rift. II: migration of isotopic and reaction fronts around cooling plutons. *Earth Planet. Sci. Lett.*, **114**, 131–48.

Knipe, R.J. and McCaig, A.M. (1994) Microstructural and microchemical consequences of fluid flow in deforming rocks, in *Geofluids: Origin, Migration and Evolution of Fluids in Sedimentary Basins* (ed. J. Parnell) Geol. Soc. Spec. Publ. **78**, 99–111.

Kohn, M.J. and Valley, J.W. (1994) Oxygen isotope constraints on metamorphic fluid flow, Townshend Dam, Vermont, USA. *Geochim. Cosmochim. Acta*, **58**, 5551–66.

Kruse, S., McNutt, M., Phipps-Morgan, J. and Roydon, L. (1991) Lithospheric extension near Lake Mead, Nevada: a model for ductile flow in the lower crust. *J. Geophys. Res.*, **96**, 4435–56.

Lasaga, A.C. and Rye, D.M. (1993) Fluid flow and chemical reaction kinetics in metamorphic systems. *Am. J. Sci.*, **293**, 361–404.

Lichtner, P.C. (1988) The quasi-stationary state approximation to coupled mass transport and fluid-rock interaction in a porous medium. *Geochim. Cosmochim. Acta*, **52**, 143–65.

Lister, C.R.B. (1972) On the thermal balance of a mid-ocean ridge. *Geophys. J. Roy. Astron. Soc.*, **26**, 515–35.

224

REFERENCES

McCaig, A.M. (this volume) The geochemistry of volatile fluid flow in shear zones.

McCaig, A.M. and Knipe, R.J. (1990) Mass-transport mechanisms in deforming rocks: recognition using microstructural and microchemical criteria. *Geology*, **18**, 824–7.

McCaig, A.M., Gong, L.Y. and Wayne, D.M. (1993) Mechanisms of permeability enhancement in carbonate mylonites, in *Geofluids '93 Extended Abstracts*.

McCaig, A.M., Wayne, D.M., Marshall, J.D. *et al.* (1995) Isotopic and fluid inclusion studies of fluid movement along the Garvanie Thrust, central Pyrenees: reaction fronts in carbonate mylonites. *Amer. J. Sci.*, **295**, 309–43.

McKenzie, D.P. (1984) The generation and compaction of partially molten rock. *J. Petrol.*, **25**, 713–65.

McKenzie, D.P. (1985) The extraction of magma from the crust and mantle. *Earth Planet. Sci. Lett.*, **74**, 81–91.

Nabelek, P.I. and Labotka, T.C. (1993) Implications of geochemical fronts in the Notch Peak contact-metamorphic aureole, Utah, USA. *Earth Planet. Sci. Lett.*, **119**, 539–59.

Ogata, A. (1964) Mathematics of dispersion with linear adsorption isotherm. *U.S. Geol. Surv. Prof. Pap.*, **411-H**, 9pp.

Oliver, N.H.S. (1996) Review and classification of structural controls on fluid flow during regional metamorphism. *J. Metam. Geol.*, **14**, 477–92.

Oliver, N.H.S., Rawling, T.J., Cartwright, I. and Pearson, P.J. (1994) High-temperature fluid-rock interaction and scapolitization in an extension-related hydrothermal system, Mary Kathleen, Australia. *J. Petrol.*, **35**, 1455–91.

Rye, R.O., Schuiling, R.D, Rye, D.M. and Jansen, J.B.H. (1976) Carbon, hydrogen and oxygen isotope studies of the regional metamorphic complex at Naxos, Greece. *Geochim. Cosmochim. Acta*, **40**, 1031–49.

Scott, D. and Stevenson, D. (1984) Magma solitons. *Geophys. Res. Lett.*, **11**, 1161–4.

Skelton, A.D.L. (1996) The timing and direction of metamorphic fluid flow in Vermont, USA. *Contrib. Mineral. Petrol.*, **125**, 75–84.

Skelton, A.D.L., Bickle, M.J. and Graham, C.M. (1997) Fluid flux and reaction rate from advective diffusive carbonation of mafic sill margins in the Dalradian, SW Scottish Highlands. *Earth Planet. Sci. Lett.*, **146**, 527–39.

Skelton, A.D.L., Graham, C.M. and Bickle, M.J. (1995) Lithological and structural controls on regional 3-D fluid flow patterns during greenschist facies metamorphism of the Dalradian of the SW Scottish Highlands. *J. Petrol.*, **36**, 563–86.

Taylor, B.E. and Bucher-Nurminen, K. (1986) Oxygen and carbon isotope and cation geochemistry of metasomatic carbonates and fluids – Bergell aureole, Northern Italy. *Geochim. Cosmochim. Acta*, **50**, 1267–79.

Taylor, H.P. and Forester, R.W. (1971) Low-^{18}O igneous rocks from the intrusive complexes of Skye, Mull, and Ardnamurchan, western Scotland. *J. Petrol.*, **12**, 465–97.

Todd, C.S. and Evans, B.W. (1993) Limited fluid-rock interaction at marble-gneiss contacts during Cretaceous granulite-facies metamorphism, Seward Peninsula, Alaska. *Contrib. Mineral. Petrol.*, **114**, 27–41.

Tullis, J., Yund, R. and Farver, J. (1996) Deformation-enhanced fluid distribution in feldspar aggregates and implications for ductile shear zones. *Geology*, **24**, 63–66.

Urai, J.L., Spears, C.J., Zwart, H.J. and Lister, G.S. (1986) Weakening of rock salt by water during long-term creep. *Nature*, **324**, 554–7.

Walther, J.V. and Orville, P.M. (1982) Volatile production and transport in regional metamorphism. *Contrib. Mineral. Petrol.*, **79**, 252–7.

Wickham, S.M. and Peters, M.T. (1992) Oxygen and carbon isotope profiles in metasediments from Lizzies Basin, East Humboldt Range, Nevada: constraints on mid-crustal metamorphic and magmatic volatile fluxes. *Contrib. Mineral. Petrol.*, **112**, 46–65.

Wiggins, C. and Spiegelman, M. (1995) Magma migration and magmatic solitary waves in 3-D. *Geophys. Res. Lett.*, **22**, 1289–92.

Wood, B.J. and Walther, J.V. (1986) Fluid flow during metamorphism and its implications for fluid-rock ratios, in *Fluid-Rock Interactions during Metamorphism* (eds J.V. Walther and B.J. Wood), Advances in Physical Geochemistry, Springer-Verlag, New York, **5**, 89–108.

Yardley, B.W.D. and Lloyd, G.E. (1995) Why metasomatic fronts are really metasomatic sides. *Geology*, **23**, 53–6.

Yardley, B.W.D., Bottrell, S.H. and Cliff, R.A. (1991a) Evidence for a regional-scale fluid loss event during mid-crustal metamorphism. *Nature*, **349**, 151–4.

Yardley, B.W.D., Rochelle, C.A., Barnicoat, A.C. and Lloyd, G.E. (1991b) Oscillatory zoning in metamorphic minerals: an indicator of infiltration metasomatism. *Mineralog. Mag.*, **55**, 357–65.

CHAPTER NINE

The geochemistry of volatile fluid flow in shear zones

Andrew M. McCaig

9.1 Introduction

Faults and shear zones are known to be important conduits for fluid flow in the lithosphere at metamorphic grades ranging from very low (e.g. Gieskes *et al.*, 1990; Moore and Vrolijk, 1992) to amphibolite (Beach, 1976; Brodie 1980; Dipple *et al.*, 1990) and even eclogite (Früh-Green, 1994). At other times, faults act as seals in sedimentary rocks (e.g. Knipe, 1992, 1993), or show little or no evidence for significant fluid–rock interaction (Kerrich *et al.*, 1980). Fluid flow through shear zones is probably an important part of the crustal fluid-flow budget (Fyfe *et al.*, 1978), is important in localizing mesothermal gold deposits (Mikuchi and Ridley, 1993; Robert *et al.*, 1995), and may even influence the Earth's climate (Kerrick and Caldeira, 1993). Fluid–rock interaction and rock deformation are both affected by many variables, so it is not surprising that when they occur together the results can be heterogeneous and unpredictable. Important questions which can be raised concerning fluid flow in faults and shear zones include the following.

- Under what circumstances does fluid flow focus along faults and shear zones?
- Does flow occur mainly during or after deformation events?
- How much fluid can flow through shear zones, and are such fluxes important in global budgets of fluid flow in the crust?
- How deep do surface-derived fluids penetrate in the crust, and is this influenced by deformation?
- What are the mechanical effects of fluid flow, and does fluid movement influence the strength of faults or of the lithosphere as a whole?
- What are the mechanisms of permeability enhancement or reduction during deformation, and how pervasive is deformation-enhanced fluid flow?

The last two questions have been comprehensively reviewed in recent years. Fluid flow during earthquakes and in low-grade fault zones has been extensively

Deformation-enhanced Fluid Transport in the Earth's Crust and Mantle. Edited by M.B. Holness. Published in 1997 by Chapman & Hall, London. ISBN 0 412 75290 5.

discussed by Sibson (1990, 1994) and Muir-Wood and King (1993), and was recently the subject of a thematic set of papers in the *Journal of Geophysical Research* (Hickman et al., 1995). General aspects of fluid flow during deformation have been addressed by Etheridge *et al.* (1983, 1984), Carter *et al.* (1990) and Oliver (1996), while Holness and Graham (1995) and Graham *et al.* (this volume) have discussed the influence of deformation on grain-scale porosity and permeability. Finally, Knipe and McCaig (1994) reviewed the effects of various deformation mechanisms on permeability and fluid flow, and established criteria for recognizing the pervasiveness or otherwise of fluid flow on a grain scale.

In view of the above, this review will concentrate on geochemical aspects of fluid flow in shear zones deformed predominantly by ductile processes (i.e. those containing mylonite in the sense of White *et al.* (1980)). Mechanical aspects of flow will be discussed only briefly, except where they influence the geochemical evolution of fault rocks, or large-scale patterns of fluid flow in the crust. The main part of this review examines the various geochemical methodologies which have been applied to shear zones, concentrating on examples where good field control and the availability of suitable geochemical tracers allow quantitative information on fluid fluxes and flow patterns to be obtained. Methodologies which can be applied in less ideal circumstances are also discussed. The information which can be obtained from geochemical studies concerning shear zone permeability, and the type, source and circulation patterns of fluid is reviewed, together with the influence of fluid flow mechanisms on alteration patterns. Finally, an attempt is made to answer the questions posed above.

9.2 Deformation in shear zones

Ideal shear zones are defined as planar zones of localized ductile strain, and show only simple shear parallel to their margins (Ramsay and Graham, 1970). In practice, much more complex strain patterns are seen, often including a component of pure shear flattening across the shear zone (e.g. Law *et al.*, 1986). Such patterns may result from strain of the wall rocks, displacement gradients within the shear zone, or partitioning of strain into domains within the zone (Knipe, 1990).

Shear zones are by definition zones of localized deformation, and hence must have been softer than surrounding rocks at the time of deformation. A variety of mechanisms contribute to softening (White *et al.*, 1980), including geometrical effects due to the development of crystallographic preferred orientations and reduction in grain size due to recrystallization. The nature of deformation within shear zones depends on temperature, effective pressure (i.e. the difference between lithostatic pressure and internal pore fluid pressure), deviatoric stress, strain rate and lithology (mineralogy and grain size). At low temperatures (<200 °C), ductile shear zones only form in unconsolidated sediments and weak materials such as halite and clays. In crystalline rocks, low temperature deformation is dominated by fracture and, in extreme cases, cataclastic flow. Ductile flow (or quasi-plastic flow; Sibson, 1986) occurs when continued deformation does not lead to work-hardening (Nicolas and Poirier, 1976). Work-hardening during dislocation creep is caused by build-up of dislocation tangles within mineral lattices, which can be removed by diffusion-controlled processes such as recovery and recrystallization. Hence, increase in temperature favours ductile deformation by dislocation creep as well as other mechanisms more directly dependent on diffusion rates. However, the transition from work-hardening and fracture to ductile flow occurs at very different temperatures in different rock-forming minerals (normally in the order sericite–calcite–quartz–feldspar–amphibole–olivine–pyroxene). Hence, different deformation mechanisms ranging from fracture to dislocation creep and diffusional mass-transfer generally occur simultaneously in typical polymineralic mylonites even under constant conditions of temperature, effective pressure and strain rate. Even monomineralic mylonites may show domainal fabrics with stress orientations and strain rates varying between different grains and grain aggregates (Knipe, 1989). In addition, variation of parameters such as temperature, strain rate, grain size, mineralogy (due to metamorphic reactions or phase changes) and especially fluid pressure may lead to either permanent changes in, or cyclical switching of, deformation mechanisms in mylonites (White et al., 1980; Brodie and Rutter, 1985, 1987; Knipe, 1989). The net effect is that mylonitic deformation mechanisms, and resulting textures, may vary rapidly both in time and space, and observed textures in natural specimens may not reflect the most important events in the history of a shear zone.

Fine grain size and continuous recrystallization generally lead to a much higher degree of isotopic equilibrium between fluid and minerals in mylonites than in coarser metamorphic rocks at comparable temperatures (Kerrich et al., 1984; Fricke et al., 1992). The recrystallized minerals in a mylonite may approach equilibrium with a fluid phase, while unrecrystallized phases, such as feldspar, may not. This is a reversal of the patterns of low temperature mineral alteration normally seen in granitic rocks in the absence of ductile deformation (Criss et al., 1987), or where the extent of deformation is small (Cartwright et al., 1993).

In natural shear zones, metamorphic reactions and deformation generally occur together. However, it is often difficult to tell whether fluid access and deformation occurred as a result of deformation, or if deformation localized in the shear zone because of metamorphism (Beach, 1980; Rubie, 1983; Brodie and Rutter, 1985; McCaig, 1984, 1987).

Knipe and McCaig (1994) and Oliver (1996) have reviewed the significance of various deformation mechanisms and parameters for permeability and fluid flow. Most ductile deformation mechanisms are unlikely to increase permeability significantly unless they are accompanied by dilatant pore opening (Walker et al., 1990; Ree, 1994; Zhang et al., 1994) or fracture. Effective pressure probably has the most influence on permeability, since low effective pressures (i.e. high pore fluid pressures) can promote dilatancy and fracture at relatively low deviatoric stresses even in rocks deforming in the ductile regime at relatively high temperatures (see section 9.8).

9.3 Causes and effects of fluid flow in shear zones

It is important to make a distinction between observing fluid flow and observing the geochemical effects of fluid flow. Fluid flow invariably occurs down hydraulic gradients, which can be produced by a variety of processes including thermal expansion and buoyancy, topography, overpressuring due to compaction of pore space or liberation of metamorphic fluids, and seismic or dilatancy pumping (Fyfe *et al.*, 1978; Etheridge *et al.*, 1983, 1984; Oliver, 1986; Garven *et al.*, 1993; Ferry, 1994; Oliver, 1996). Alteration occurs when a moving fluid is out of equilibrium with the rock, either because of changes in pressure or temperature along the flow path, or because fluid moves from one lithology to another. Alteration may be either absent or limited in extent if the moving fluid reaches equilibrium with the rocks along the flow path or if fluid–rock reaction is slow compared with the velocity of fluid movement (Bickle, 1992; Knipe and McCaig, 1994). It is quite possible for large volumes of fluid to pass through a particular rock leaving no measurable geochemical signature. Conversely, diffusion can lead to alteration of a rock without any advection of fluid. Interpretation of alteration patterns in shear zones requires a good understanding of the theory of diffusive-advective mass-transport (see section 9.5).

The total amount of fluid which passes through a shear zone is a function of the time-integrated hydraulic gradient and the time-integrated permeability. Focusing of fluid flow into a shear zone is a result of higher average permeability in the shear zone than in its wall rocks (sections 9.7 and 9.8). It follows that variations in the intensity of alteration within and between shear zones may result from variations in average permeability, variations in average hydraulic gradient, or variations in the extent to which fluid was out of equilibrium with the rock at a particular site. In the next section, various methods of constraining the extent of alteration are discussed, while the significance of alteration patterns is discussed in section 9.5.

9.4 Geochemical techniques applied to shear zones

In this section, various geochemical methodologies which have been applied to shear zone alteration, and some of the particular problems associated with defining the extent of alteration in shear zones, are discussed.

Table 9.1 summarizes previous work on fluid–rock interaction in shear zones. As in all studies of alteration, the composition of a mylonite must be compared with an assumed protolith in order to determine whether changes in rock composition have occurred through fluid–rock interaction. In small shear zones the protolith can often be established with some certainty, since layers in the wall rocks can often be traced into and across the shear zone (e.g. McCaig, 1984; Sinha et al., 1986; Gilotti, 1989). However, the inhomogeneous nature of many mylonites leads to particular problems:

- Relics of unaltered material or porphyroclasts inherited from the protolith may be present. Whole-rock analyses may represent an average of altered and unaltered material, or of layers derived from different protoliths. Wall-rock alteration may also make unaltered protolith rocks difficult to find (McCaig et al., 1990).
- Many mylonites contain abundant vein material precipitated directly from a fluid (McCaig, 1987; Cox and Etheridge, 1989; McCaig and Knipe, 1990; Knipe and McCaig, 1994). This material may be present on a variety of scales, and may be difficult to detect if it occurs only within certain mineral grains or in specific microstructural sites. In some cases, veins are subject to recrystallization and breakdown into individual minerals during ductile flow (Knipe and McCaig, 1994).
- Volume increase or decrease during alteration can lead to passive concentration or dilution of components, so that whole-rock alteration must be interpreted with care (Gresens, 1967; O'Hara, 1988).
- Local diffusive redistribution of material by pressure solution or due to metamorphic reactions (Knipe and Wintsch, 1985; Wintsch et al., 1995) may complicate alteration patterns and lead to erroneous inferences of large-scale mass-transfer if analytical volumes are poorly selected. For example, Tempest (1991) showed that mylonitization of an initially homogeneous granodiorite can lead to domains of widely different mineralogy and chemistry (section 9.4.2).

9.4.1 Major and trace element analysis

This is the most universally applied analytical method, but suffers to the greatest degree from the problems described above. A pioneering study was that of Beach (1976) on amphibolite facies shear zones in the Lewisian of NW Scotland. He used a correlation matrix approach to show that alkalis had been added to the mylonites and Fe, Ca and Mg removed. Graphical techniques have been preferred by most subsequent authors. Problems of volume change can be resolved using the following equation (Gresens, 1967) to relate initial (rock A) and final (rock B) compositions:

$$X_n = 100[f_v(g^B/g^A)C^B_n - C^A_n]$$ (9.1)

where X_n is the actual mass gain or loss of component n in wgt%, f_v is the volume factor for the change, g^B and g^A are the specific gravities of the two rocks, and C_n is the mass fraction of component n in rock A or B. A variety of plots have been

Table 9.1 Summary of geochemical studies on shear zones
Shear zone type – A: Steep reverse

Reference	Location and lithology	Fluid type	Metasomatic effects	Fluid flux
Beach, 1976, 1980	Lewisian gneiss, NW Scotland.	? Mantle fluids (Beach and Tarney, 1987).	Increase in K, Na, loss of Fe, Ca, Mg.	Unknown.
Kerrich et al., 1980	Granite, Mieville, Switzerland.	Unknown.	Minor hydration only.	Undetectable.
Kerrich et al., 1984	(a) Uranium-mineralized amphibolite facies shear zones in gneiss, Bahia, Brazil.	Meteoric formation waters from underthrust sediments.	Lowering of $\delta^{18}O$; increase in Na, Fe^{3+}, U, REE; decrease in Si, K.	High.
	(b) Au-bearing shear zones in greenstones, Yellowknife.	Reduced metamorphic fluids.	Increase in $\delta^{18}O$, increase in Fe^{2+}/Fe^{3+}.	W/R 3-30 based on down-T flow.
	(c) Amphibolite facies mylonites, Grenville front, Ontario.	Metamorphic fluids; meteoric water in late veins.	Local albitization; little change in $\delta^{18}O$ except in late veins.	Low, except for albitite.
Winchester and Max, 1984	Prograde shear zones in amphibolite facies basic and silicic gneisses, Co. Mayo, Ireland.	? Metamorphic fluids.	Complex lithology-dependent major element changes.	Unknown
McCaig, 1984; McCaig et al., 1990	Alpine shear zones in Variscan basement, Pyrenees.	Hypersaline brine (cf. Tempest, 1991), probably formation water from low $\delta^{18}O$ values.	Mineralogical and chemical changes linked to progressive strain; lowering of $\delta^{18}O$; muscovitization and albitization of granite.	1.25×10^5 m³/m², based on muscovitization reaction.
Marquer et al., 1985; Fourcade et al., 1989	Grimsel granodiorite, Aar Massif, Alps.	~1m Cl brine (Dipple and Ferry, 1992b) (?)	Increase in K, Mg; decrease in Ca, Na in highest strained rocks; 1 to 2 permil increase in $\delta^{18}O$ (Fourcade et al., 1989).	1.1×10^4 m³/m² (Dipple and Ferry, 1992b).
McCaig, 1987	Shear banded phyllonite, Merens fault, Pyrenees.	? Brine (cf. Tempest, 1991).	Introduction of Mg, Fe, Si into dilatant shear bands.	Unknown.
Losh, 1989	Alpine shear zones in Neouvielle granodiorite, Pyrenees.	Hypersaline brine (Henderson and McCaig, 1996); mixing with high level fluid.	Increase and decrease in $\delta^{18}O$.	Unknown.
Romer, 1990	Caledonide shear zones in Proterozoic basement, N. Sweden.	?	Introduction of radiogenic Pb isotopes.	?

Reference	Location / rock type	Fluid source	Chemical changes	Water/rock ratio
Tobisch et al., 1991	(a) Tonalite, Foothills Terrane, Sierra Nevada, USA and Lachlan Foldbelt, SE Australia.	Modified seawater or connate water (low $\delta^{18}O$ values).	Increase in Si, Ca, Fe, Sr; decrease in Na, K, Mg, Ba, Rb, and Fe^{2+}/Fe^{3+}; decrease in $\delta^{18}O$. Volume increase.	Minimum water/rock ratios 0.1 to 10.
	(b) Adamellite, Lachlan Foldbelt.	As above.	Increase in Sr, decrease in K, Ba, Pb, Rb, Si, Al; decrease in $\delta^{18}O$. Volume loss.	
Arreaza and Persoz, 1992	Greenschist/amphibolite shear zones in granitoids, E. Gotthard Massif, Alps.	Uncertain.	In greenschist facies increase in K, Rb, Fe and trace elements; loss of Na, Sr. In amphibolite facies loss of Si, Na; gain of some trace elements.	?
Früh-Green, 1994	Eclogitic shear zones in quartz diorite, Sesia Zone, W Alps.	Metamorphic fluids.	Increase in Si, H_2O; decrease in K, Na, Ca, Al; volume increase in $\delta^{18}O$.	?
Marquer et al., 1994	Alpine shear zones in feldspathic gneiss, Aiguilles Rouges Massif, Alps.	Uncertain.	Isovolumetric and isochemical except for Ca. No change in $\delta^{18}O$ of quartz.	Small amounts of fluid, then influx of Ca-rich fluid.

B: Thrust

Reference	Location / rock type	Fluid source	Chemical changes	Water/rock ratio
Dostal et al., 1980; Jamieson and Strong, 1981	Biotite mylonite in metamorphic sole of St Anthony ophiolite complex, NW Newfoundland.	Unknown.	Changes in major and trace element ratios, including LREE enrichment.	Unknown.
Knipe and Wintsch, 1985	Moine Thrust Zone, NW Scotland.	Unknown.	Transfer of Na, K, Ca between layers.	Small, largely diffusive mass-transfer.
Sinha et al., 1986, 1988	Brevard Fault ultramylonites in granitoids, Appalachians.	Unknown.	Increase in Ca, Fe; decrease in Si, K, Na.	Minimum water/rock of 250 for up T flow.
Vocke et al., 1987	Blueschist/greenschist thrusting of granite gneiss, Suretta nappe, Switzerland.	? Metamorphic.	Gain of Mg, K, Fe^{3+}; loss of Na, Sr, Fe^{2+} and HREE.	Unknown.
O'Hara, 1988; O'Hara and Haak, 1992	Mylonitized granites, Rector Branch thrust, N. Carolina.	Brines and surface-derived fluids (O'Hara and Haak, 1992).	Volume loss; loss of K, Na, Si, Ca, Al, Sr, Rb.	Minimum water/rock ratios of 10^2 to 10^3.
Gillotti, 1989	Mylonitized basaltic dykes, Seve-Koli thrust, Norway.	Unknown.	Isochemical except for K increase at very high strains.	Low except at high strain.

Table 9.1 Summary of geochemical studies on shear zones (*continued*)
B: Thrust (continued)

Reference	Location and lithology	Fluid type	Metasomatic effects	Fluid flux
Dipple et al., 1990	Hunts Brook fault zone, Connecticut. Amphibolite facies mylonites in granodiorite gneiss.	Metamorphic fluid, down T flow (Dipple and Ferry, 1992b).	Increase in Si, K; decrease in Na, Ca.	4×10^4 m²/m² (Dipple and Ferry, 1992b).
Barovich and Patchett, 1992	Harquahala thrust, Arizona; mylonite in granite.	Unknown.	No consistent major element changes; Sr isotopes mobile, Nd, Hf isotopes immobile.	Low.
Burkhard et al., 1992; Burkhard and Kerrich, 1990	Glarus Thust, Swiss Alps.	Formation/metamorphic waters.	Lowering of $\delta^{18}O$ and increase in $^{87}Sr/^{86}Sr$ in carbonate mylonites.	3 to 6×103 (Bowman et al., 1994).
Newman and Mitra, 1993	Mylonite, Linville Falls fault, N. Carolina.	Formation/metamorphic waters?	Loss of Si; muscovitization; 20–80% volume reduction.	
Mattey et al., 1994	Eclogite facies shear zone in granulite, Bergen Arcs, Norway.	Hydrous metamorphic fluid in eclogite facies; N_2/CO_2-rich fluid in late retrogression.	Si increase in eclogitization; Si, K, Ba, Rb enriched in post-kinematic regression.	Unknown.
McCaig et al., 1995a,b; Banks et al., 1991, 1994	Gavarnie Thrust, Pyrenees. Carbonate mylonites and fluid inclusions in quartz veins.	Hypersaline formation waters, probably evaporitic (Banks et al., 1994).	Increase in $^{87}Sr/^{86}Sr$ and lowering of $\delta^{18}O$ in carbonate mylonites. Mixing of Ca/Na, Sr, and Pb isotopes in fluid inclusions.	>1.8–3.0×10^3 m³/m².
O'Hara et al., 1995	Shear zone in Blue Ridge province, S. Appalachians.	Ca–Mg brines of variable salinity, probably carbonate formation waters.	Loss of Si, alkalis. Changes in $\delta^{18}O$.	
Crespo-Blanc et al., 1995	Diablarets Thrust, Swiss Alps.	Formation/metamorphic waters.	Lowering of $\delta^{18}O$ and $\delta^{13}C$.	Unknown.

C: Extensional

Reference	Location and lithology	Fluid type	Metasomatic effects	Fluid flux
Brodie, 1980	Ultramylonite in peridotite, Finero, Italian Alps.	Estimated 1m Cl brine (Dipple and Ferry, 1992b).	Increase in K/Si and decrease in Mg/Si.	$\sim 1.4 \times 10^4$ m³/m² (Dibble and Ferry, 1992b).
Kerrich and Hyndman, 1986	Mylonite and chlorite breccia, Bitteroot Lobe/Sapphire block detachment, Montana.	Magmatic/metamorphic in mylonites, meteoric in cht breccias.	Lowering of $\delta^{18}O$ in cht breccias and in less resistant minerals in mylonites.	Unknown.

234

Reference	Location/Description	Fluid	Chemical changes	Flux
Kerrich and Rehrig, 1987; Kerrich, 1988	Mylonite and chlorite breccia in granite, Pichacho core complex, Arizona.	Magmatic/metamorphic or highly evolved formation waters in both mylonite and cht breccias.	Increase in $\delta^{18}O$, Fe, Mg in cht breccias.	Unknown.
Glazner and Bartley, 1991	Mylonites, central Mojave core complex, California.	Meteoric?	Decrease in Si; 20–70% volume loss.	Large fluxes in hydrothermal circulation
Selverstone et al., 1991	High grade extensional mylonite in convergent regime, Tauern window, Austrian Alps.	Up T flow of metamorphic fluid from underthrust units.	Decrease in Si, Ca, Na, Sr; increase in Mg, Fe; 60% volume increase.	$>10^6$ m^3/m^2.
Fricke et al., 1992	Mylonites in granitoids, Ruby Mountains core complex, Nevada.	Meteoric waters infiltrating mylonites during ductile deformation.	Lowering of $\delta^{18}O$; resetting of δD.	Unknown.
Morrison, 1994	Whipple Mountains core complex, California.	Meteoric waters in upper plate and late fractures. Metamorphic waters in mylonites at higher grades.	Lowering of $\delta^{18}O$ where meteoric waters involved.	Unknown.

D: Strike slip

Reference	Location/Description	Fluid	Chemical changes	Flux
Federowich et al., 1991	Tartan Lake gold deposit, Proterozoic Trans-Hudson orogen, Manitoba. Reverse then strike-slip movement on shear zones in gabbros and volcanics.	Metamorphic/magmatic fluids, 5–10 wgt% salinity.	Uncertain.	Unknown.
Evans and Chester, 1995	Cataclasite in San Gabriel fault zone California.	Meteoric?	Increase in Ti, Fe, Ca, Mg, P; decrease in Si, Na, K; 37% volume loss.	Unknown.

E: Reviews and theoretical treatments

Reference	Location/Description	Fluid	Chemical changes	Flux
Kerrich, 1986	General review of oxygen isotope and whole rock alteration in shear zones.	Variable, mainly metamorphic.	Variable.	Variable.
Dipple and Ferry, 1992	Modelling of metasomatic changes due to flow along temperature gradients.	Unknown, salinities assumed 5–10%.	Changes in major element and isotopic ratios.	Typically 2×10^4 m^3/m^2.
Mikucki and Ridley, 1993	Review of Archean lode-gold deposits, greenschist to granulite grades.	Metamorphic fluids equilibrated with felsic rocks.	Various.	?

used to illustrate the above equation (e.g. Marquer, 1989). Gresens plots (Gresens, 1967; Marquer *et al.*, 1985) use f_v as one axis and gains or losses of components as the other. The isocon plot of Grant (1986), in which the mass fractions of various components in the altered and unaltered rocks are plotted against one another, allows errors due to combining several analyses into representative initial and final compositions to be shown. Figure 9.1 shows isocon plots for alteration of granodiorite in shear zones cutting the Bassies granodiorite in the Pyrenees (Tempest, 1991). Volume factors can be estimated from the slope of 'isocons' passing through components assumed to be immobile, and a volume increase of about 20% can be inferred from the data in Figure 9.1(a). The dangers associated with mylonite heterogeneity are illustrated by Figure 9.1(b), which shows that if a different sample from the same hand-specimen is used to represent the mylonite composition, no volume change can be inferred. Studies using isocon plots include that of O'Hara (1988), who inferred volume loss of 60% during mylonitization on a thrust in the Appalachians, based on the isocons for TiO_2, Fe_2O_3, MnO and Zr. This requires loss of SiO_2, Al_2O_3, MgO and alkalis from the shear zone.

Major element data can also be used to calculate parameters related to specific reactions which are observed in alteration. For example, McCaig *et al.* (1990) used a muscovitization parameter $M = (Na + K)/Al(-)$ to track the alter-

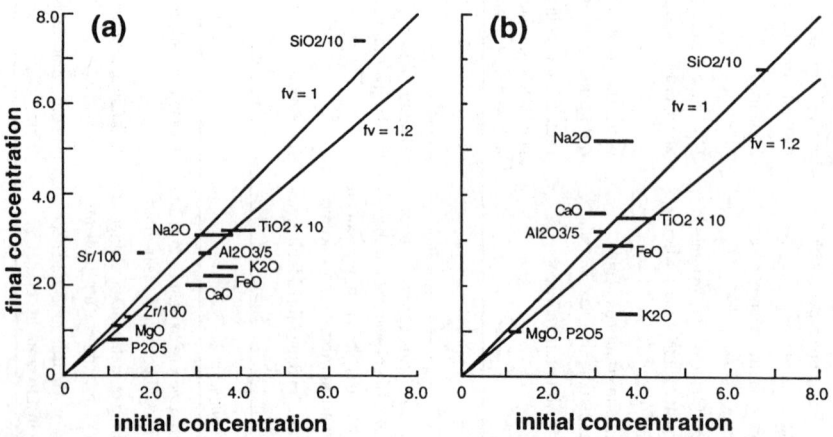

Figure 9.1 Isocon diagrams for transition of granodiorite to albitized mylonite, Bassiès granodiorite, Pyrenees (Tempest, 1991). Bars indicate standard deviation in mean granodiorite data, based on five analyses. Similar or greater uncertainties must exist in the single sample mylonite data. (a) Comparison of granodiorite and bulk mylonite analyses suggests a volume factor (f_v) of 1.2 based on an isocon passing through immobile components TiO_2, Al_2O_3, Zr and P_2O_5. On this basis, SiO_2 and Sr are gained and K_2O, FeO and CaO are lost. (b) Comparison with analysis of a thin section chip from the same mylonite sample suggests no significant volume change, but increase of Na_2O and CaO, with decrease of K_2O. See text for discussion.

236

ation of feldspars to muscovite in shear zones from the Pyrenees (see Figure 9.8). This parameter is based on the molar proportions of Na, K, Al and Ca in the analysis, with $Al(-) = Al-2Ca$ to correct for Al contained in anorthite. A pure alkali feldspar rock has $M = 1.0$, while a pure muscovite rock has $M = 0.33$. Albitization reactions were tracked in a similar way using Na/K ratios. This approach works well in mineralogically simple lithologies.

9.4.2 Mineralogy and mineral chemistry

The use of changes in mineral assemblages to track the progress of metasomatic reactions has been popularized by Ferry (1986). He suggested using (i) the modal abundance of reactant or product phases, (ii) the changing ratio of the abundance of two phases, or (iii) the changing chemical composition of mineral solid solutions, to measure reaction progress.

This approach has been used qualitatively by McCaig (1984) to show that a change in bulk rock chemistry accompanied the mylonitization and retrogression of diorite in shear zones from the Pyrenees, and quantitatively by Gilotti (1989) in her study of mylonitized dykes in the basal thrust zone of the Särv-Koli nappe in the Scandinavian Caledonides. However, quantitative use of the method in mylonites can be difficult for the reasons of inhomogeneity outlined above. For example, Tempest (1991) used a combination of point-counting on scanning electron microscope (SEM) photos and electron microprobe analyses to calculate the compositions of four different domains within a single thin section of mylonitized granodiorite. When he attempted to reconstruct the composition of the bulk rock from the proportions of the domains, significant differences were present compared with an analysis of the thin-section chip by X-ray fluorescence (XRF), particularly for K_2O and FeO. Possible reasons were inaccurate point-counting of minor phases such as Fe-oxides, and poor correspondence between the area of the domains in thin-section and their volumes in the thin-section chip. When the analysis of the thin-section chip was compared with another XRF analysis of a larger specimen taken from the same hand sample, even greater differences occurred (Figure 9.1). This example illustrates that great care must be taken in defining reaction volumes in mylonites for either modal or whole rock analysis, and shows how time-consuming modal analysis of mylonites with heterogeneous grain sizes and domainal fabrics is likely to be. An additional problem in some cases could be caused by variations in mineral composition reflecting lack of equilibrium (McCaig and Knipe, 1990; Knipe and McCaig, 1994). The conclusion is that quantifying bulk reaction progress by point-counting is only likely to be successful in very simple mineral assemblages (for example the muscovitized mylonites described by McCaig et al., 1990), or very homogeneous mylonites lacking fine banding or domainal fabrics. The approach can, however, be used for comparing domains in a mylonite or assessing directions of chemical change between overprinting fabrics (cf. McCaig, 1984, 1987).

9.4.3   Isotopic analysis

Oxygen (and where appropriate carbon) isotopic analysis has frequently been used to constrain shear zone alteration (Kerrich *et al.*, 1984; Kerrich, 1986; Burkhard and Kerrich, 1988, 1990; Losh, 1989; McCaig *et al.*, 1990; Fourcade *et al.*, 1989; Fricke *et al.*, 1992; Burkhard *et al.*, 1992; Crespo-Blanc *et al.*, 1995). Fine grain size means that most oxygen isotope studies of mylonites are based on analysis of whole rocks rather than separated minerals. These data need to be treated with care, since changes in mineralogy between wall rocks and their sheared equivalent can lead to changes in bulk $\partial^{18}O$ without significant fluid–rock interaction, and various minerals in the mylonite may not have been in isotopic equilibrium with each other or the fluid (Cartwright *et al.*, 1993). Ideally, separated minerals such as quartz should show isotopic shifts in parallel with those shown by the whole rocks (e.g. McCaig *et al.*, 1990). Where fine grain size precludes separation of minerals, the existence of isotopic change can be established either by comparison of syntectonic vein quartz with quartz grains in the protolith, or by point-counting of the mineralogy of the mylonite and reconstruction of individual mineral compositions from known fractionation factors (e.g. Losh, 1989).

Hydrogen isotopes have only rarely been analysed in mylonites (Burkhard *et al.*, 1992; Fricke *et al.*, 1992). Problems of mineral separation and weathering of fine grained phyllosilicates mean that ∂D analysis of syntectonic fluid inclusions (Fricke *et al.*, 1992) may be a more reliable way of identifying the source of fluid in shear zones.

Radiogenic isotopes such as Sr, Pb and Nd are not fractionated significantly between the solid and fluid phase, so that in contrast to oxygen isotopes, temperature effects are not important. Unlike oxygen, these trace elements have highly variable concentrations in both rocks and fluids, and isotopic ratios change with time through radioactive decay. Using radiogenic isotopes as tracers of fluid–rock interaction only works when both the initial rock composition immediately before alteration and the composition of the incoming fluid can be constrained. A good example is that of Sr in carbonates where a uniform initial ratio in equilibrium with seawater is relatively insensitive to radioactive decay because of low Rb/Sr ratios. Where shear zones in carbonates are infiltrated by Sr-rich fluids derived from more pelitic rock types, significant metasomatic effects have been described (Burkhard *et al.*, 1992; McCaig *et al.*, 1995a). In pelitic and granitic rocks, Rb/Sr ratios and hence $^{87}Sr/^{86}Sr$ ratios at any particular time tend to be rather variable even within a single rock type, making protolith compositions before alteration hard to define. Hence most studies have been directed towards dating the age of mylonitization (e.g. Sinha *et al.*, 1988), although metasomatic effects have also been reported (e.g. Barovich and Patchett, 1992; Wayne, 1995).

Pb isotopes have also been used mainly to date deformation, particularly using zircons (Wayne *et al.*, 1992; Getty and Gromet, 1992). In some cases, systematic changes in age-corrected Pb isotopic ratios between mylonites and their protoliths have been reported (Romer, 1990; Wayne, 1995). Interpretation is complicated by the possibility of selective dissolution of phases with unusual Pb isotopic compositions such as metamict zircons (Wayne, 1995; Wayne *et al.*, 1992). Banks *et al.* (1991) presented Pb and Sr isotopic data from fluid inclusions from quartz veins associated with the Gavarnie thrust in the Pyrenees. Further data (McCaig *et al.*, 1995b) suggest that separate local fluid sources supplied fluid into carbonate mylonites along the thrust, and later into the quartz veins.

Other isotope systems such as Nd and Hf seem to be relatively undisturbed during shear zone alteration (Barovich and Patchett, 1992), although rare earth element mobility has been reported in some cases (Dostal *et al.*, 1980; Vocke *et al.*, 1987).

9.5 Spatial variations in alteration and their interpretation

In order for alteration to occur through fluid–rock interaction, a fluid which is out of equilibrium with the mylonite must move into or through the shear zone. There are two possible reasons for this lack of equilibrium (Dipple and Ferry, 1992b):

- flow of fluid along temperature or pressure gradients;
- introduction of fluid which previously equilibrated with other lithologies along the flow path or is derived from some 'exotic' fluid reservoir such as meteoric or sea water.

Flow of fluid along a temperature gradient can significantly affect oxygen and hydrogen isotopic ratios and major element compositions, but will not alter isotopic ratios which show only small temperature effects on mineral–fluid fractionations such as $^{87}Sr/^{86}Sr$ (Dipple and Ferry, 1992a, 1992b). Dipple and Ferry (1992a) showed that $\partial^{18}O$ in carbonates or quartzo-feldspathic rocks is lowered for flow up-temperature, and increased for flow down-temperature, assuming initially isotopically homogeneous rock, and fluid in equilibrium with the rock at the start of the flow path. In both cases, the difference between final and initial rock compositions increases with increasing flux, and decreases gradually along the flow path towards higher temperatures as the size of equilibrium fractionations between rock and fluid decreases. Moderate to high fluid fluxes are required to produce significant shifts in $\partial^{18}O$. For example, Dipple and Ferry (1992b) calculated that an integrated flux of 10^4 m^3/m^2 flowing down a 25 °C/km temperature gradient could account for an increase in average $\partial^{18}O$ of 1.7‰ in quartzo-feldspathic mylonites at Grimsel, Switzerland. This is higher than most values calculated in regional metamorphic rocks (Graham *et al.*, this volume) but comparable with values calculated for hydrothermal systems around shallow intrusions (Norton and Taylor, 1979).

Dipple and Ferry (1992b) used standard thermodynamic data sets to calculate fluid compositions in equilibrium with a variety of amphibolite facies

239

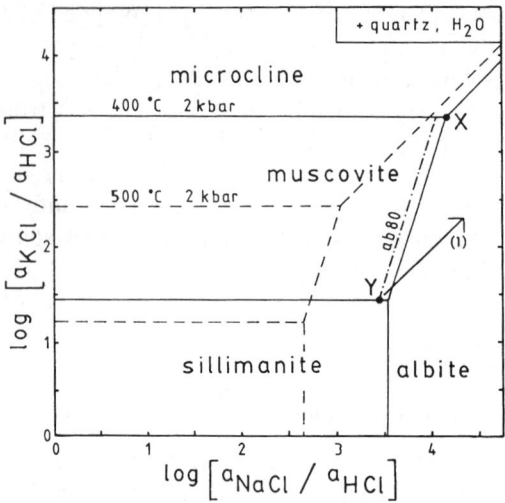

Figure 9.2 Activity diagram for fluid compositions in (metastable) equilibrium with granitic and aluminosilicate-bearing assemblages at 400 and 500 °C in shear zones from the Pyrenees. Fields of paragonite (between albite and muscovite) and pyrophyllite (replacing sillimanite at 400 °C) were omitted because these phases were not observed. Line marked ab80 indicates the position of the albite–muscovite reaction line for plagioclase of this composition. Flow down temperature of fluid in equilibrium with a granite involves increase in a_{KCl}/a_{HCl} and a_{NaCl}/a_{HCl} ratios and increase in Na/K ratios, with opposite changes occurring in the rock in a closed system. Reproduced from McCaig *et al.* (1990) with permission from Springer-Verlag.

mylonitic assemblages, together with the changes in fluid cation ratios resulting from perturbation in pressure and temperature. From this they were able to calculate the effects on rock composition which would result from flow of fluid of a specified chloride molality up or down specified temperature or pressure gradients. Significant effects are produced only for moderate to high chloride molalities (≥ 1), high fluxes (2×10^4 m³/m²) and for flow along temperature (not pressure) gradients. Flow down-temperature should result in gain of Si and increase in K/Na and K/Ca ratios, with the reverse effects for flow up-temperature. This is illustrated qualitatively by the activity diagram in Figure 9.2. Fluid in equilibrium with a typical granitic assemblage at point X will have an a_{KCl}/a_{NaCl} ratio defined by equilibrium with the solid minerals at 400 °C. Fluid in equilibrium with the same assemblage at 500 °C has lower ratios of a_{KCl}/a_{HCl} and a_{NaCl}/a_{HCl} and a higher a_{KCl}/a_{NaCl} ratio. Flow of fluid up-temperature must result in transfer of K to the fluid and Na to the rock, as well as reaction of muscovite to feldspars. Dipple and Ferry (1992b) modelled a number of shear zones showing metasomatic changes, finding examples of both up- and down-temperature flow and typically calculating fluxes of 2×10^4 m³/m². It should be noted, however, that the model did not always account for all the metasomatic changes which occurred, for example the gain of Mg in the mylonite at Grimsel. This suggests either that flow along a temperature gradient was not the only source of metasomatic effects, or that deficiencies exist in the thermodynamic database used.

Change in SiO_2 content in mylonites has also been used to infer flow up or down-temperature gradients (Sinha *et al.*, 1986; O'Hara, 1988; Selverstone *et al.*, 1991; Dipple and Ferry, 1992b). Because quartz solubility increases with increasing temperature, SiO_2 content should increase for down-temperature flow, and decrease for up-temperature flow. SiO_2 gains and losses are complicated by local redistribution of SiO_2 through pressure solution and quartz veining (Kerrich *et al.*, 1977; Etheridge *et al.*, 1984), but do not depend on chloride molality.

Introduction of an exotic fluid into a flow path within a homogeneous lithology should lead to a new lithological assemblage in chemical and isotopic equilibrium with the incoming fluid. In ideal advective mass-transport, a sharp front (Figure 9.3) between fully altered and unaltered rock will move down the flow path at a velocity less than that of the fluid (Bickle, 1992; Graham *et al.*, this volume, and references therein). The distance (Z_{GF}) moved by the step is given by the relationship:

$$J_{int} \approx Z_{GF}K_v \qquad (9.2)$$

where J_{int} is the integrated flux (m^3/m^2) and K_v is the volumetric solid–fluid partition coefficient for the particular tracer. This means that tracers with higher concentrations in the fluid than in the rock (e.g. ∂D, Br) will move much greater distances than tracers which are concentrated in the rock (e.g. Sr in dilute fluids, C in an aqueous fluid moving through carbonates). In practice, diffusion, hydrodynamic dispersion and kinetic limitations on fluid–rock equilibrium result in continuous changes in rock isotopic composition in the direction of fluid flow rather than sharp fronts (Lassey and Blattner, 1988; Bowman *et al.*, 1994; Graham *et al.*, this volume). Profiles of isotopic composition measured in the flow direction can be modelled in terms of the dimensionless Peclet (*Pe*) and Damköhler (N_D) numbers (Bickle, 1992), where:

$$Pe = \omega_0\phi h/D_{eff} \qquad (9.3)$$

and

$$N_D = \kappa h/\omega_0 \qquad (9.4)$$

Pe is a measure of the relative importance of advection, expressed by the fluid pore flow velocity (ω_0), and diffusion, expressed by the effective diffusion constant D_{eff}. ϕ is porosity and h is an arbitrary length scale for mass-transport. N_D relates the linear-kinetic exchange constant between fluid and rock (κ) to the fluid flow velocity. Sharp isotopic fronts will occur in the direction of fluid flow if both *Pe* and N_D are high (>100), while if either or both are low (<10), more gentle profiles will result. For advective-diffusive flux the breadth of the front in real (as opposed to dimensionless) space depends only on D and time, is independent of the advection distance, and is restricted to a few metres or tens of

241

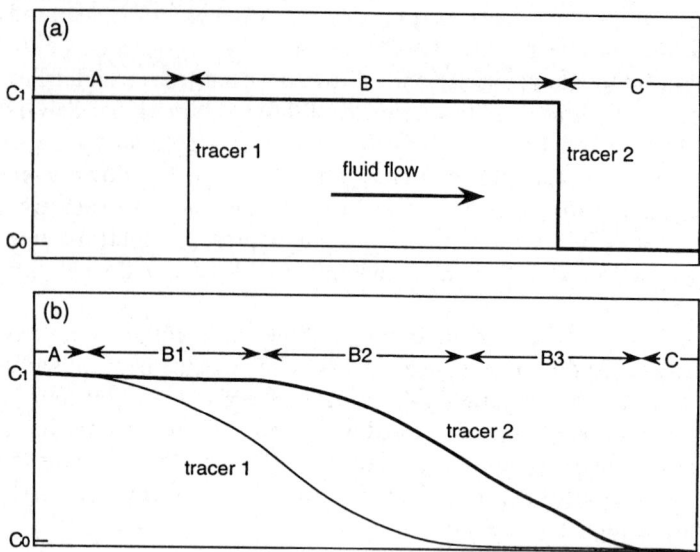

Figure 9.3 Movement of idealized isotopic fronts along a shear zone. C_0 and C_1 are the initial and final compositions of the rock (for two independent isotopic tracers) before and after reaching equilibrium with an incoming fluid. (a) Isotopic profiles for pure advective transport with no diffusion or dispersion, and perfect equilibrium between fluid and rock at all points. Sharp compositional steps move in the direction of fluid flow at rates determined by the rock–fluid partition coefficients (K_y) for each tracer (see text). In region A of the flow path, the rock is fully altered for each tracer, and conventional fluid–rock ratio calculations would give W/R = ∞. In region C, the rock is completely unaltered and W/R of zero would be calculated, despite the fact that exactly the same amount of fluid has passed through all parts of the flow path. In region B, different values of W/R would be calculated using each tracer. (b) Schematic geochemical profiles for the same tracers, but at low Peclet and/or Damköhler numbers. Geochemical fronts are broadened due to diffusion, dispersion or kinetically limited exchange, and hence the profiles for the two tracers may overlap. Regions A and C show the same geochemical characteristics as in the case of pure advection (diagram a), but region B now contains zones where one tracer is fully altered and the other partially altered (B1 and B3), and a zone of partial alteration for both isotopes (B2). Conventional water–rock ratio calculations would give a range of values of W/R which might be incorrectly interpreted to reflect variable fluid fluxes.

metres at metamorphic temperatures (Fyfe *et al.*, 1978; Bickle, 1992). For kinetically limited exchange, the breadth of the front increases with increasing flow velocity and hence advection distance. If hydrodynamic dispersion is believed to be important, a combined diffusion-dispersion coefficient

$$D = D_{eff} + a\omega_0 \qquad (9.5)$$

can be used, where a is the dispersivity with units of length (Bowman *et al.*, 1994; Domenico and Schwartz, 1990). For dispersive flow, the breadth of the front probably depends on both the scale of mass-transfer (through variations in a) and

242

the flow velocity through equation 9.5, but according to Bowman *et al.* (1994) there is no firm theoretical basis for assessing values of dispersivity and hence the importance of dispersion in fluid flow in different circumstances is uncertain.

In metamorphic rocks, the above theory has mostly been applied to relatively small displacements of isotopic fronts across the boundaries of carbonate layers or dykes (e.g. Bickle and Baker, 1990; Cartwright, 1994; Skelton *et al.*, 1995; Graham *et al.*, this volume). The difficulty in shear zones is that large fluid fluxes can lead to front displacements of several kilometres, and sampling must be on a correspondingly large scale. On the other hand, the presence of a planar fluid conduit, easily recognized in the field, greatly simplifies the planning of a sampling programme. Both the possibilities and the difficulties of applying geochemical front theory to shear zones are exemplified by the Glarus thrust in the Swiss Alps, which places Permian Verrucano sandstones on top of Cretaceous carbonates and then Tertiary flysch (Figure 9.4). A thin sliver of mylonitized carbonate, probably derived from Cretaceous carbonates, marks the thrust plane over a large area of the Swiss Alps. Burkhard and Kerrich (1988, 1990) and Burkhard *et al.* (1992) measured oxygen and carbon isotopes in a series of profiles across the mylonite zone (Figure 9.5). A continuous increase in both minimum and average $\partial^{18}O$ values occurs from south to north along the shear zone (Figure 9.5), with values at the northern end of the traverse being closest to those in undeformed Helvetic carbonates ($\partial^{18}O = 25.2 \pm 2‰$; Burkhard and Kerrich, 1988). In the most southerly outcrops, $\partial^{13}C$ values are lowered and $^{87}Sr/^{86}Sr$ ratios increased relative to marine carbonates, with these differences dying out northwards (Burkhard *et al.*, 1992). These changes were interpreted to reflect influx and northward flow of fluid which had equilibrated with Verrucano or basement lithologies at or beyond the southern end of the transect (Burkhard *et al.*, 1992). Bowman *et al.* (1994) used a numerical model to generate theoretical curves through the data shown in Figure 9.5 (Figure 9.6). They assumed that rapid isotopic exchange between calcite and the fluid would lead to high values of N_D, and so attempted to fit the curves to the data by varying *Pe* (where *Pe* includes hydrodynamic dispersion according to equation 9.3). Since the scale of front broadening was too large to be explained by diffusion, this effectively involved varying the dispersivity (*a*). Bowman *et al.* (1994) concluded that the data could best be fit by values of *Pe* between 1 and 10, corresponding to values of dispersivity between 500 m and 5 km. Fluid fluxes in the range 3 to 6×10^3 m^3/m^2 were estimated.

Similar alteration patterns to those seen in the Glarus thrust have been described from other Helvetic thrusts (Crespo-Blanc *et al.*, 1995), and McCaig *et al.* (1995a,b) found similar results for Sr isotopes along the Gavarnie Thrust in the Pyrenees. In the latter case a minimum of 1.8 to 3×10^3 m^3/m^2 of hypersaline brine containing up to 2000 ppm Sr was inferred to have been expelled southwards out of Triassic redbeds beneath the thrust along a mylonite zone in Cretaceous carbonates. The major uncertainty when using Sr isotopes is in estimating K_v. This was done by comparing Sr contents of quartz-hosted fluid

243

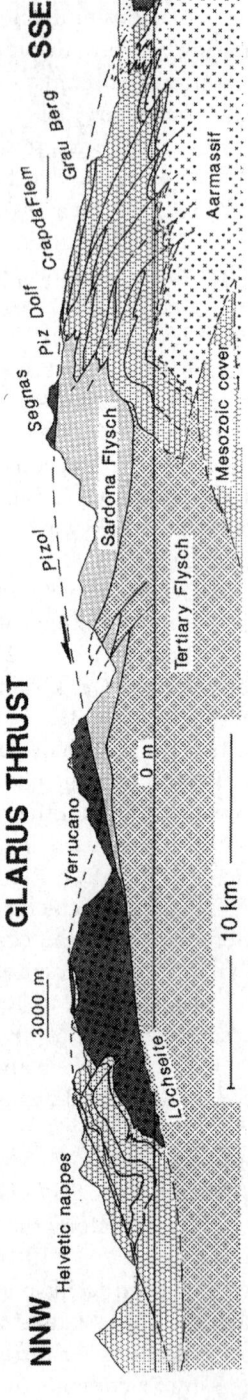

Figure 9.4 Geological cross-section through the Glarus Thrust with horizontal scale corrected (M. Burkhard, pers. comm., 1996). Reproduced from Burkhard *et al.* (1992) with permission from Springer-Verlag. Localities where samples of carbonate mylonite were collected for stable isotopic analysis (Figure 9.5) are shown.

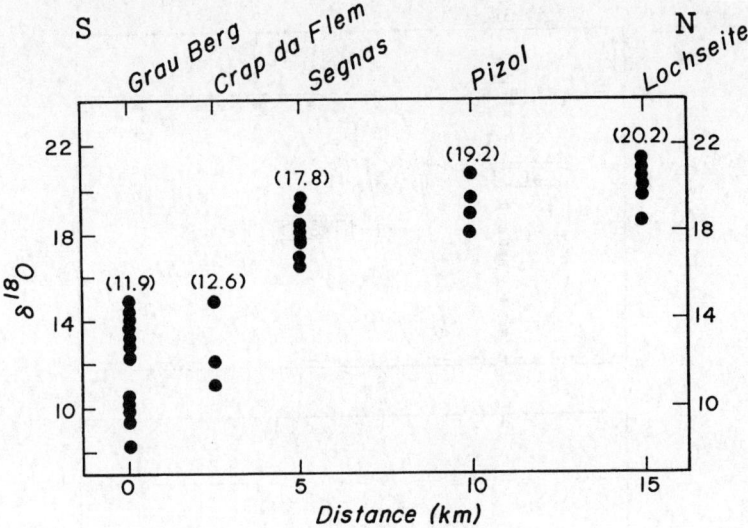

Figure 9.5 Oxygen isotope data from carbonate mylonites along the Glarus Thrust (Burkhard and Kerrich, 1990) summarized by Bowman *et al.* (1994), reproduced with permission from *American Journal of Science*. For localities see Figure 9.4, but note that due to projection onto a different section line, distances do not correspond exactly to those in Figure 9.4 (M. Burkhard, pers. comm., 1996). The systematic northward increase in both average and minimum $\delta^{18}O$ values is inferred to have resulted from northward flow of low $\delta^{18}O$ fluid along the mylonite.

inclusions with Sr contents of coexisting vein calcite, and independently by comparing the length scales of Sr and O isotopic transport (McCaig *et al.*, 1995b).

Figure 9.6 illustrates some major problems in attempting to model fluid–rock interaction in shear zones affected by large fluxes of fluid.

- The kilometric scale of mass-transfer makes it extremely difficult to sample accurately the shape of the isotopic profile in the direction of flow (see also McCaig *et al.*, 1995b).
- At any particular site, $\partial^{18}O$ may take a range of values (see also McCaig *et al.*, 1995b). Should curves be fitted to the average values, or to the most altered, and in the latter case how can collection of the most altered samples be ensured?
- Bowman *et al.* (1994) were obliged to use an unaltered rock $\partial^{18}O$ value of +20‰ in order to get a reasonable fit to the data. Use of the value of +25‰ given by Burkhard *et al.* (1992) would require extremely low values of *Pe* to fit the curves since significant reductions in $\partial^{18}O$ compared with undeformed carbonates occur even at the northern end of the traverse. This may be due to direct influx of fluid into the mylonite from the underlying flysch at various points along the flow path (Burkhard *et al.*, 1992), a possibility not accounted for in the one-dimensional flow model.

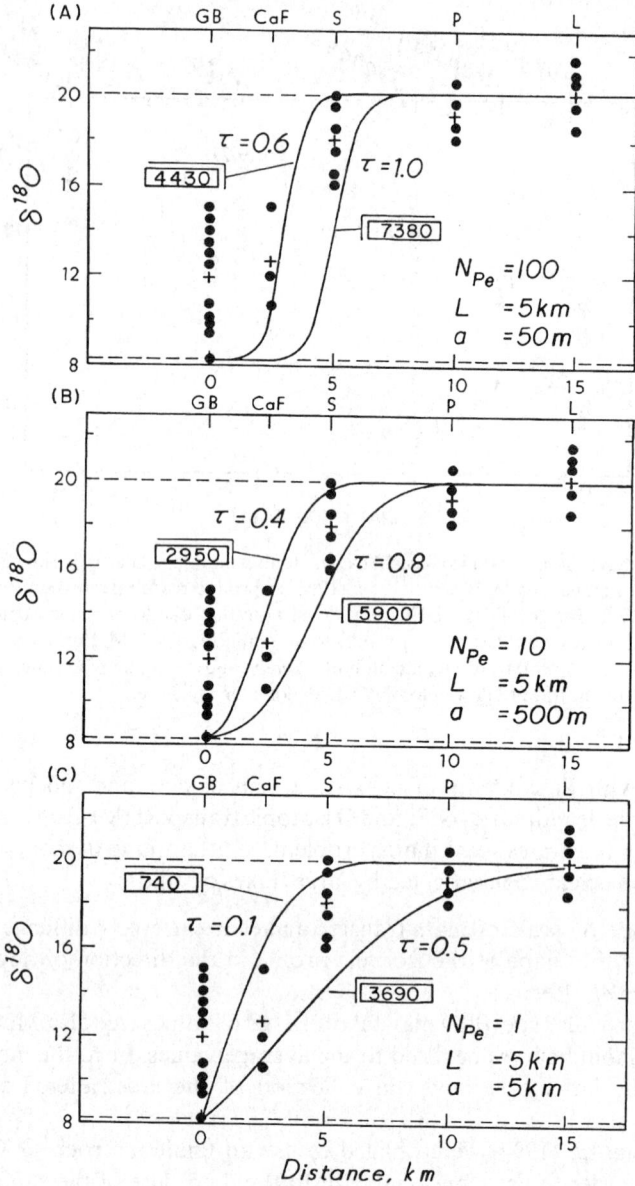

Figure 9.6 Fitting of curves calculated (reproduced from Bowman *et al.* (1994) with permission from *American Journal of Science*) for variable Peclet number (N_{Pe}) and dimensionless time (τ) to the data in Figure 9.5. Numbers in boxes are minimum fluid flux in m³/m². Crosses are average values for each site. Variations in N_{pe} were assumed to result from variations in dispersivity [a]. Fluid–rock equilibrium was assumed at all times ($N_D = \infty$). Fluid flow is from left to right. Use of $N_{pe} = 10$ gives a reasonable fit to the minimum values in the southern half of the profile, but no models give a good fit to the data at the northern two localities.

- The calcmylonite layer may be made up largely or even entirely by vein material (Burkhard *et al.*, 1992). It is not clear how this would affect the model.
- Diffuse isotopic fronts can be modelled either by varying Pe or N_D, and although there are subtle differences in profile shape, these cannot normally be resolved. Although most authors agree that diffusion is unlikely to be important where mass-transport and front-broadening are on a kilometric scale, the relative importance of kinetic and hydrodynamic dispersion is less clear. In either case, porous medium flow is unlikely to be an appropriate mechanism for fluid flow in shear zones (Bickle, 1992, and see below).

Despite these provisos, it is clear that spatially constrained sampling can lead to a much better understanding of isotopic changes in shear zones than studies at a single sample site. In particular, the fluid flow direction, sites of fluid input into the mylonite, initial fluid composition and fluid flux can all be constrained far better than is possible with spatially unconstrained samples. The same is true for whole-rock chemical change. Figure 9.7 shows the distribution of muscovitization in granite mylonites in a shear zone from the Aston Massif, central Pyrenees (McCaig *et al.*, 1990). Field mapping shows that muscovitization of feldspars dies out upwards within about 125 m of the contact between granite gneiss and underlying sillimanite gneiss. McCaig *et al.* (1990) used the activity diagram shown in Figure 9.2 to show that a fluid in (metastable) equilibrium with sillimanite gneiss at point Y would react to produce muscovite from feldspars on entering a granite according to reactions of the form:

$$3 \text{ microcline} + 2HCl(aq) = \text{muscovite} + 2KCl(aq) + 6SiO_2 \qquad (9.6)$$

Equilibrium fluid compositions at points X and Y (Figure 9.2) were calculated from thermodynamic data (assuming a chloride molality of 1) and combined with XRF data to calculate the minimum water–rock ratio (W/R) required for the alteration. W/R is equivalent to K_v in equation 9.1 (see Skelton *et al.*, this volume, equation 8.6), and was combined with the length scale of muscovitization (125 m) to give a fluid flux of about $10^5 \text{ m}^3/\text{m}^2$ (McCaig *et al.*, 1990). Subsequent fluid inclusion work (Tempest, 1991) on related shear zones suggests that chloride molalities were in fact in the range 5–7, so that the fluid flux estimate should be reduced to 1.5 to $2 \times 10^4 \text{ m}^3/\text{m}^2$. Muscovitization was preceded in the shear zones by albitization of K-feldspar, as illustrated by the profile across the margin of a minor muscovitized shear zone shown in Figure 9.8. Figure 9.2 shows that while muscovitization could be explained by down-temperature fluid flow as modelled by Dipple and Ferry (1992b), down-temperature flow would lead to replacement of albite by K-feldspar, not the reverse. Simultaneous muscovitization and albitization can only be explained by influx of a low K/Na, low KCl/NaCl fluid such as that at point Y in Figure 9.2, in agreement with the model described above. Figure 9.8 also shows that $\partial^{18}O$ values were reduced in the mylonites and their immediate wall rocks compared with fairly uniform

247

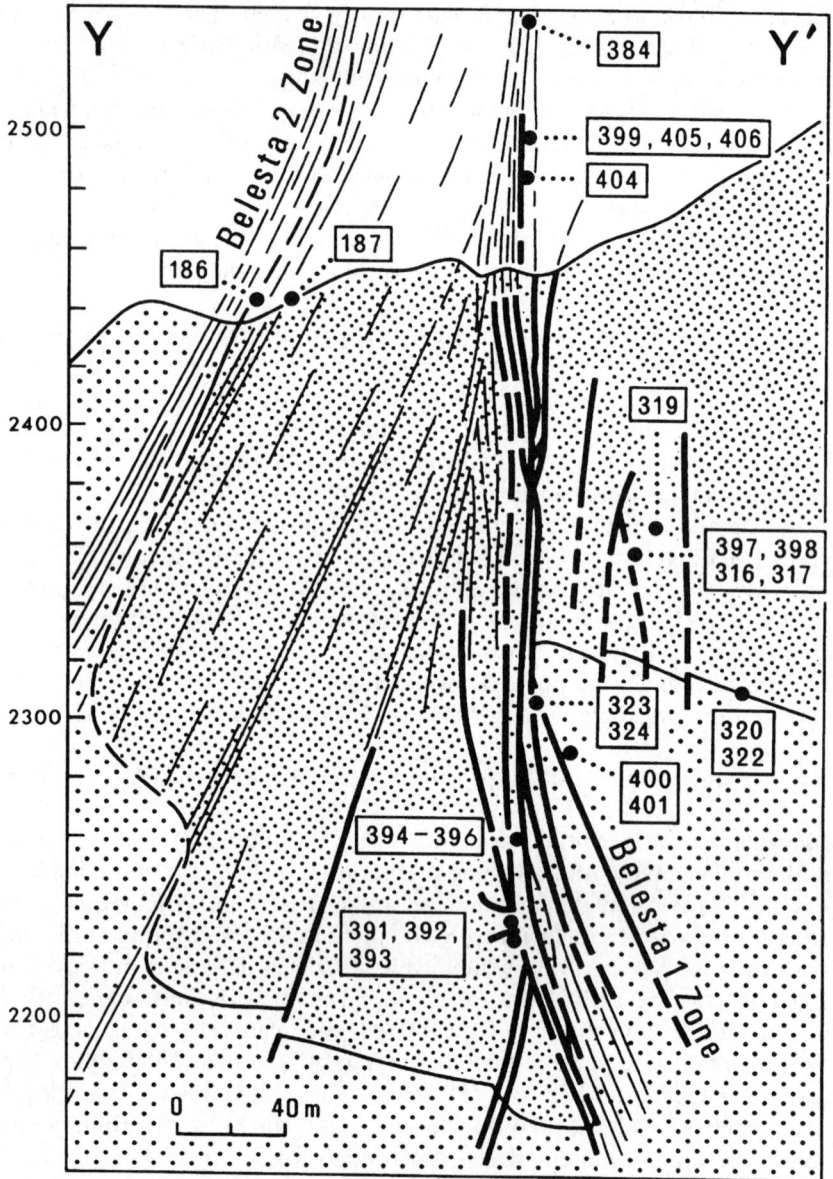

Figure 9.7 North–south cross-section showing two shear zones (Belesta-1 and -2 zones) cutting the contact between granite gneiss (heavy stipple) and sillimanite gneiss (light stipple) in the Aston Massif, central Pyrenees. Y is in the south. Heavy lines show mapped zones of muscovitization of granite within the shear zones. Thin lines show mylonites in which feldspars are preserved. Muscovitization dies out in the granitic mylonites about 120 m above the contact with sillimanite gneiss. Numbers are sample numbers of McCaig *et al.* (1990). Reproduced from McCaig *et al.* (1990) with permission from Springer-Verlag.

248

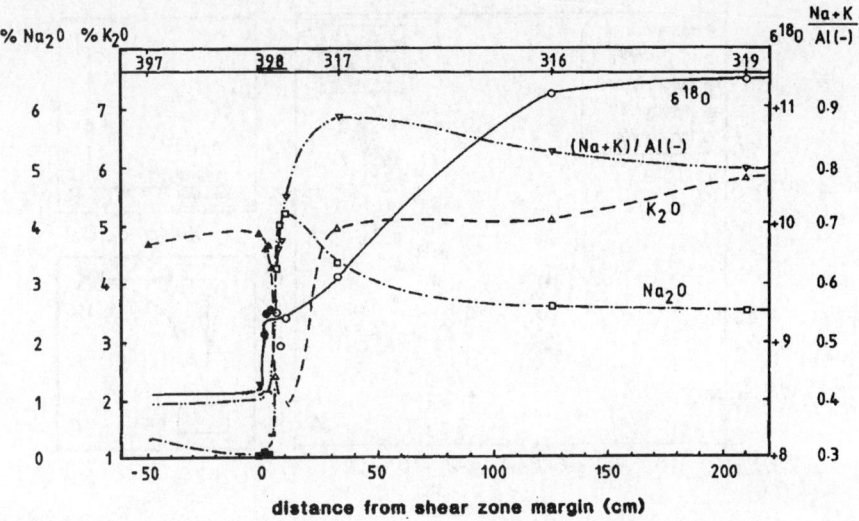

Figure 9.8 Geochemical profile across the margin of a 1 m thick muscovitized shear zone in granite gneiss 40 m north of the Belesta-1 zone in Figure 9.7. (Na + K)/Al(-) is a muscovitization parameter (*M*) based on whole rock analyses normalized to cation concentrations. Al(-) = Al – 2Ca, and corrects for Al contained in plagioclase such that (Na + K)/Al(-) = 1.0 in a pure feldspar rock and 0.33 in a pure muscovite rock. Note the low values of *M*, Na_2O and $\delta^{18}O$ in the mylonite, and high values of Na_2O and low values of K_2O in the immediate wall rock of the shear zone due to albitization of K-feldspar. Reproduced from McCaig *et al.* (1990) with permission from Springer-Verlag.

values of +11 to +12 ‰ in both granitic and pelitic gneisses. These $\partial^{18}O$ reductions occurred in all the mylonites, irrespective of degree of muscovitization, and reflect the much longer length-scale of oxygen isotope transport due to the low value of solid–fluid K_v for oxygen isotope exchange compared with W/R for the muscovitization reaction (cf. Graham *et al.*, this volume). In shear zone wall rocks, in contrast, the length-scales of oxygen isotope transport and muscovitization are similar (Figure 9.8). This is because the wall-rock alteration is a 'metasomatic side' (Yardley and Lloyd, 1995) rather than a metasomatic front. The profiles in Figure 9.8 are probably largely diffusive rather than advective in origin, although with a pinned boundary (Bickle and Baker, 1990) due to flow of low $\partial^{18}O$ fluid up the shear zone.

In most cases, lack of outcrop or the large length-scale of mass-transport precludes sampling of geochemical fronts, particularly in steep shear zones where upward flow is suspected. Most studies of mylonites show isotopic values shifted relative to wall rocks, but rather variable at a particular sample site (Figure 9.6). In this case, some information about flow paths and fluid compositions can be gained by plotting isotope correlation plots. Figure 9.9 shows data from the Glarus thrust plotted in terms of $\partial^{18}O$ and $\partial^{13}C$ (Burkhard *et al.*, 1992). Calc-mylonites show a trend of decreasing $\partial^{18}O$ at fairly constant $\partial^{13}C$, with a few of the most altered samples showing lower $\partial^{13}C$ values, similar to calcite in

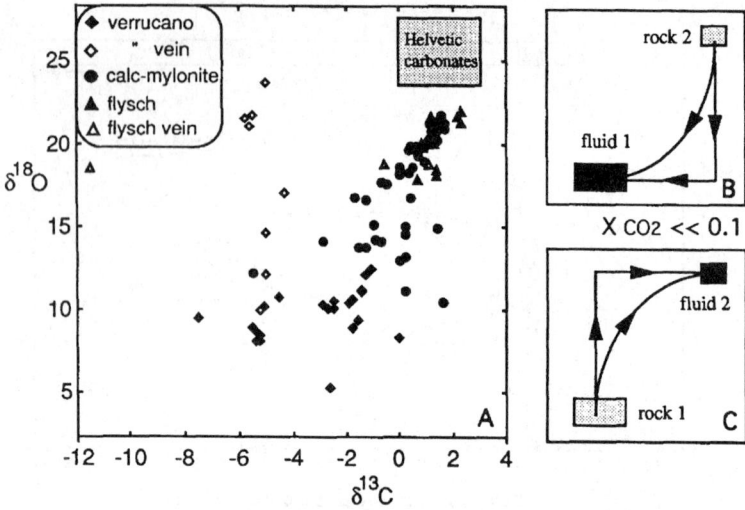

Figure 9.9 Stable isotope correlation plot, reproduced from Burkhard *et al.* (1992) with permission from Springer-Verlag, for carbonate mylonites, carbonate in Verrucano and flysch lithologies, and veins associated with the Glarus Thrust (Figure 9.4). Values in calc-mylonites can be interpreted in terms of progressive exchange of unaltered carbonates (rock 2) with a low CO_2 fluid (fluid 1) equilibrated with the Verrucano. Veins in the Verrucano (rock 1) can be interpreted in terms of influx of a low CO_2 fluid derived from unaltered Helvetic carbonates (fluid 2). Diagrams of this sort allow the CO_2 content of the fluid and the fluid movement direction, but not the water–rock ratio, to be estimated (Baumgartner and Rumble, 1988).

the Verrucano (Figure 9.6). This trend can be interpreted in terms of progressive interaction of a low X_{CO_2} fluid, with a composition in equilibrium with calcite in the Verrucano, with carbonates along the thrust (cf. Baumgartner and Rumble, 1988). A different trend is seen in calcite veins within the Verrucano, which retain low $\partial^{13}C$ values as $\partial^{18}O$ increases towards values similar to undeformed Helvetic carbonates. This trend can also be interpreted in terms of influx of a low X_{CO_2} fluid, but this time from unaltered limestones into the Verrucano. This pattern of fluid flow along the mylonite with limited escape of fluid into hangingwall veins can be inferred without any knowledge of the isotopic profile within the mylonites. Note that it is not possible to quantify alteration in terms of fluid–rock ratio from this type of plot (e.g. Rye and Bradbury, 1988), since the location of particular samples within the trend is a function of position along the flow path and the degree of overlap of the isotopic profiles for O and C, as well as fluid flux (Baumgartner and Rumble, 1988; McCaig *et al.*, 1995a).

Correlation plots have been used in other examples of shear zone alteration. For example, McCaig *et al.* (1995a) interpreted variations in the slope of arrays in a $\partial^{18}O$ versus $^{87}Sr/^{86}Sr$ plot in terms of variable salinities and hence variable K_v values for Sr at different sampling sites within the Gavarnie thrust (K_v for oxygen varies only slightly with H_2O content of the fluid). McCaig *et al.* (1990)

used a plot of $\partial^{18}O$ against a muscovitization parameter calculated from whole rock data (cf. Figure 9.8) to illustrate that shifts in oxygen isotopic values occurred in the absence of muscovitization in the mylonites.

Another method which can be used to constrain flow paths in shear zones is to plot the isotopic compositions of veins against those of the local matrix (Figure 9.10). McCaig *et al.* (1995a) showed that carbonate veins in mylonites and carbonate wall rocks of the Gavarnie thrust were consistently more radiogenic in $^{87}Sr/^{86}Sr$ ratios than the adjacent matrix. They inferred unidirectional flow along the mylonite zone such that fluid always flowed from more altered to less altered rocks. In hangingwall phyllites, veins were less radiogenic than the adjacent matrix, but showed values similar to the carbonate mylonites. This was interpreted to reflect limited escape of fluid from the thrust into hangingwall vein systems (see also Burkhard *et al.*, 1992). Late quartz veins which formed at

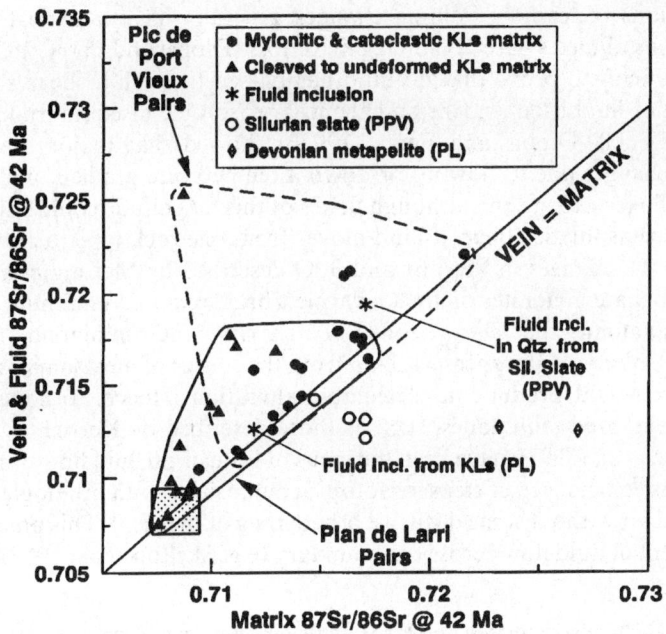

Figure 9.10 $^{87}Sr/^{86}Sr$ values of veins plotted against those of the adjacent matrix for Cretaceous carbonates (Kls) beneath the Gavarnie Thrust, Pyrenees, and for other rock types in the footwall and hangingwall of the thrust (McCaig *et al.*, 1995a). Carbonate mylonites along the thrust are enriched in ^{87}Sr relative to undeformed carbonates due to flow of fluid derived from lithologies containing more radiogenic strontium along the thrust. In the carbonates, veins are almost invariably similar to, or more radiogenic than, the matrix, suggesting highly organized unidirectional flow down the isotopic composition gradient in the mylonites. In hangingwall Devonian phyllites, escape of relatively unradiogenic fluids from the thrust zone carbonates leads to veins which are less radiogenic than the adjacent matrix. PL and PPV are two different localities. (See McCaig *et al.*, 1995a for details.)

the same time as a faulted culmination which deforms the mylonites show a different trend in fluid inclusion Pb isotopes from either the carbonate mylonites or the carbonate veins (McCaig et al., 1995b). This probably reflects lateral or downward movement of fluid out of Devonian phyllites in the hangingwall of the thrust during growth of the culmination, compared with simple expulsion of fluid during formation of the mylonites.

There does not seem to be any consistent pattern of volume change in shear zones. If elements such as Al and Ti are assumed immobile, volume changes can be estimated using Gresens or isocon plots as described above. Apparent volume increase may in some cases be related to introduction of vein material (McCaig, 1987), but more commonly is associated with gain in SiO_2 (Tempest, 1991). Some shear zones appear to have remained isovolumetric despite large changes in major element chemistry (McCaig, 1984; McCaig et al., 1990), while others have undergone volume loss (Sinha et al., 1986, 1988; O'Hara, 1988; Selverstone et al., 1991; Newman and Mitra, 1993; O'Hara et al., 1995). Assumptions concerning volume changes generally have a critical effect on calculations of fluxes based on SiO_2 gain or loss (Dipple and Ferry, 1992a).

It is difficult to prove that no fluid has passed through a shear zone, since detection of fluid is dependent on chemical or isotopic disequilibrium between fluid and rock. Dipple and Ferry (1992b) showed that major element and isotopic changes due to flow up or down a temperature gradient will be negligible for fluxes $<10^3$ m^3/m^2, although fluxes of this magnitude could still produce sizeable metasomatic effects if fluid moves from one rock type to another. For example, the changes in $^{87}Sr/^{86}Sr$ and $\partial^{18}O$ described by McCaig et al. (1995a) from carbonate mylonites on the Gavarnie Thrust required minimum transport distances along the shear zone of 1.5 km, and minimum fluxes of 2.85×10^3 m^3/m^2. At distances >1.5 km from the source of metasomatic fluid, the same flux would produce no detectable alteration. Hence, to conclude that unmetasomatized fault zones, such as those described by Kerrich et al. (1980) and Smith et al. (1991), have been the locus of little or no fluid flow, is risky. The best test is coincidence of steps in isotopic composition with lithological boundaries (with or without some diffusive broadening of the step). This precludes any component of fluid flow across the boundary (e.g. Skelton et al., 1995).

9.6 Fluid chemistry, origin and large-scale circulation patterns

9.6.1 Fluid chemistry

Table 9.1 summarizes the types of fluid inferred to have caused alteration in shear zones in various environments. The most common fluid in compressional regimes appears to be a calcic brine with salinities in the 15–35 wt% range. The strongest evidence for this comes from fluid inclusions in syntectonic veins (Banks et al., 1991; O'Hara and Haak, 1992; Tempest, 1991; Bennett and Barker, 1992; O'Hara et al., 1995; Henderson and McCaig, 1996), particularly where the kinematics of vein opening can be related to the direction of movement on the shear zone.

Dipple and Ferry (1992b) calculated salinities in the 3–10 wt% range on the basis of the Cl content of apatite and mica in the Hunts Brook fault zone, Connecticut. These values were consistent with those required to explain changes in K and Na content of the mylonites for fluxes fixed with reference to Si gain, using a model of fluid flow down a temperature gradient. In fact, mass-transport of cations such as Na, K and Ca on any but a very local scale requires extremely high fluid fluxes unless brines with a significant Cl content are involved. A related argument can be applied to thrust zones in the Alps and Pyrenees where Sr and O isotopes have been transported on a similar length scale to each other (Burkhard et al., 1992; McCaig et al., 1995a,b), implying similar K_v values for Sr and O. Dilute fluids such as meteoric water, seawater and the likely products of metamorphic dehydration reactions do not contain enough Sr for transport on scales greater than a few metres.

Hypersaline brines in shear zones could result either from passive concentration of Cl through retrograde hydration reactions (Bennett and Barker, 1992), or introduction of evaporitic formation waters derived from sedimentary basins. Formation of brines by retrograde concentration is hard to reconcile with evidence for very large fluxes (e.g. Dipple and Ferry, 1992b; McCaig et al., 1990). Fluid inclusion crush-leach data from both thrust faults and steep shear zones cutting granodiorites in the Pyrenees (Banks et al., 1994) reveals high Br/Cl ratios which are most easily explained by progressive evaporation of seawater with halite precipitation. These data suggest penetration of sedimentary formation waters ultimately derived from seawater into retrograde shear zones at temperatures of around 400 °C and depths of at least 10 km. A similar conclusion was reached by McCaig et al. (1990) to explain low values of $\partial^{18}O$ in related shear zones cutting gneisses in the Aston Massif, Pyrenees (see Figure 9.8).

Stable isotope data suggest that shear zones in continental extensional environments contain a mixture of meteoric and metamorphic fluids. Higher grade (350–600 °C) mylonites beneath detachment faults normally show evidence for influx of only small amounts of fluid, of metamorphic origin (Kerrich, 1988; Smith et al., 1991; Morrison, 1994). Upper plate rocks and chlorite breccias along detachment faults often show large ^{18}O depletions suggesting meteoric water circulation, but enrichment in ^{18}O indicative of large fluxes of metamorphic water also occurs (Kerrich, 1988). Penetration of meteoric water into lower plate mylonites during ductile shearing has been demonstrated by Fricke et al. (1992) in the Ruby Mountains, but in other cases meteoric water penetration into lower plate rocks appears to occur during late brittle overprinting (Morrison, 1994).

Magmatic waters appear to be involved in skarn formation in some high grade shear zones related to granite emplacement (Rothstein et al., 1994), but are normally difficult to distinguish from deep-seated metamorphic fluids. The latter seem to be the typical fluids involved in shear zone hosted Archean and

Proterozoic mesothermal lode gold deposits (Fedorowich et al., 1991; Peters, 1993; Mikuchi and Ridley, 1993; Fayek and Kyser, 1995), and are characterized by low salinity, X_{CO_2} in the range 0.1 to 0.3, and sometimes the presence of CH_4, although high salinity fluids are occasionally involved (Ettner et al., 1994). Equilibration with felsic crustal rocks in the middle to deep crust is suggested by cation ratios (Mikuchi and Ridley, 1993) and Pb isotope data (McNaughton et al., 1993), although Robert et al. (1995) have attributed high boron content in gold-quartz veins in the Abitibi belt to derivation from underthrust greywackes in the Pontiac subprovince.

9.6.2 Fluid circulation patterns

The geochemical evidence cited above and in section 9.5 places important constraints on large-scale patterns of fluid flow. In compressional regimes, the presence of fluids originating as either sedimentary formation waters or products of metamorphic dehydration is normally attributed to underthrusting of sedimentary rocks beneath shear zone networks in the middle crust (Lobato et al., 1983; Sinha et al., 1986, 1988; Selverstone et al., 1991; Marquer and Burkhard, 1992; Robert et al., 1995). In either case, large reservoirs of fluid can be emplaced at depth and overpressured due to increase in overburden and/or metamorphic dehydration reactions. Such fluid is expected to move upwards through shear zones or upwards and outwards along major thrust faults (Oliver, 1986). The role of shear zones is mainly to provide high permeability conduits for fluid escape towards the high-level (<5 km) hydrostatic regime, although the rate of fluid escape and any related mineralization may be strongly influenced by fault-valve behaviour and the earthquake cycle (Sibson et al., 1988; Sibson, 1994; Robert et al., 1995; Henderson and McCaig, 1996; Odling, this volume).

In some cases the scale of overthrusting does not appear to be sufficient to emplace large volumes of unmetamorphosed sediments beneath steeply-dipping shear zones, and other mechanisms such as seismic or dilatancy pumping have been invoked to bring high-level fluids into the shear zones. Direct downward movement of fluid into the ductile regime has been suggested by Losh (1989) and Upton et al. (1995). In the Aston Massif of the Pyrenees, evidence for upward movement of low $\partial^{18}O$ fluids in shear zones (Figure 9.7 and section 9.5) led McCaig (1988) and McCaig et al. (1990) to infer seismic pumping of fluid down a shallow decollement, with subsequent movement upwards through a hanging-wall shear zone network.

In continental extensional regimes, the typical pattern appears to be one of mixing of an upper plate circulation system consisting of meteoric or formation waters, with smaller volumes of metamorphic water moving up low angle extensional detachments. Fluid circulation in the upper plate is driven by thermal convection or topography-driven flow in an essentially hydrostatic regime, and penetration of this fluid into the ductile mylonites appears to occur mainly during fracturing events.

The patterns of alteration discussed in section 9.5 can also place important constraints on models for large-scale fluid circulation. For example, several authors have inferred up-temperature flow on the basis of volume loss or chemical changes in shear zones (Lobato *et al.*, 1983; Sinha *et al.*, 1988; O'Hara, 1988; Selverstone *et al.*, 1991; Dipple and Ferry, 1992b; Newman and Mitra, 1993). These studies clearly indicate either downward flow, or upward flow in a regime with inverted thermal gradients due to overthrusting. More direct information on flow directions can be gained from displaced geochemical fronts. Although geochemical front models are strictly one-dimensional, by studying variably oriented lithological contacts in a restricted area it is possible to define a single flux vector (Skelton *et al.*, 1995; Graham *et al.*, this volume). In shear zones, the flux vector is invariably sub-parallel to the shear zone margins, reducing the problem by one dimension, but on the other hand lithological boundaries within the mylonites also tend to be in this orientation. So far, no unique flux vectors have been published from shear zone studies, although directions of flow have been constrained (Bowman *et al.*, 1994; McCaig *et al.*, 1990, 1995a,b). Vein-matrix studies (Figure 9.10) also allow constraints to be placed on fluid movement directions, and can potentially be used to distinguish between unidirectional fluid expulsion and dilatancy pumping, as discussed in section 9.5.

9.7 Permeability in shear zones

Rough estimates of shear zone permeability can be obtained by applying Darcy's Law, if the fluid flux, the hydraulic gradient, and the fluid viscosity are known (Ferry, 1989; McCaig *et al.*, 1990). Because the time-scales of both deformation and fluid flow are generally poorly known, fluid flux cannot be precisely calculated from the integrated fluxes yielded by geochemical calculations. Hydraulic gradients are also uncertain. McCaig *et al.* (1990) estimated permeabilities in the range 10^{-15} to 10^{-18} m^2 for greenschist facies shear zones in the Pyrenees, while Dipple and Ferry (1992b) calculated values in the range 10^{-15} to 10^{-17} m^2 for the five amphibolite facies shear zones they studied. These average permeabilities are greater than those normally associated with prograde or contact metamorphism, but still much less than typical sandstones (Dipple and Ferry, 1992b; McCaig *et al.*, 1990).

An alternative approach is to estimate the extent to which permeability was enhanced in the shear zone compared with undeformed wall rocks. If the duration of the event and the hydraulic gradient are the same in each case, then the ratio of the characteristic transport distances for the same metasomatic front in the wall rock and the shear zone is equal to the ratio of the permeabilities. McCaig *et al.* (1990, 1995b) used this method to suggest that average permeabilities were about three orders of magnitude greater in shear zones than in their wall rocks, both for retrograde shear zones cutting granitic gneisses, and for prograde mylonites in carbonates.

255

9.8 Implications of permeability structure and fluid flow mechanisms for shear zone geochemistry

The subject of permeability enhancement during deformation has been reviewed recently by Knipe and McCaig (1994) and Oliver (1996). It is unlikely that dihedral angles of volatile fluids are sufficiently low in most crustal rocks to create high intrinsic permeabilities at low porosities (Holness, this volume), and in any case great care should be used in applying such models to deforming rocks (Holness and Graham, 1995). More likely, permeability is enhanced through generation of porosity or cracks of various sizes during deformation at high pore fluid pressures (Beach, 1980; Etheridge *et al.*, 1983, 1984; Cox and Etheridge, 1989; McCaig and Knipe, 1990; Oliver *et al.*, 1990; Zhang *et al.*, 1994; Knipe and McCaig, 1994; Holness and Graham, 1995; Oliver, 1996). Knipe and McCaig (1994) used microchemical evidence to infer pervasive fluid flow at a grain scale in carbonate mylonites from the Gavarnie thrust in the Pyrenees, but non-pervasive flow in contemporaneous mylonites cutting pelitic gneisses in the Aston Massif, Pyrenees. Even in the carbonate mylonites, the presence of deformed veins, often with isotopic compositions slightly different from the adjacent matrix (Figure 9.10) suggests rapid, transient flow events alternating with more pervasive flow. It is most unlikely that models based on constant velocity, one-dimensional flow through a porous medium (Baumgartner and Rumble, 1988; Lassey and Blattner, 1988; Bowman *et al.*, 1994) will quantitatively reproduce the patterns of alteration produced by transient, localized flow in a mylonite, even if they represent a useful end-member for qualitative analysis.

Both McCaig *et al.* (1995b) and Bowman *et al.* (1994) attempted to explain some of the complexities of natural alteration patterns in terms of heterogeneous fluid flow within different layers in the mylonite. Figure 9.11 illustrates how a variety of permeabilities, and hence integrated fluxes, in different layers in a mylonite can lead to a wide variety of rock isotopic compositions at a single sample site. Different isotopic profiles in the direction of fluid flow are developed in each layer in the mylonite. The shape of the profiles is determined by the Peclet and Damköhler numbers, which will vary between different layers because of variations in fluid flow velocity (section 9.5). The positions of the profiles depend on the integrated fluid flux, so high permeability layers are likely to be upstream of any geochemical front and show highly altered isotopic values, while low permeability layers may be completely unaltered. Even if flow was confined within each layer, with no mass-transfer across layer boundaries, a variety of isotopic values would be recorded at each site. If the layers can communicate with each other chemically, either by advection or diffusion, then more complex effects are possible (Cartwright, 1994). Figure 9.8 demonstrates that oxygen isotopic anomalies can penetrate between 0.5 and 1 m into the undeformed wall rock of a greenschist facies shear zone. If diffusion is the dominant mechanism for mass-

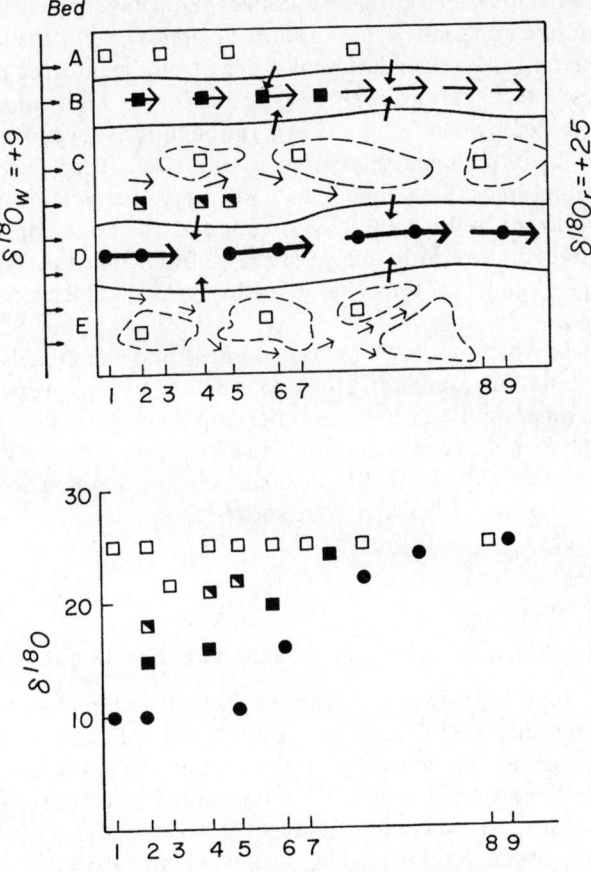

Figure 9.11 Qualitative model to explain dispersion during advective flow and the variability of isotopic compositions at single sites in Figures 9.5 and 9.6 (reproduced from Bowman *et al.*, 1994 with permission of *American Journal of Science*). Variable permeabilities in adjacent layers in the mylonite lead to variable transport distances for the isotopic signature of an infiltrating fluid. Hence adjacent layers at a sampling site may be fully or partially altered, or completely unaltered in isotopic composition. The lower diagram shows isotope datasets which might be collected by sampling different layers at sites 1 to 9. See text for discussion.

transfer, then some of the high $\partial^{18}O$ signature of the wall rock will also have diffused into the shear zone, even if this part of the profile has been swept away by advective flow up the shear zone parallel to the boundary. In the case of the variable-flux shear zone model in Figure 9.11, diffusion out of a high flux layer such as D at point 4 will lead to lower than expected (more altered) values of $\partial^{18}O$ in adjacent low flux layers such as C and E, while diffusion the other way will lead to higher than expected values of $\partial^{18}O$ in the high flux layer D. In this way, profiles in both low and high flux layers will be smeared

257

out to give lower model Peclet or Damköhler numbers when curves are fitted to the data, but the apparent values of both Pe and N_D will depend on sample positions relative to layer boundaries, as much as on the intrinsic properties of the layer. These effects are probably responsible for the high values of dispersivity estimated by Bowman *et al.* (1994), although it is questionable whether they represent hydrodynamic dispersion in the strict sense. They are more analogous to the crack flow model with diffusion perpendicular to cracks which was developed by Bickle (1992), except that the cracks are replaced by high flux zones of finite width. Bickle (1992) showed that this type of model produces profiles similar to those modelled using Damköhler numbers for all but very low N_D.

Another effect which needs to be incorporated into the models is the transience of high fluid flux events. In effect, two flow systems of very different Pe and/or N_D are superimposed on one another; rapid advective flow into veins at potentially high Pe and low N_D, and slow flow in between veining events at low Pe and high N_D. Finally, 2- or 3-D flow models are required to account for multiple fluid influx into, or escape from, shear zones.

9.9 Conclusions

Some of the questions posed in the introduction can now be answered:

- Syntectonic fluid flow always appears to focus into faults and shear zones under metamorphic conditions. Lack of fluid–rock interaction in shear zones does not necessarily imply low permeability; fluid may not have been available, or may have already reached equilibrium with the rock. There are no documented examples where less fluid has flowed through a shear zone than its wall rocks, although there may be cases when pervasive post-tectonic flow is superimposed on shear zones and wall rocks alike (Oliver, 1986). The action of faults as seals in sedimentary rocks reflects the much higher permeability of porous sediments than metamorphic rocks, rather than any intrinsic difference in process in the fault zones.
- Most fluid flow and alteration appears to be synchronous with deformation, although alternation of deformation and flow events (Tobisch *et al.*, 1991) cannot be ruled out. Permeability is mainly enhanced by dynamic processes such as crack development, and probably returns to near-normal values between deformation events. Occasionally, permanent enhancement of permeability is suggested by post-kinematic metasomatic assemblages (Mattey *et al.*, 1994).
- Considerable fluxes of fluid, comparable to those documented in high-level hydrothermal systems, have been documented in shear zones. The total flux may well be limited by the availability of fluid rather than the permeability. On the other hand, the presence of cavity-fill veins at all times in the deformation history of some shear zones suggests that high fluid pressures and signifi-

cant fluid volumes were present even at the end of their active lifetimes. This suggests that average permeabilities were not high enough to allow efficient drainage of fluid out of the shear zone channelway. The most important hydrogeologic role of shear zones is to provide reasonably permeable pathways through otherwise effectively impermeable blocks of crust, and to introduce a strong permeability anisotropy into large volumes of crust.

- Most information is gained by sampling which is well constrained spatially, and in particular by sampling for appropriate tracers along the flow path and across lithological contacts within the shear zone. In practice, this makes the quantitative study of large-displacement shear zones very difficult. Considerable information about flow directions and patterns, the nature of the fluid, and whether flow was up- or down-temperature can, however, be gained from spatially unconstrained sampling or single profiles across a shear zone, particularly if several tracers are used simultaneously, and appropriate use is made of isotope correlation plots and vein-matrix plots.

- Fluid–rock interaction in shear zones is complicated by complex flow patterns on a grain scale, and variable permeabilities in both space and time. On the other hand, continuous recrystallization results in a closer approach to equilibrium between the fluid and deformable phases than in undeformed rocks at the same temperature. Incorporating realistic flow paths and mechanisms into geochemical models is a challenge for the future, although sampling difficulties and problems of non-uniqueness mean that such models will always be difficult to test using field data.

- There is growing evidence for the involvement of sedimentary (often evaporitic) formation waters in ductile shear zones at depths of 10 to 15 km in the continental crust. In some cases this can be attributed to deep burial of sediments by thrust sheets, but this is not always a viable explanation, and more speculative mechanisms such as seismic pumping have been invoked for moving fluid downwards across the brittle–ductile transition and against the normal hydraulic gradient. In extensional regimes, surface-derived fluids circulate freely in the brittle upper crust, but penetrate only a short distance into ductile rocks, mixing with upward-moving metamorphic fluids.

Questions concerning mechanical aspects of fluid movement in shear zones have been addressed in recent reviews (Knipe and McCaig, 1994; Oliver, 1996). The presence of even small amounts of fluid in shear zones has profound mechanical consequences – such zones are probably in a more or less continuous state of incipient brittle failure due to high pore fluid pressures (e.g. Sleep and Blanpied, 1992). This means that both the majority of slip and the majority of fluid flow in greenschist facies shear zones probably occurs during brittle fracture events, even when microstructural evidence reflects mainly the long periods of ductile creep in between fracture events. This clearly has important implications for rheological models for the lithosphere (e.g. Ord and Hobbs, 1989).

259

References

Arreaza, C.M. and Persoz, F.P. (1992) The Medel-Cristallina intrusive (Eastern Gotthard Massif) Alpine deformation and chemical alteration. *Schweizerische Mineralogische und Petrographische Mitteilungen*, **72**, 179–96.

Banks, D.A., Davies, G.R., Yardley, B.W.D. *et al.* (1991) The chemistry of brines from an Alpine thrust system in the Pyrenees: An application of fluid inclusion analysis to the study of fluid behaviour in orogenesis. *Geochimica et Cosmochimica Acta*, **55**, 1021–30.

Banks, D.A., Henderson, I.C.H., McCaig, A.M. and Yardley, B.W.D. (1994) Origin of the fluid that lubricated Alpine deformation of the Pyrenees. *Geological Society America Abstracts with Programmes*, **26**, A280.

Barovich, K.M. and Patchett, P.J. (1992) Behavior of isotopic systematics during deformation and metamorphism – A Hf, Nd, and Sr isotopic study of mylonitized granite. *Contributions to Mineralogy and Petrology*, **109**, 386–93.

Baumgartner, L.P., and Rumble, D. (1988) Transport of stable isotopes. I: Development of a kinetic continuum theory for stable isotope transport. *Contributions to Mineralogy and Petrology*, **98**, 417–30.

Beach, A. (1976) The interrelations of fluid transport, deformation, geochemistry and heat flow in early Proterozoic shear zones in the Lewisian Complex. *Philosophical Transactions of the Royal Society of London*, **A280**, 569–604.

Beach, A. (1980) Retrogressive metamorphic processes in shear zones with special reference to the Lewisian complex. *Journal of Structural Geology*, **2**, 257–63.

Beach, A. and Tarney, J. (1978) Major and trace element patterns established during retrogressive metamorphism of granulite facies gneisses, NW Scotland. *Precambrian Research*, **7**, 325–48.

Bennett, D.J. and Barker, A.J. (1992) High salinity fluids – the result of retrograde metamorphism in thrust zones. *Geochimica et Cosmochimica Acta*, **56**, 81–95.

Bickle M.J. (1992) Transport mechanisms by fluid flow in metamorphic rocks: Oxygen and strontium decoupling in the Trois Seigneurs Massif – a consequence of kinetic dispersion? *American Journal of Science*, **292**, 289–316.

Bickle M.J. and Baker J. (1990) Migration of reaction and isotopic fronts in infiltration zones: assessments of fluid flux in metamorphic terrains. *Earth and Planetary Science Letters*, **98**, 1–13.

Bowman, J.R., Willett, S.D. and Cook, S.J. (1994) Oxygen isotopic transport and exchange during fluid flow: one-D models and applications. *American Journal of Science*, **294**, 1–55.

Brodie, K.H. (1980) Variations in mineral chemistry across a shear zone in phlogopite peridotite. *Journal of Structural Geology*, **2**, 265–72.

Brodie, K.H. and Rutter, E.H. (1985) On the relationship between deformation and metamorphism with special reference to the behaviour of basic rocks, in *Kinetics, Textures and Deformation* (eds A.B. Thompson and D.C. Rubie), *Advances in Geochemistry*, **4**, 138–79.

Brodie, K.H. and Rutter, E.H. (1987) The role of transiently fine-grained reaction products in syntectonic metamorphism: natural and experimental examples. *Canadian Journal of Earth Sciences*, **24**, 556–64.

Burkhard, M. and Kerrich, R. (1988) Fluid regimes in the deformation of the Helvetic nappes, Switzerland, as inferred from stable isotopic data. *Contributions to Mineralogy and Petrology*, **99**, 416–29.

Burkhard, M. and Kerrich, R. (1990) Fluid-rock interactions during thrusting of the Glarus nappe evidence from geochemical and isotopic data. *Schweizerische Mineralogische und Petrographische Mitteilungen*, **70**, 77–82.

Burkhard, M., Kerrich, R., Maas, R. and Fyfe, W.S. (1992) Stable and Sr-isotope evidence for fluid advection during thrusting of the Glarus nappe (Swiss Alps). *Contributions to Mineralogy and Petrology*, **112**, 293–311.

Carter, N.L., Kronenburg, A.K., Ross, J.V. and Wiltschko, D.V. (1990) Control of fluids on deformation in rocks, in *Deformation Mechanisms, Rheology and Tectonics* (eds R.J. Knipe and E.H. Rutter), Geological Society of London Special Publications, **54**, 1–13.

Cartwright, I. (1994) The two dimensional pattern of metamorphic fluid flow at Mary Kathleen, Australia. Fluid focussing, transverse dispersion, and implications for modelling fluid flow. *American Mineralogist*, **79**, 526–35.

Cartwright, I., Valley, J.W. and Hazelwood, A.M. (1993) Resetting of oxybarometers and oxygen isotope ratios in granulite facies orthogneisses during cooling and shearing, Adirondack Mountains, New York. *Contributions to Mineralogy and Petrology*, **113**, 208–25.

Cox, S.F., and Etheridge, M.A. (1989) Coupled grainscale dilatancy and mass-transfer during deformation at high fluid pressures; Examples from Mt. Lyell, Tasmania. *Journal of Structural Geology*, **11**, 147–62.

Crespo-Blanc, A., Masson, H., Sharp, Z. *et al.* (1995) A stable and Ar^{40}/Ar^{39} isotope study of a major thrust in the Helvetic Nappes (Swiss Alps) – evidence for fluid-flow and constraints on Nappe kinematics. *Geological Society of America Bulletin*, **197**, 1129–44.

Criss, R.E., Gregory, R.T. and Taylor, H.P., Jr. (1987) Kinematic theory of oxygen isotope exchange between minerals and water. *Geochimica et Cosmochimica Acta*, **51**, 1099–1108.

Dipple, G.M. and Ferry, J.M. (1992a) Fluid flow and stable isotopic alteration in rocks at elevated temperatures with applications to metamorphism. *Geochimica et Cosmochimica Acta*, **56**, 3539–50.

Dipple, G.M. and Ferry, J.M. (1992b) Metasomatism and fluid flow in ductile fault zones. *Contributions to Mineralogy and Petrology*, **112**, 149–64.

Dipple, G.M., Wintsch, R.P. and Andrews, M.S. (1990) Identification of the scales of differential element mobility in a ductile fault zone. *Journal of Metamorphic Geology*, **8**, 645–61.

Domenico, P.A., and Schwartz, F.W. (1990) *Physical and Chemical Hydrogeology*, John Wiley & Son, New York.

Dostal, J., Strong, D.F. and Jamieson, R.A. (1980) Trace element mobility in the mylonite zone within the ophiolite aureole, St. Anthony Complex, Newfoundland. *Earth and Planetary Science Letters*, **49**, 188–92.

Etheridge, M.A., Wall, V.J. and Cox, S.F. (1984) High fluid pressure during regional metamorphism and deformation: implications for mass transport and deformation mechanisms. *Journal of Geophysical Research*, **89**, 4344–58.

Etheridge, M.A., Wall, V.J. and Vernon, R.H. (1983) The role of the fluid phase during regional metamorphism and deformation. *Journal of Metamorphic Geology*, **1**, 205–26.

Ettner, D.C., Bjorlykke, A. and Andersen, T. (1994) A fluid inclusion and stable-isotope study of the Proterozoic Bidjovagge Au-Cu deposit, Finnmark, Northern Norway. *Mineralium Deposita*, **29**, 16–29.

Evans, J.P. and Chester, F.M. (1995) Fluid-rock interaction in faults of the San Andreas system: Inferences from San Gabriel fault rock geochemistry and microstructures. *Journal of Geophysical Research*, **100**, 13007–20.

Fayek, M. and Kyser, T.K. (1995) Characteristics of auriferous and barren fluids associated with the Proterozoic Contact Lake lode gold deposit, Saskatchewan, Canada. *Economic Geology and the Bulletin of the Society of Economic Geologists*, **90**, 385–406.

Fedorowich, J., Stauffer, M. and Kerrich R. (1991) Structural setting and fluid characteristics of the Proterozoic Tartan Lake gold deposit, Trans-Hudson Orogen, Northern Manitoba. *Economic Geology and the Bulletin of the Society of Economic Geologists*, **86**, 1434–67.

Ferry, J.M. (1986) Reaction progress: a monitor of fluid-rock interaction during metamorphic and hydrothermal events, in *Fluid-rock interactions during metamorphism* (eds J.V. Walther, and B.J. Wood), 60–88.

Ferry, J.M. (1989) Contact metamorphism of roof pendants at Hope Valley, Alpine County, California, USA. *Contributions to Mineralogy and Petrology*, **101**, 402–17.

261

Ferry, J.M. (1994) A historical review of metamorphic fluid flow. *Journal of Geophysical Research*, **99**, 15487–98.

Fourcade, S., Marquer, D. and Javoy, M. (1989) $^{18}O/^{16}O$ variations and fluid circulation in a deep shear zone: the case of the alpine ultramylonites from the Aar Massif (Central Alps, Switzerland). *Chemical Geology*, **77**, 119–32.

Fricke, H.C., Wickham, S.M. and O'Neil, J.R. (1992), Oxygen and hydrogen isotope evidence for meteoric water infiltration during mylonitization and uplift in the Ruby Mountains-East Humboldt Range core complex, Nevada. *Contributions to Mineralogy and Petrology*, **111**, 203–21.

Früh-Green, G.L. (1994) Interdependence of deformation, fluid infiltration and reaction progress recorded in eclogitic metagranitoids (Sesia Zone, Westem Alps). *Journal of Metamorphic Geology*, **12**, 327–43.

Fyfe, W.S., Price, N.J. and Thompson, A.B. (1978) *Fluids in the Earth's Crust*, Elsevier, New York.

Garven G., Ge, S., Person, M.A. and Sverjensky, D.A. (1993) Genesis of stratabound ore deposits in the midcontinent basins of North America, 1, role of groundwater flow. *American Journal of Science*, **293**, 497–568.

Getty, S.R., and Gromet, L.P. (1992) Geochronological constraints on ductile deformation, crustal extension and doming about a basement-cover boundary, New England, Appalachians. *American Journal of Science*, **292**, 359–97.

Gieskes, J.M., Vrolijk, P. and Blanc, G. (1990) Hydrogeochemistry of the North Barbados Accretionary Complex transect: Ocean Drilling Program, Leg 110. *Journal of Geophysical Research*, **95**, 8809–18.

Gilotti, J.A. (1989) Reaction progress during mylonitization of basaltic dikes along the Särv thrust, Swedish Caledonides. *Contributions to Mineralogy and Petrology*, **101**, 30–45.

Glazner, A.F. and Bartley, J.M. (1991) Volume loss, fluid-flow and state of strain in extensional mylonites from the Central Mojave Desert, California. *Journal of Structural Geology*, **13**, 587–94.

Graham, C.M., Skelton, A.D.L., Bickle, M.J. and Cole, C. (this volume) Lithological, structural and deformation controls on fluid flow during regional metamorphism.

Grant, J.A. (1986) The isocon diagram – a simple solution to Gresens' equation for metasomatic alteration. *Economic Geology*, **18**, 1976–82.

Gresens R.L. (1967) Composition-volume relationships of metasomatism. *Chemical Geology*, **2**, 47–65.

Henderson, I.H.C. and McCaig, A.M. (1996) Fluid pressure cycling in brittle-ductile shear zones, Pyrenees, *Tectonophysics*, **262**, 321–48.

Hickman, S., Sibson, R.H. and Bruhn, R. (1995) Introduction to special section: Mechanical involvement of fluids in faulting. *Journal of Geophysical Research*, **100**, 12831–40.

Holness, M.B. (this volume) The permeability of non-deforming rock.

Holness, M.B. and Graham, C.M. (1995) P-T-X effects on equilibrium carbonate-H_2O-CO_2-NaCl dihedral angles: constraints on carbonate permeability and the role of deformation during fluid infiltration. *Contributions to Mineralogy and Petrology*, **119**, 301–13.

Jamieson, R.A. and Strong, D.F. (1981) A metasomatic mylonite zone within the ophiolite aureole, St. Anthony complex, Newfoundland. *American Journal of Science*, **281**, 264–81.

Kerrich, R. (1986) Fluid transport in lineaments. *Philosophical Transactions of the Royal Society of London*, **A317b**, 219–51.

Kerrich, R. (1988) Detachment zones of Cordilleran metamorphic core complexes: thermal, fluid and metasomatic regimes. *Geologische Rundschau*, **77**, 157–82.

Kerrich, R. and Hyndman, D. (1986) Thermal and fluid regimes in the Bitteroot lobe Sapphire block detachment zone, Montana: Evidence from $^{18}O/^{16}O$ and geological relations. *Geological Society of America Bulletin*, **97**, 147–55.

Kerrich, R. and Rehrig, W. (1987) Fluid motion associated with Tertiary mylonitization and detachment faulting: $^{18}O/^{16}O$ evidence from the Picacho metamorphic core complex, Arizona. *Geology*, **15**, 58–62.

Kerrich, R., La Tour, T., and Willmore, L. (1984), Fluid participation in deep fault zones: Evidence from geological, geochemical and $^{18}O/^{16}O$ relations. *Journal of Geophysical Research*, **89**, 4331–43.

Kerrich, R., Allison, I, Barnett, R.S. *et al.* (1980) Microstructural and chemical transformations accompanying deformation of granite in a shear zone at Mieville, Switzerland, with implications for stress corrosion cracking and superplastic flow. *Contributions to Mineralogy and Petrology*, **73**, 221–42.

Kerrich, R., Fyfe, W.S., Gorman, B.E. and Allison, I. (1977) Local modification of rock chemistry by deformation. *Contributions to Mineralogy and Petrology*, **65**, 183–90.

Kerrick, D.M. and Caldeira, K. (1993) Paleoatmospheric consequences of CO_2 released during early Cenozoic regional metamorphism in the Tethyan orogen. *Chemical Geology*, **108**, 201–30.

Knipe, R.J. (1989) Deformation mechanisms – recognition from natural tectonites. *Journal of Structural Geology*, **11**, 127–46.

Knipe, R.J. (1990) Microstructural analysis and tectonic evolution in thrust systems: examples from the Assynt region of the Moine Thrust Zone, NW Scotland, in *Deformation Processes in Minerals, Ceramics and Rocks* (eds D.J. Barber and P.G. Meredith), The Mineralogical Society Series, **1**, 228–59.

Knipe, R.J. (1992) Faulting processes and fault seal, in *Structural and Tectonic Modelling and its Application to Petroleum Geology* (ed. R.M. Larsen), NPF Special Publications, **1**, 325–43.

Knipe, R.J. (1993) The influence of fault zone processes on fluid flow and diagenesis, in *Diagenesis and basin development* (eds E.D. Horbury and A.G. Robinson), American Association of Petroleum Geologists Studies in Geology, **36**, 135–54.

Knipe, R.J. and McCaig, A.M. (1994) Microstructural and microchemical consequences of fluid flow in deforming rocks, in *Geofluids: Origin, Migration, and Evolution of Fluids in Sedimentary Basins* (ed. J. Parnell), Geological Society Special Publication, **78**, 99–111.

Knipe, R.J. and Wintsch, R.P. (1985) Heterogeneous deformation, foliation development and metamorphic processes in a polyphase mylonite, in *Kinetics, Textures and Deformation* (eds A.B. Thompson and D.C. Rubie), *Advances in Geochemistry*, **4**, 180–210.

Lassey, K.R. and Blattner P. (1988) Kinetically controlled oxygen isotope exchange between fluid and rock in one-dimensional advective flow. *Geochimica et Cosmochimica Acta*, **52**, 2169–75.

Law, R.D., Casey, M. and Knipe, R.J. (1986) Kinematic and tectonic significance of microstructures and crystallographic fabrics within quartz mylonites from the Assynt and Eriboll regions of the Moine thrust zone, NW Scotland. *Transactions of the Royal Society of Edinburgh: Earth Sciences*, **77**, 99–123.

Lobato, L.M., Forman, J.M.A., Fazikawa, K. *et al.* (1983) Uranium in overthrust Archean basement, Bahia, Brazil. *Canadian Mineralogist*, **21**, 647–54.

Losh, S. (1989) Fluid-rock interaction in an evolving ductile shear zone and across the brittle–ductile transition, central Pyrenees, France. *American Journal of Science*, **289**, 601–48.

McCaig, A.M. (1984) Fluid-rock interaction in some shear zones from the Pyrenees. *Journal of Metamorphic Geology*, **2**, 129–41.

McCaig, A.M. (1987) Deformation and fluid-rock interaction in metasomatic dilatant shear bands. *Tectonophysics*, **135**, 121–32.

McCaig, A.M. (1988) Deep fluid circulation in fault zones. *Geology*, **16**, 867–70.

McCaig, A.M. and Knipe, R.J. (1990) Mass-transport mechanisms in deforming rocks: Recognition using microstructural and microchemical criteria. *Geology*, **18**, 824–7.

McCaig, A.M., Wickham, S.M. and Taylor H.P. Jr. (1990) Deep fluid circulation in alpine shear zones, Pyrenees, France: field and oxygen isotope studies. *Contributions to Mineralogy and Petrology*, **106**, 41–60.

McCaig, A.M., Wayne, D.M., Marshall, D. M. *et al.* (1995a) Isotopic and fluid inclusion studies of fluid movement along the Garvarnie Thrust, Central Pyrenees: Reaction fronts in carbonate mylonite. *American Journal of Science*, **295**, 309–43.

McCaig, A.M., Wayne, D.M. and Banks, D.A. (1995b) Fluid movement during thrusting in the south Pyrenean basin: Isotopic and fluid inclusion studies. *Abstract supplement No. 1 to Terra Nova*, **7**, 198.

McNaughton, N.J., Groves, D.I. and Witt, W.K. (1993) The source of lead in Archean lode gold deposits of the Menzies-Kalgoorlie-Kambalda Region, Yigarn Block, Western Australia. *Mineralium Deposita*, **28**, 495–502.

Marquer, D. (1989) Transfert de matière et déformation des granitoides: aspects méthodologiques. *Schweizerische Mineralogische und Petrographische Mitteilungen*, **69**, 15–35.

Marquer, D. and Burkhard, M. (1992) Fluid circulation, progressive deformation and mass transfer processes in the upper crust: the example of basement-cover relationships in the External Crystalline Massifs, Switzerland. *Journal of Structural Geology*, **14**, 1047–57.

Marquer, D., Gapais, D. and Capdevila, R. (1985) Comportement chimique et orthogneissification d'une granodiorite en faciès schistes verts (Masif de l'Aar, Alpes Centrales), *Bulletin Mineralogique*, **108**, 209–21.

Marquer, D., Petrucci, E. and Iacumin, P. (1994) Fluid advection in shear zones – evidence from geological and geochemical relationships in the Aiguilles-Rouges Massif (Western Alps, Switzerland). *Schweizerische Mineralogische und Petrographische Mitteilungen*, **74**, 137–48.

Mattey, D., Jackson, D.H., Harris, N.B.W. and Kelley, S. (1994) Isotopic constraints on fluid infiltration from an eclogite-facies shear zone, Holsenoy, Norway. *Journal of Metamorphic Geology*, **12**, 311–25.

Mikuchi, E.J. and Ridley, J.R. (1993) The hydrothermal fluid of archean lode-gold deposits at different metamorphic grades – compositional constraints from ore and wallrock alteration assemblages. *Mineralium Deposita*, **28**, 469–81.

Moore, C. and Vrolijk, P. (1992) Fluids in accretionary prisms. *Reviews in Geophysics*, **30**, 113–35.

Morrison, J. (1994) Meteoric water-rock interaction in the lower plate of the Whipple Mountain metamorphic core complex, California. *Journal of Metamorphic Geology*, **12**, 827–40.

Muir-Wood, R. and King, G.C.P. (1993) Hydrological signatures of earthquake strain. *Journal of Geophysical Research*, **98**, 22035–68.

Nicolas, A. and Poirier J.P. (1976) *Crystalline Plasticity and Solid State Flow in Metamorphic Rocks*, John Wiley & Sons, London.

Newman, J. and Mitra, G. (1993) Lateral variations in mylonite zone thickness as influenced by fluid-rock interactions, Linville Falls fault, North Carolina. *Journal of Structural Geology*, **15**, 849–63.

Norton, D. and Taylor, H.P. Jr. (1979) Quantitative simulation of the hydrothermal circulation systems of crystallising magmas on the basis of transport theory and oxygen isotope data: an analysis of the Skaergaard intrusion. *Journal of Petrology*, **20**, 421–86.

Odling, N.E. (this volume) Fluid flow in fractured rocks at shallow levels in the Earth's crust: an overview.

O'Hara, K. (1988) Fluid flow and volume loss during mylonitization – An origin for phyllonite in an overthrust setting, North Carolina, USA. *Tectonophysics*, **156**, 21–36.

O'Hara, K. and Haak (1992) A fluid inclusion study of fluid pressure and salinity variations in the footwall of the Rector Branch thrust, North Carolina, USA. *Journal of Structural Geology*, **14**, 579–89.

O'Hara, K.D., Kirschner, D.L. and Moecher, D.P. (1995) Petrologic constraints on the source of fluid during mylonitization in the Blue Ridge province, North Carolina and Virginia, USA. *Journal of Geodynamics*, **19**, 271–87.

Oliver, J. (1986) Fluids expelled tectonically from orogenic belts: Their role in hydrocarbon migration and other geological phenomena. *Geology*, **14**, 99–102.

Oliver, N.H.S. (1996) Review and classification of structural controls on fluid flow during regional metamorphism. *Journal of Metamorphic Geology*, **14**, 477–92.

Oliver, N.H.S., Valenta, R.K. and Wall, V.J. (1990) The effect of heterogeneous stress and strain on metamorphic fluid flow, Mary Kathleen, Australia, and a model for large-scale fluid circulation. *Journal of Metamorphic Geology*, **8**, 311–31.

Ord, A. and Hobbs, B.E. (1989) The strength of the continental crust, detachment zones, and the development of plastic instabilities. *Tectonophysics*, **158**, 269–89.

Peters, S. G. (1993) Formation of oreshoots in mesothermal gold-quartz vein deposits – examples from Queensland, Australia. *Ore Geology Reviews*, **8**, 277–301.

Ramsay, J.G. and Graham, R.H. (1970) Strain variation in shear belts. *Canadian Journal of Earth Sciences*, **7**, 786–813.

Ree, J.H. (1994) Grain boundary sliding and development of grain boundary openings in experimentally deformed octachloropropane. *Journal of Structural Geology*, **16**, 403–18.

Robert, F., Boullier, A.-M. and Firdaous, K. (1995) Gold-quartz veins in metamorphic terranes and their bearing on the role of fluids in faulting. *Journal of Geophysical Research*, **100**, 12861–79.

Romer, R.L. (1990) Lead mobilization during foreland metamorphism in orogenic belts – examples from Northern Sweden. *Geologische Rundschau*, **79**, 693–707.

Rothstein, D.A., Karlstrom, K.E., Hoisch, T.D. and Morrison, J. (1994) Synkinematic contact metasomatism – implications for the timing of pluton emplacement and regional deformation in the Scanlon shear zone, South Eastern California. *Journal of Metamorphic Geology*, **12**, 709–21.

Rubie, D.C. (1983) Reaction enhanced ductility: the role of solid-solid univariant reaction in deformation of the crust and mantle. *Tectonophysics*, **96**, 331–52.

Rye, D.M. and Bradbury, H.J. (1988) Fluid flow in the crust: An example from a Pyrenean thrust ramp. *American Journal of Science*, **288**, 197–235.

Selverstone, J., Morteani, G. and Staude, J.M. (1991). Fluid channeling during ductile shearing: transformation of granodiorite into aluminous schist in the Tauern window, Eastern Alps. *Journal of Metamorphic Geology*, **9**, 419–31.

Sibson, R.H. (1986) Brecciation processes in fault zones: Inferences from earthquake rupturing. *Pure and Applied Geophysics*, **124**, 159–75.

Sibson, R.H. (1990) Conditions of fault-valve behaviour, in *Deformation Mechanisms, Rheology and Tectonics*, (eds R.J. Knipe and E.H. Rutter), Geological Society Special Publication, **54**, 15–28.

Sibson, R.H. (1994) Crustal stress, faulting, and fluid flow, in *Geofluids: Origin, Migration, and Evolution of Fluids in Sedimentary Basins*, (ed J. Parnell). Geological Society Special Publication, **78**, 69–84.

Sibson, R.H., Robert, F. and Poulsen, K.H. (1988) High-angle reverse faults, fluid pressure cycling and mesothermal gold-quartz deposits. *Geology*, **16**, 551–5.

Sinha, A.K., Hewitt, D. and Rimstidt, D. (1986) Fluid interaction and element mobility in the development of ultramylonites. *Geology*, **14**, 883–6.

Sinha, A.K., Hewitt, D. and Rimstidt, D. (1988) Metamorphic petrology and strontium isotope geochemistry associated with the development of mylonites: An example from the Brevard fault zone, N Carolina. *American Journal of Science*, **288-A**, 115–47.

Skelton, A.D.L., Graham, C.M. and Bickle, M.J. (1995) Lithological and structural controls on regional 3D-fluid flow patterns during greenschist facies metamorphism of the Dalradian of the SW Scottish Highlands. *Journal of Petrology*, **36**, 563–86.

Sleep, N.H. and Blanpied, M.L. (1992) Creep, compaction and the weak rheology of major faults. *Nature*, **359**, 687–92.

Smith, B.M., Reynolds, S.J., Day, H.W. and Bodnar, R.J. (1991) Deep-seated fluid involvement in ductile-brittle deformation and mineralization, South Mountains metamorphic core complex, Arizona. *Geological Society of American Bulletin*, **103**, 559–69.

Tempest, S.A. (1991) Fluid-rock interaction in ductile shear zones, central-eastern Pyrenees. Ph.D. thesis, Leeds University, 202 pp.

Tobisch, O.T., Barton, M.D., Vemon, R.H. and Paterson, S.R. (1991) Fluid-enhanced deformation: Transformation of granitoids to banded mylonites, Western Sierra Nevada, California and southeastern Australia. *Journal of Structural Geology*, **13**, 1137–56.

Upton, P., Koons, P.O. and Chamberlain, C.P. (1995) Penetration of deformation-driven meteoric water into ductile rocks – isotopic and model observations from the Southern Alps, New Zealand. *New Zealand Journal of Geology and Geophysics*, **38**, 535–43.

Vocke, R.D., Hanson, G.N. and Grunenfelder, M. (1987) Rare earth element mobility in the Roffen gneiss, Switzerland. *Contributions to Mineralogy and Petrology*, **95**, 145–54.

White, S. H., Burrows, S.E., Carreras, J. *et al.* (1980) On mylonites in ductile shear zones. *Journal of Structural Geology*, **2**, 175–87.

Walker, A.N., Rutter, E.H. and Brodie, K.H. (1990) Experimental study of grain size sensitive flow of synthetic, hot pressed calcite rocks, in *Deformation Mechanisms, Rheology and Tectonics*, (eds Knipe, R.J. and Rutter, E.H.), Geological Society Special Publication, **54**, 259–82.

Wayne, D.M. (1995) Rb-Sr redistribution during amphibolite-grade mylonitization: An example from the Hope Valley shear zone, Massachusetts, U.S.A. *Journal of Geodynamics*, **19**, 351–77.

Wayne, D.M., Sinha A.K. and Hewitt, D.A. (1992) Differential response of zircon U-Pb systematics to metamorphism across a lithological boundary: an example fom the Hope Valley Shear Zone, Southeastern Massachusetts, USA. *Contributions to Mineralogy and Petrology*, **109**, 408–20.

Wintsch, R.P., Christoffersen, R. and Kronenberg, A.K. (1995) Fluid-rock reaction weakening of fault zones. *Journal of Geophysical Research*, **100**, 13021–32.

Winchester, J.A. and Max, M.D. (1984) Element mobility associated with syn-metamorphic shear zones near Scotchport, NW Mayo, Ireland. *Journal of Metamorphic Geology*, **2**, 1–11 .

Yardley, B.W.D. and Lloyd, G.E. (1995) Why metasomatic fronts are really meatsomatic sides. *Geology*, **23**, 53–6.

Zhang, S., Patterson, M.S., Cox, S.F. (1994) Porosity and permeability evolution during hot isostatic pressing of calcite aggregates. *Journal of Geophysical Research*, **99**, 15741–60.

CHAPTER TEN

Segregation veins: evidence for the deformation and dewatering of a low-grade metapelite

S.L. Brantley, D.M. Fisher, P. Deines, M.B. Clark and G. Myers

10.1 Introduction

Veins in sedimentary and metamorphic rocks document local or long-distance mass transfer of mineral components. Many workers have argued that vein formation occurs by mineral precipitation associated with through-flowing fluids: for example, precipitation from continuous, long-distance upward fluid flow along open fracture networks (Walther and Orville, 1982; Yardley, 1984; Ferry and Dipple, 1991) or episodic precipitation from flowing fluids driven by pressure variations (e.g. fault valve behaviour; Sibson, 1994). Such models of vein formation necessarily require large volumes of fluid, due to the relatively low solubility of most vein-forming minerals in metamorphic fluids.

Etheridge *et al.* (1984) argued that the observation of crack-seal textures in some veins may indicate local pumping of fluid from the matrix to the newly opened fracture, followed by draining of the fluid (after mineral precipitation in the fracture) due to the opening of a new fracture. This model assumes that fluid flow is local, and that the fluid passes episodically from fracture to fracture through the matrix. Such a model is similar to the so-called mechanism of 'dilational pumping' described by Yardley (1984) and does not require large volumes of fluid because it reuses the same fluid. Etheridge *et al.* (1984) have also suggested convection cells as a means of reusing fluids to transport vein-forming minerals.

Although many workers implicitly or explicitly assume that vein growth documents such local or long-distance fluid flow, Elliott (1973) presented a model for mineral segregation by diffusion of locally derived mineral components. Diffusive transport of vein minerals requires minimal fluid and therefore

Deformation-enhanced Fluid Transport in the Earth's Crust and Mantle. Edited by M.B. Holness. Published in 1997 by Chapman & Hall, London. ISBN 0 412 75290 5.

relieves the large fluid volume problem. In a series of papers, we have documented a set of well-exposed quartz veins whose textures, distribution, and chemical features are consistent with such a model for local diffusion of silica from wall rock to vein (Fisher and Brantley, 1992; Fisher et al., 1995; Clark et al., 1995). In this chapter, we present new isotope and fluid inclusion data and argue that the fluids which filled these fractures may have also been locally derived. Diffusion-driven quartz precipitation such as we have documented may be a realistic model for vein formation in many metapelites. Such a diffusion mechanism requires substantially lower volumes of fluid than the fluid flow models.

In the following sections, we review our earlier work describing the geometry and textures of a set of segregation veins developed as episodic fractures in a thick turbidite sequence after underplating and incorporation into an accretionary wedge at 8–12 km depth. Emphasizing vein texture and distribution in combination with local variations in wall-rock composition, we review the evidence for a diffusive mass transfer model wherein Si is predominantly derived from local reactions in the rock matrix. We then present oxygen isotope and fluid inclusion evidence that is consistent with locally derived fluids. Finally, we review models of vein formation, and argue that positive feedback between vein growth and deformation is manifested in power-law vein thickness distributions. Based on this model, the veins therefore document the deformation and dewatering of the metapelites within the sequence, and suggest that external fluids from the high fluid-pressure decollement lying below may not have crossed this low permeability barrier.

10.2 Geologic setting

The Kodiak accretionary complex consists of a set of north-east-trending turbidite belts that decrease in age trenchward (Figure 10.1). Veins discussed in this paper are restricted to the Maastrichtian Kodiak formation, an 80-km wide sequence of folded and imbricated turbidites accreted in the late Cretaceous or early Tertiary. Early deformation (D_1) consisted of progressive dewatering and lithification related to underthrusting, and resulted in bedding-perpendicular calcite and quartz veins restricted to sandstone layers. Later deformation (D_2) involved folding, imbrication, slaty cleavage development, lower greenschist facies metamorphism, and two regionally extensive vein sets distinguished by orientation. A third deformation (D_3) folded the accreted package into a broad regional anticline that exposed the deepest D_2 structural levels in the core (known as the central belt) and progressively higher structural levels on either limb (known as the landward and seaward belts respectively).

In this paper, we focus on the D_2 quartz veins which developed at the same time as the cleavage and are oriented roughly perpendicular to the sub-horizontal cleavage and bedding in the central belt. The central belt of the Kodiak formation is an example of a sub-horizontal shear zone and may have

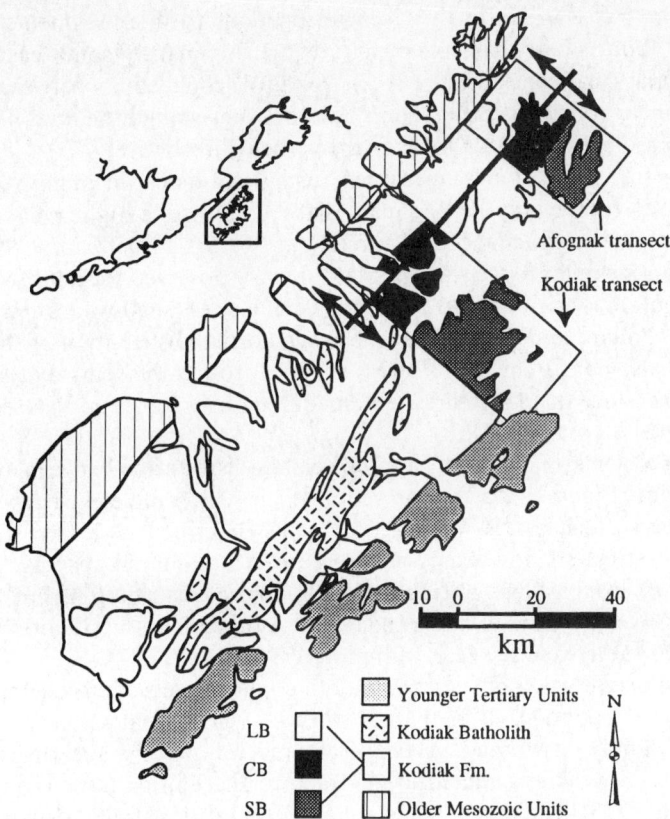

Figure 10.1 Location map of Kodiak and Afognak islands in south-west Alaska. The Afognak and Kodiak transects include the seaward (SB), central (CB) and landward (LB) belts of the Kodiak Formation.

separated packages of underplated material. During vein formation, the sub-horizontal cleavage presumably acted as a barrier to pervasive upward fluid flow. We have examined the D_2 veins along two transects: the Afognak transect along the north-east coast of Afognak island, and the Kodiak transect between Kodiak and Afognak islands. Three-dimensional characterization of the D_2 veins is possible because of the near-continuous cliff and wave-cut platform exposures along the coastlines of Afognak and Kodiak islands (Figure 10.1).

10.3 Vein geometry

The vein network, documented over a 90 km × 15 km area within a package of at least 1–5 km thickness (Fisher and Byrne, 1992), is composed of nearly vertical, ~1 m long veins with a median thickness of <1 mm and a spacing ranging from

269

0.5 to 3 cm. Veins widen in their central portions to thick segments (median thickness about 1 cm) that collectively define dominantly south-east-dipping brittle ductile shear zones (Figure 10.2). Very little connectivity between veins of different shear zones was observed in outcrop, suggesting little fluid flow could have occurred between shear zones in the vertical direction.

Spacing of shear zones measured using scanlines in many locations throughout a transect across Afognak Island comprises a log-normal distribution with a median spacing of 514 mm (Figure 10.2(e); Fisher *et al.*, 1995). This distribution is similar to the distribution typically observed for joint sets, where the distribution is attributed to the presence of a stress shadow around an open crack (e.g. Pollard and Segall, 1987). The log-normal distribution of shear zone spacing is also consistent with a stress shadow around the veins that lie within shear zones, since these fractures remained open over long periods (see discussion below).

Spacing of veins, in contrast, has a power-law distribution between at least 10 and 400 mm (Figure 10.2(f)). For veins which seal completely between fracturing events, such as the majority of veins described here (see discussion below), no stress shadow exists and there is no minimum spacing. For this reason, vein spacing does not define a log-normal distribution, but rather is power-law over an order of magnitude. Such a power-law distribution has been observed for other vein sets (e.g. Manning, 1994).

The lack of distortion of vein tips in both conjugate sets of en echelon veins is consistent with a model (Figure 10.3) whereby veins initiated and propagated first as randomly distributed, vertical fractures, followed by shearing localized along zones of weakness related to the fracture distribution (type II en echelon sets, Beach, 1975). The tips of veins are approximately parallel for both south-east-dipping and less abundant north-west-dipping en echelon sets. The brittle–ductile shear zones dip moderately to the south-east (trenchward) and contain sigmoidal veins which indicate south-east side-down shear.

Figure 10.2 Geometry of one shear zone in the D$_2$ veins in the Kodiak Formation. White zones are quartz veins and grey shading represents the slate wallrock. In (b), (c) and (d) individual quartz crystals in the vein are shown as they appear under crossed nicols. (b) Continuous crack-seal banding dominates the texture of the tips and thin parts of the veins. Banding is due to planar layers of chlorite inclusions engulfed by quartz during vein growth. Banding documents the multiple fracture and sealing episodes. Planar bands show that the fracture closed by sealing due to quartz precipitation rather than collapse. (c) Closer to the central, widest sections of the veins, discontinuous crack seal banding is observed. In this texture, inclusion bands are discontinuous, and are strictly confined to those quartz crystals which grew the fastest. The texture indicates multiple fracturing and partial resealing episodes. (d) Euhedral growth textures dominate the central, widest quartz portions of the veins. Although these veins typically show a few early bands of continuous, then discontinuous crack seal events, the majority of quartz growth results in euhedral crystals. Crystal terminations are marked by pressure solution selvedges along the contact with the wall rock, indicating that closure occurred by collapse rather than sealing. Occasional dark residues that formed at the vein margin are found engulfed in the euhedral quartz growth, indicating a few

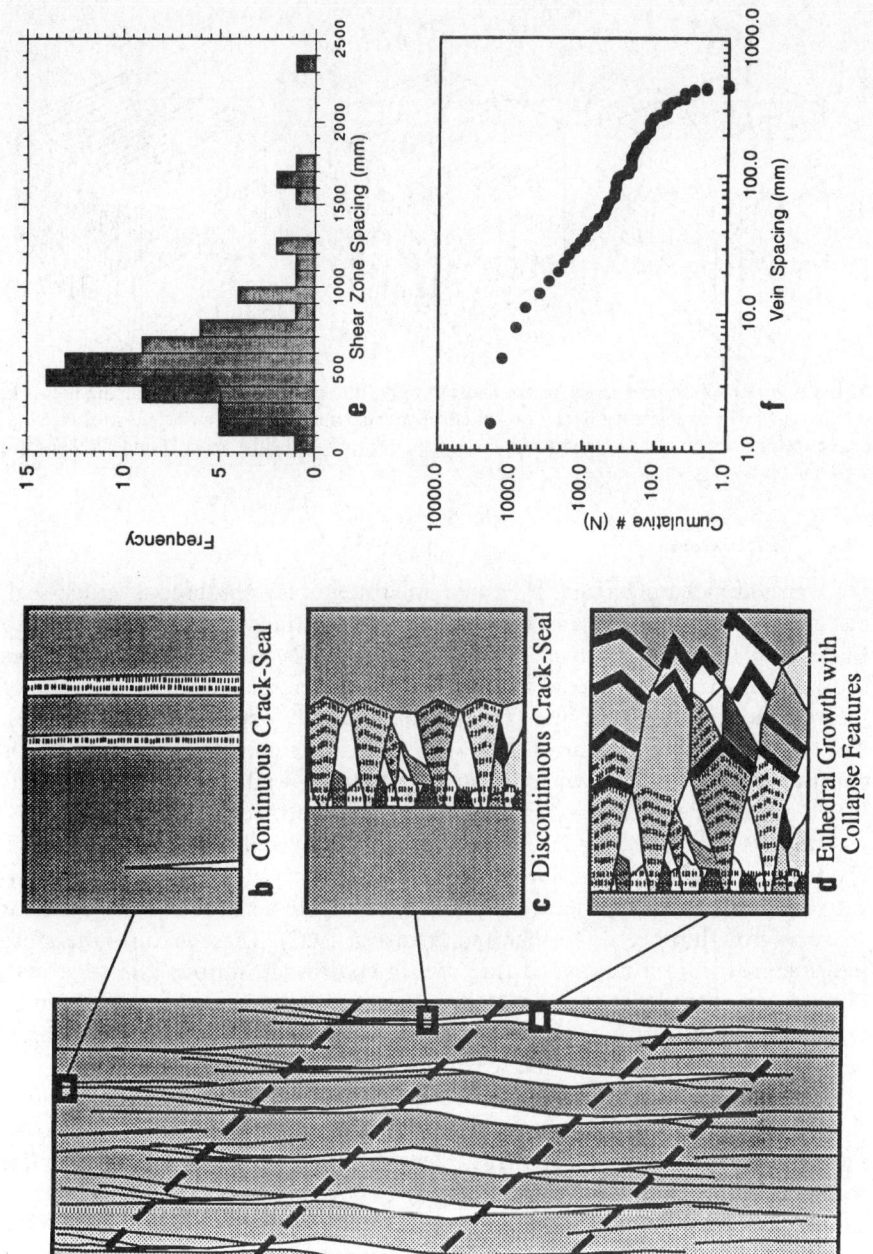

collapse events during the vein (fracture) lifetime. (e) Histogram showing the spacing between shear zones observed throughout the Afognak transect (Fisher *et al.*, 1995). (f) Cumulative number, *N*, of vein spacings observed throughout the Afognak transect (Fisher *et al.*, 1995).

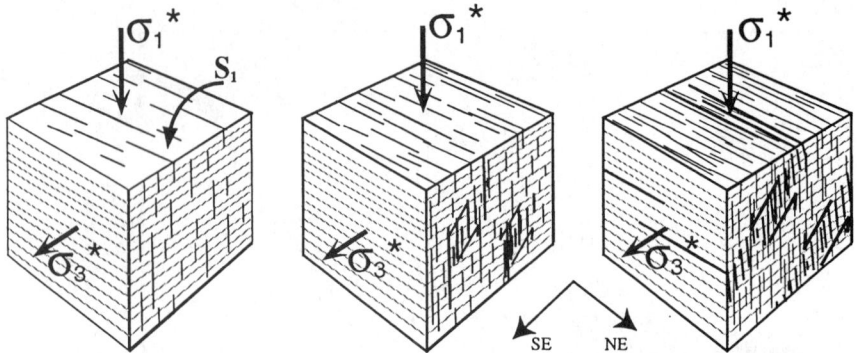

Figure 10.3 Block diagram showing, from left to right, the evolution of the D_2 fracture network. $\sigma_1{}^*$ and $\sigma_3{}^*$ arrows represent orientations of effective maximum and least principal stress vectors, respectively. Horizontal lines represent cleavage planes (perpendicular to $\sigma_1{}^*$) and D_2 veins are shown in black.

10.4 Vein textures

D_2 veins are characterized by three microtextures: continuous crack-seal, discontinuous crack-seal, and euhedral growth textures (Figure 10.2; see also Fisher and Byrne, 1990; Fisher and Brantley, 1992). All veins of all textures show asymmetrical, antitaxial growth, with the open crack restricted to one of the vein margins (i.e. vein quartz was added on only one side of the fracture). The direction of growth varies from vein to vein. Asymmetrical vein growth is consistent with a model wherein one of the vein–wall-rock boundaries was sealed early in the history and in many cases never refractured.

Although crack-seal textures are sometimes referred to as fibrous quartz crystal growth (e.g. Mullis, 1987), in the D_2 veins, crystals are blocky or elongate, and preferentially oriented crystals outcompete neighbouring crystals for space as growth proceeds (Fisher and Brantley, 1992). These textures therefore indicate growth in a fluid-filled fracture, in contrast to fibrous quartz growth observed in other veins and pressure shadows in the same rocks, which can be explained by diffusive mass transport through cohesive grain boundaries without influx of a bulk fluid phase (Fisher and Brantley, 1992).

Continuous crack-seal textures are characterized by quartz bands of median thickness of ~7 μm separated by laterally continuous, planar bands of chlorite inclusions. This texture results from periodic cracking followed by complete sealing of the open fracture by quartz growth (see Ramsay, 1980). Upon refracturing, small inclusions of chlorite grown in the wall rock are entrapped by the vein quartz, eventually remaining as bands of inclusions parallel to the fracture margin. Because the inclusion bands remain relatively planar, without developing noticeable crystal terminations, crack closure must occur by sealing, rather than by collapse (Fisher and Brantley, 1992). Discontinuous crack-seal textures are characterized by quartz bands of median thickness (~10 μm) separated by discontinuous bands of chlorite inclusions. Discontinuous inclusion bands reflect partial sealing of the fracture before the subsequent fracturing event. Closure was only achieved by quartz crystals favourably oriented for fast growth (c-axes perpendicular to the fracture walls). Discontinuously banded crack-seal textures also show relatively planar inclusion bands, indicating partial crack closure by sealing rather than collapse.

Euhedral growth textures are dominated by euhedral inclusion-free quartz growth (Figure 10.2). Prismatic terminations indicate that quartz grew into fluid-filled fractures, and that closure occurred by collapse of pointed crystals into wall rock rather than by sealing due to quartz precipitation. Thus, euhedral growth veins reflect relatively long duration growth into open cracks without high-frequency sealing events. Periodic collapse of these fractures is documented by irregularly spaced insoluble wall-rock residues trapped, parallel to fracture walls, within the veins. Identical insoluble wall-rock residues are also observed along euhedral terminations which jut into the wall rock. Spacing between wall-rock residues within the vein quartz is on the order of millimetres, documenting that collapse events were infrequent compared to crack-seal events.

The thickest veins reflect an evolution from continuously banded, to discontinuously banded, to euhedral growth, documenting the increasing length of time during which the fluid-filled fractures remained open. Periodically, fluid was totally expelled from the reservoirs and fractures collapsed, perhaps due to large-scale linkage of fractures. Only these collapse events document large-scale fluid expulsion from the section.

10.5 Mechanism of vein formation: textural and compositional arguments

One of four models for vein formation could explain the D_2 veins in the Kodiak Formation (Fisher and Brantley, 1992):

1. vein growth due to silica derived from a single influx of fluid;
2. vein growth due to silica precipitated from flow of fluid derived strictly from the local matrix;
3. vein growth due to silica derived from fracture-channelized long-distance fluid flow through the fracture;
4. vein growth from silica derived from diffusive transport into the fracture.

In analysing the model which assumes silica is transported as a solute in a single influx of fluid, we conclude that, for any reasonable supersaturation value, fractures would have to open to unreasonably large apertures to contain enough silica to precipitate 7 μm quartz bands (Fisher and Brantley, 1992; see also Ferry and Dipple, 1991). For this reason, model 1 is discounted.

Similarly, strong evidence argues against the reasonableness of model 2. Given a median vein spacing of about 1 cm and vein thickness of 10 μm, the ratio of volume of water to volume of rock is $\sim 10^3$ for a reasonable supersaturation (Fisher and Brantley, 1992; note, however, in that earlier publication we had no vein spacing distribution data and so all spacings in that publication were estimated). This ratio is much larger than the amount of water produced per rock volume (~0.1, the estimated water to rock volume ratio released from shales during late diagenesis; Bjorlykke, 1979). Even if this fluid is pumped back and forth from matrix to vein, such dilational pumping (e.g. Yardley, 1984) cannot explain the observed textures; for continuously banded crack-seal veins, it is realistic that each pumping event should correspond to one precipitation event. However, this is equivalent to the single pulse model for each crackseal layer, and requires very large *in situ* crack apertures per event. Furthermore, a dilational pumping model, as described by previous workers, leads to episodic fracture collapse, whereas textures of crack-seal veins indicate that crack closure occurs as a consequence of sealing.

Model 3, long-distance fluid flow, also does not explain the observed textures since this model requires continuous flow of large volumes of fluid through an ever diminishing crack aperture. Because the flow rate is proportional to the cube of the aperture (e.g. Domenico and Schwartz, 1990), fracture-channelized flow must become exponentially more difficult as the crack seals. Furthermore, the largest crystals in euhedral quartz veins often coincide with sandy or quartz-rich layers in the wall rock (Fisher and Brantley, 1992; see also Figure 10.4), from which we infer that Si concentration gradients in the fluid existed along the vertical axis of the fractures. Such gradients would not be maintained where fluid flow was the dominant mechanism of silica transport. Finally, field obser-

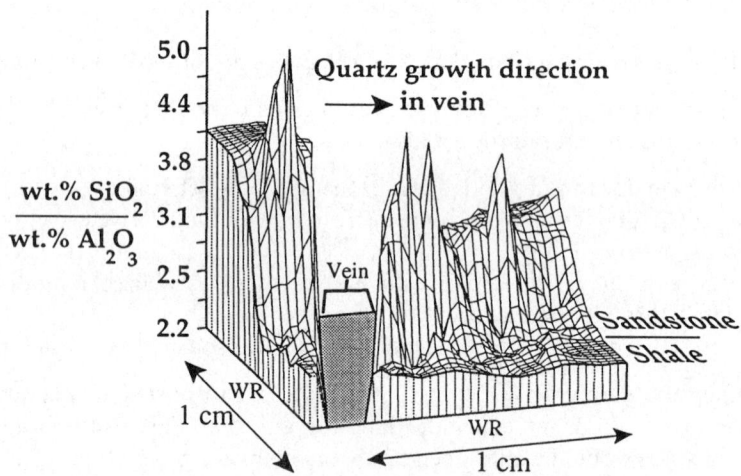

Figure 10.4 Maps of wt% SiO_2 normalized by wt% Al_2O_3 in the wall rock (WR) around a euhedral growth vein in one sample. Textures within the vein indicate growth from left to right. The low relief section of the surface in the foreground is a shale layer whereas the high relief 'ridge' is a sandy layer. The vein is represented by a gap in the data. Note the drop in SiO_2/Al_2O_3 across the vein (depleted zone on the youngest, growth side of the vein) and that the depletion is more pronounced in the sandy layer than in the shaly layer.

vations demonstrate that little vertical connectivity exists among veins (Fisher *et al.*, 1995), making continuous upward migration of large fluid volumes difficult. Specifically, the veins which define each en echelon set are frequently not connected with overlying or underlying sets and may not be connected with each other. Although fluid flow could have occurred along-strike, such flow would not have produced the cooling and depressurizing necessary to drive silica precipitation.

Based on these arguments, local diffusion of silica from matrix to veins is the most likely model to explain the observed textures in the Kodiak D_2 veins. At the time of our earlier publication (Fisher and Brantley, 1992), the best estimates of coefficients of grain-boundary diffusion were very low (Farver and Yund, 1991; Joesten, 1991), and we concluded that grain-boundary diffusive transport of silica would be prohibitively slow. Newly published data for diffusion along grain boundaries in novaculite in the presence of an interconnected pore fluid suggests that grain-boundary diffusion coefficients may, however, be orders of magnitude faster when water is present than previously determined for dry grain boundaries (Farver and Yund, 1992). Therefore, local transport of silica from matrix to vein may have occurred by grain-boundary instead of pore-fluid diffusion. The chemical potential gradient that leads to diffusion from the rock matrix into the crack could be due to the pressure drop that accompanies fracturing (if the permeability of the rock matrix was low enough to maintain pressure gradients from wall rock to fracture), due to silica-producing reactions in the wall rock, due to variations in quartz grain size, or due to the higher dislocation density for quartz grains in the matrix relative to the undeformed, larger vein quartz crystals. While a local pressure drop may be short-lived, a chemical potential gradient related to variations in chemical reaction, grain size, or strain energy would persist. Regardless of the transport mechanism and driving force, however, textural arguments require that transport was local in nature, and that for wide veins, a silica depletion zone should have developed on the growth side of the vein.

Such silica depletion zones around euhedral growth veins have been identified for six veins (Fisher *et al.*, 1995). Electron microprobe scans of silica and alumina compositions on either side of euhedral growth veins document that, for every vein thicker than several millimetres which was analysed, silica is depleted in the matrix on the youngest (most recently growing) side of the vein. For two veins, 2D maps of SiO_2/Al_2O_3 content in the matrix were completed by combining the electron microprobe measurement of the compositions of 20 μm^2 spots distributed on either side of the veins (Fisher *et al.*, 1995). Figure 10.4 shows a contour plot of SiO_2/Al_2O_3 on either side of one of the veins, based upon 140 electron microprobe measurements over an area of ~95 mm^2. The arrow indicates the direction of growth of the vein quartz. A marked depletion of silica is shown in the plot, where the wall-rock depletion was greater in a sandy layer than in a shaly layer. For this vein, and for the other vein for which a similar analysis was performed, approximately 80% of the Si in the vein could have been derived from the depleted wall rock (Fisher *et al.*, 1995). Therefore, a model of

275

local, diffusive transport of silica from wall rock to vein could explain most, if not all, of the vein quartz observed.

10.6 Oxygen isotopic measurements

Although local Si transport explains the provenance of Si in the veins, fluids could still be derived from local or long-distance transport. Oxygen isotope analysis of one sample was completed in order to investigate whether vein quartz formed from fluid whose isotopic signature demanded a deeper source. Small (0.5 mg) grains of wall rock and vein quartz were separated from the sample so as to define three transects across the vein. For each transect, a sample was taken in the wall rock along a bedding plane on both sides, and from the vein quartz between wall-rock samples.

Isotopic measurements were completed using *in situ* laser extraction techniques (Sharp, 1992). Grains were oxidized using BrF_5 while each grain was heated with a CO_2 laser. Released O_2 was combusted with graphite catalysed by Pt and the converted CO_2 was analysed using conventional mass spectrometry.

We observed that the average $\partial^{18}O$ for the vein quartz was 17.2‰ vs. SMOW and showed little variability (range 17.0–17.4‰). On average the wall rock ($\partial^{18}O$ range 12.6–14.5‰) on either side of the vein was 3.4‰ lower in $\partial^{18}O$ than the vein quartz.

The lower $\partial^{18}O$ content of the wall rock is due to the fact that, in addition to quartz, wall rock contains chlorite and mica, which tend to concentrate $\partial^{18}O$ less than quartz. If we assume that isotopic equilibrium was established between all phases present, we can use the difference in isotopic composition between the vein and the wall rock to obtain an estimate of the temperature at which the isotope record was established. This requires knowledge of the relative abundances of quartz, mica and chlorite in the wall rock as well as the oxygen isotope fraction among these minerals as a function of temperature. We can write a mass balance for the isotopic composition of the rock, δ_r:

$$\delta_r = x_q * \delta_q + x_m * \delta_m = x_c * \delta_c \qquad (10.1)$$

where x is the mole fraction of oxygen in the rock contained in the minerals quartz *(q)*, mica *(m)* and chlorite *(c)*, δ representing their oxygen isotopic composition.

The isotopic compositions of the minerals are related through fractionation factors. For the quartz–water fractionation we have used the experimentally measured fractionations reported by Matsuhisa *et al.* (1979), and for mica–water, the experimental results of O'Neil and Taylor (1969). In view of the composition of the mica in the wall rock (Fisher *et al.*, 1995), it is worth noting that the experiments showed that the substitution of Na for K does not affect the isotope fractionation between muscovite and water. No experimental work is available for the fractionation between chlorite and water and we have used the empirically derived fractionations of Savin and Lee (1988). The mineral–water fractionations were combined in order to obtain the mica and chlorite isotopic composition (δ_m and δ_c) in terms of that of quartz, δ_q. Substitution in the mass-balance equation above yields:

$$\delta_r(T) = x_q * \delta_q + x_m * [\alpha_{m-q}(T) * (\delta_q + 1000) - 1000] + x_c * [\alpha_{c-q}(T) * (\delta_q + 1000) - 1000]$$

$$(10.2)$$

Here α_{m-q} and α_{c-q} represent the fractionation factors for the muscovite–quartz and chlorite–quartz fractionations, which are a function of the temperature T (K). This expression was used to find the temperature at which the measured quartz isotopic composition (17.2‰), together with the mole fractions of oxygen in quartz ($x_q = 0.21$–0.46), mica ($x_m = 0.24$–0.55) and chlorite ($x_c = 0.24$–0.3) predicted the measured rock isotopic composition (13.8‰). The ranges in oxygen mole fractions were obtained by casting the measured rock compositions to normative compositions of quartz, mica and chlorite.

The temperature range evaluated in this manner was found to be 230–330 °C. On the basis of this temperature range, the measured quartz isotopic composition, and the quartz–water fractionation measured by Matsuhisa *et al.* (1979), we compute the range in isotopic composition of the vein fluid to be 7–11.2‰, well within the range of metamorphic fluid compositions (Sheppard, 1986). We conclude, therefore, that the isotopic composition of the vein quartz is consistent with a model where fracture fluids are in equilibrium with the slaty matrix at temperatures in the vicinity of 280 °C.

10.7 Fluid inclusions

Although compositional evidence documents that most silica was derived from local diffusive transport, and isotope evidence indicates that fracture fluids were in oxygen isotopic equilibrium with the wall rock at 280 °C, fluid in fractures may have been derived either locally or from a distant source. In a further attempt to constrain the provenance of the fluid, we examined the chemistry of fluid inclusions from the Afognak and Kodiak transects. Along both transects, fluid inclusions that we observed in D_2 veins were irregular, negative-crystal, or rounded and elliptical in shape, and ranged in size from <1 µm to ~15 µm. Larger and more abundant inclusions were observed in the euhedral growth portions of the veins compared to the crack-seal portions. Characteristics of inclusions indicated primary rather than secondary entrapment: in some cases inclusions were aligned parallel to euhedral crystal faces, while others were distributed randomly. Inclusions were two-phase (L/V) at room temperature, with a vapour bubble that occupied ~15–30% of the volume of the inclusion.

Along the Afognak transect, we measured homogenization temperatures (T_H) of fluid inclusions in four samples, and temperature of last melting (T_{LM}) in three samples. T_H (±0.1 °C) was measured using the cycling technique of Roedder (1984). Several of the same samples were also cooled to around –300 °C and gradually warmed to check for the presence of CH_4 or CO_2. No changes were observed in the inclusions near the expected melting temperatures for these components, and we concluded that all the observed inclusions are H_2O-rich, with minor or absent CH_4 and CO_2. Because of the small size of inclusions, the temperature of first melting of ice could not be measured. Upon continued warming, ice crystals could be observed. The samples were heated until the ice crystals could no longer be observed, then cooled again to see if the ice crystals grew back. The temperature at which no ice crystals grew back immediately upon cooling (i.e. the vapour bubble decreased in size gradually, rather than decreasing abruptly) was taken as the temperature of last melting (T_{LM}).

Along the Kodiak transect, similar observations indicate the presence of H_2O-rich inclusions (without CO_2 or CH_4) in six of ten samples. However, in contrast to the Afognak transect where only H_2O-rich inclusions were observed, D_2 veins in four of ten samples from the Kodiak transect contained inclusions with varying proportions of CO_2 in addition to the H_2O-rich inclusions. The shapes, sizes, vapour bubble volume proportions, homogenization, and melting temperatures were consistent for all observed H_2O-rich inclusions, regardless of whether CO_2–H_2O inclusions were also present. The H_2O-rich inclusions from the Kodiak transect were similar in every respect to the inclusions from the north-east coast of Afognak Island. There was no evidence for CO_2 or CH_4 in the H_2O-rich inclusions, and the evidence for the simultaneous entrapment of the H_2O-rich and CO_2–H_2O inclusions (where present) was not definitive.

Figure 10.5(a) shows the histogram of homogenization temperatures we obtained for 104 inclusions from the Afognak transect, both within euhedral-growth and discontinuous crack-seal portions of veins. Homogenization temperatures were statistically indistinguishable from texturally different portions of the vein and the data sets were therefore combined. We observed a median homogenization temperature of approximately 159 °C (with a range of 106–333 °C). Several inclusions yielded anomalously high temperatures (>300 °C) that may reflect decrepitation of inclusions, secondary inclusions, or anomalous fluids. Along the Kodiak transect, we observed similar T_H values (124–294 °C) in the six samples containing only H_2O-rich inclusions (Figure 10.5(b)).

Our measured temperatures of last melting from the Afognak transect (Figure 10.5(c)) ranged from –3 to –0.9 °C corresponding to approximately 4.7 to 1.5 equivalent wt% NaCl. Similar values were found along the Kodiak transect (between –5.0 and 0.0 °C, consistent with salinities between 7.9 and 0.0 equivalent wt% NaCl). In a cross plot of T_H vs. T_{LM} for the Afognak transect (Figure 10.5(d)) there appears to be a positive correlation, with the lowest homogenization temperatures corresponding to the lowest temperatures of last melting (highest wt% NaCl). In addition, the four anomalously high homogenization temperatures corresponded to inclusions with the lowest NaCl content.

Combining both the Afognak and Kodiak transects, a total of four of 14 samples contained CO_2–H_2O inclusions. In the CO_2–H_2O inclusions, measurements were made of the melting temperature of CO_2 ($T_m^{CO_2}$) (Figure 10.5(e)), the melting temperature of clathrate ($T_m^{clathrate}$) (Figure 10.5(f)), the melting temperature of the water phase ($T_m^{H_2O}$) (Figure 10.5(g)), and the decrepitation temperature (T_d) (Figure 10.5(h)). There is a depression of the $T_m^{CO_2}$ which is attributed to the presence of 5–15 mol% CH_4 in the CO_2-rich phase. Observed ranges in $T_m^{clathrate}$ in the 10% of CO_2–H_2O inclusions where clathrate was observed are attributed to the presence of dissolved salts in the range of 0.8 to 7.5 equivalent wt% NaCl. However, this is a conservative estimate due to the opposing effects of CH_4 and NaCl. The distribution of CO_2 and H_2O in individual inclusions varies widely within a sample. All four CO_2-containing samples show a strongly peaked distribution with many inclusions containing about 30 mol% CO_2; however, three samples show a weak second peak at ~80 mol% CO_2.

The volumetric proportions of CO_2 suggest that three D_2 samples crystallized from a heterogeneous mixture, and one sample from a homogeneous mixture of CO_2–H_2O. Based on well-constrained evidence from other fluid inclusion studies, the distribution in mol% CO_2 could best be explained by mixing of multiple primary fluids or by pressure and temperature fluctuations during growth (see Myers, 1987). Because of the observed close distribution of all fluid inclusion types within three-dimensional arrays in individual veins, however, multiple primary fluid mixing is unlikely. Possibly, temperature drops or pressure drops of about 100–200 MPa across the CO_2–H_2O solvus caused trapping of dissimilar fluids (note, however, that the critical curve is nearly vertical in P–T space for CO_2-rich fluids). The decrepitation pressure and temperature of one sample from the central belt trapped below the CO_2–H_2O solvus can be used to estimate the pressure and temperature of vein formation: 260 ± 40 MPa (minimum of 10.5 km depth) and 269 °C. This temperature is consistent with the isotopic calculations.

Myers (1987) investigated the fluid inclusion chemistry measured in veins associated with the earlier, lower grade underthrusting regime (D_1). She observed $CH_4 \pm H_2O \pm NaCl \pm CO_2$ fluids in inclusions associated with D_1, as compared to the $H_2O \pm CO_2 \pm NaCl \pm CH_4$ fluids in D_2. These two fluid chemistries exemplify the two most common fluid types found in low-grade metamorphic rocks: methane zone fluids (CH_4–H_2O mixtures with 1 to >90 mol% methane) and water zone fluids (H_2O–CO_2–CH_4–N_2 mixtures with 80 to >99 mol% H_2O, <10 mol% CO_2 and <1 mol% CH_4; Mullis, 1987). These latter fluids typically contain dissolved chlorides and may contain more CH_4 or N_2 if liberated from rocks with concentrated organic matter, such as black shales, or more CO_2 where carbonate is present. Fluids from low-grade metamorphic rocks evolve with increasing temperature and pressure from the methane to the water zone. Mullis (1987) argues that the transition from the methane to the water zone occurs because:

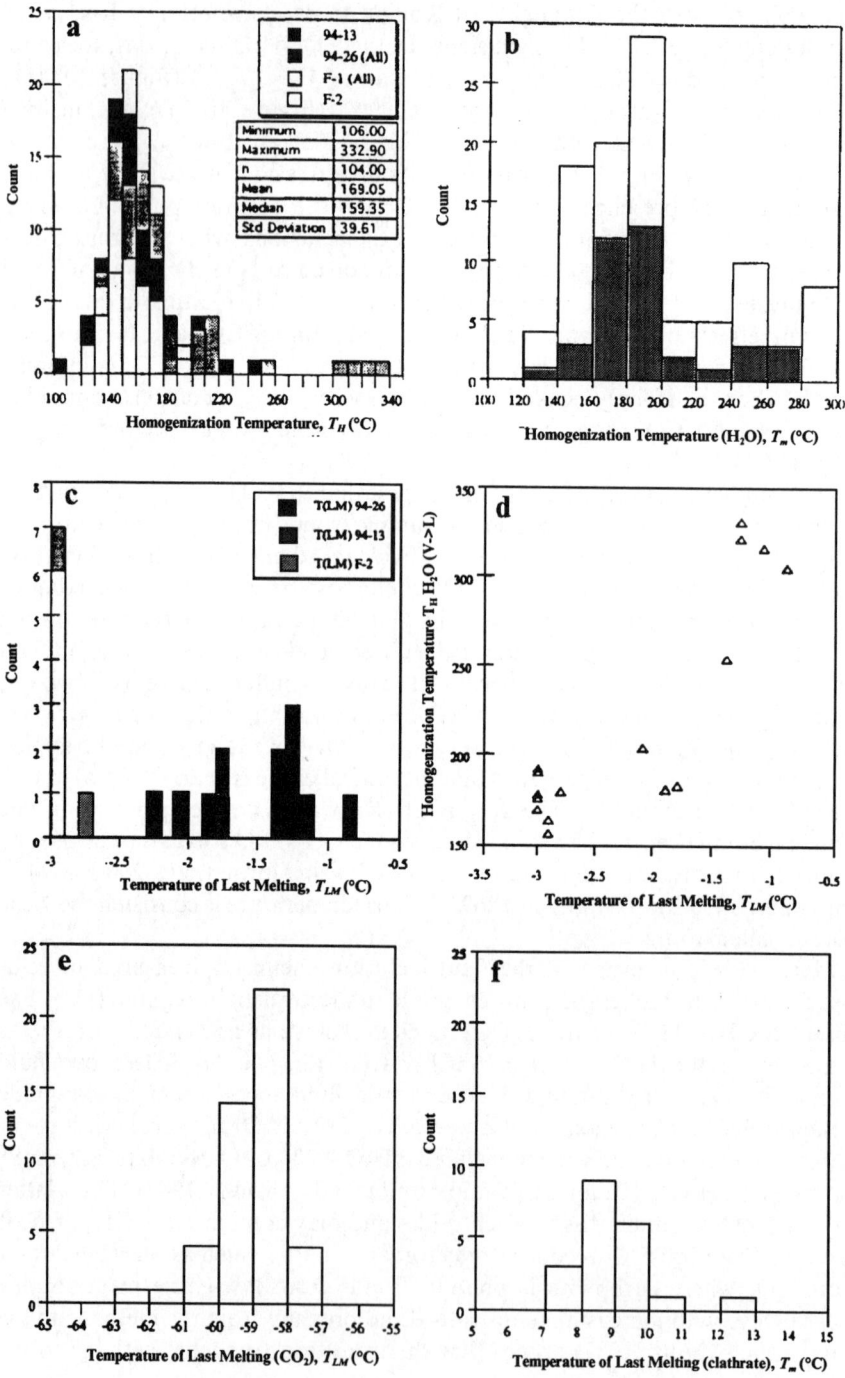

280

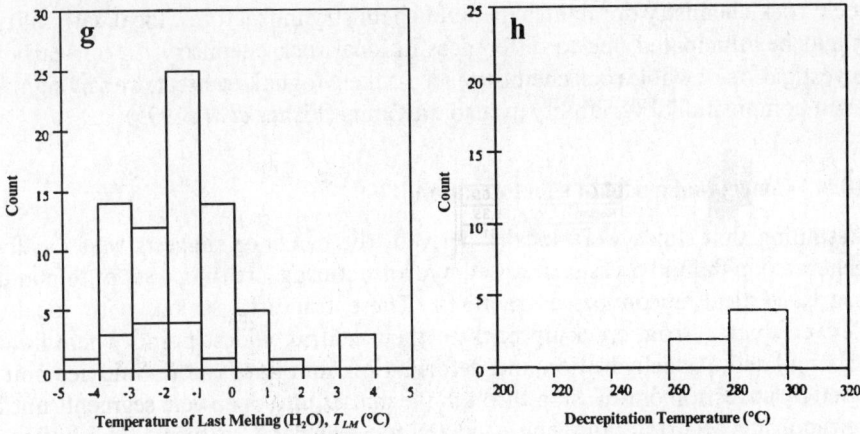

Figure 10.5 Histogram of homogenization temperature (T_H) measurements in four samples from the Afognak transect (94-13, 94-26, F-1, F-2). Inclusions from euhedral growth and crack-seal growth vein sections are all included. (b) Histogram of homogenization temperature (T_H) measurements for H_2O-rich inclusions in samples from the Kodiak transect. Shaded portions represent samples with H_2O- and CO_2-rich inclusions whereas unshaded portions represent samples with only H_2O-rich inclusions. (c) Histogram of temperatures of lasting melting (T_{LM}) measured for three samples (94-13, 94-26, F-2) from the Afognak transect. (d) T_H plotted vs. (T_{LM}) for several inclusions from the Afognak transect. Histograms from the Kodiak transect showing (e) the melting temperature of CO_2 ($T_m^{CO_2}$), (f) the melting temperature of clathrate ($T_m^{clathrate}$), (g) the melting temperature of the water phase ($T_m^{H_2O}$), and (h) the decrepitation temperature (T_d).

- CH_4 production due to thermal cracking of kerogen decreases as grade increases;
- H_2O production increases due to dehydration and water-releasing recrystallization reactions; and
- redox reactions between C–O–H fluids and Fe-silicates or oxides may cause CH_4 oxidation to CO_2. In some cases, CO_2 will precipitate as a carbonate, or decarbonation reactions will release CO_2.

Mullis (1987) also notes that marked differences in CO_2, CH_4, or N_2 content have often been correlated with differences in host rock lithology.

Based upon such observations, the evolution of fluids from $CH_4 \pm H_2O \pm NaCl \pm CO_2$ to $H_2O \pm CO_2 \pm NaCl \pm CH_4$ in the D_1 to D_2 transition probably documents the increasing temperature and pressure of metamorphism. A simple explanation for the variability in NaCl, CO_2, and CH_4 fluid composition observed for veins from Kodiak and Afognak islands is that local dehydration and decarbonation reactions dominated fluid chemistry: if long-distance fluid transport along the decollement had introduced fluids to the system, these fluids would have become well-mixed during transport. In contrast, where

281

local rock chemistry dominates the fluid chemistry in fractures, local variability might be anticipated due to differences in local rock chemistry. In our earlier investigation of whole rock chemistry across the Afognak transect, we saw significant compositional variability in turbidite units (Fisher *et al.*, 1995).

10.8 Conceptual model of vein formation

Assuming that fluids were locally derived, the evidence suggests that locally generated high fluid pressures caused hydrofracturing, creating a set of thin and nearly vertical unconnected fractures. These fractures sealed, with locally derived quartz, from crack-tip back to their central, widest point, where fluid accumulated. As dehydration and deformation continued, more fluid accumulated by advection or diffusion through the matrix into open vein segments until refracturing occurred, allowing a new crack-seal band to form. Each refracturing event would correspond to a pressure pulse within the fluid. If the pulse represented a large pressure drop of 100–200 MPa, it could be related to the varying proportions of CO_2–H_2O observed in the fluid inclusions.

Because of the lack of vertical interconnectivity, resealing of fractures must have repeatedly segregated fluids back into lens-like reservoirs in the central portions of veins (Figure 10.6). Where neighbouring fluid lenses were aligned coincident with the overall stress regime, zones of weakness would have developed that localized strain as brittle–ductile shear zones. Fluid would be stored in

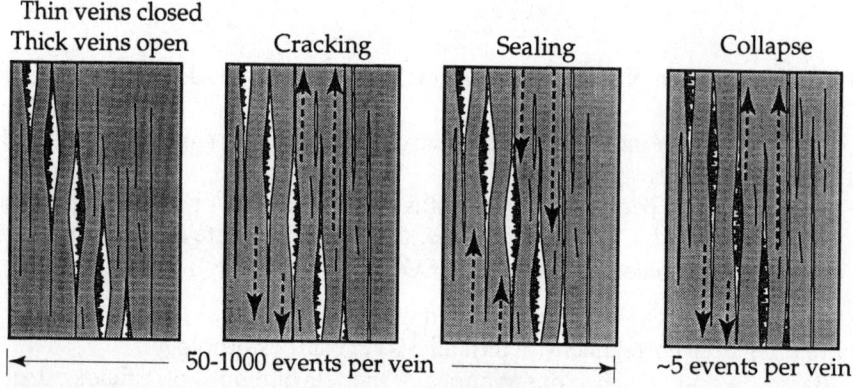

Figure 10.6 Schematic diagram of the fracturing and resealing model. For most of the life of the vein, water is only present in the widest central part – the shear zones. Upon build-up of water pressure, cracking occurs, opening some of the fracture tips (not all tips are open simultaneously or completely). Quartz precipitation occurs from tip back to central zone, leaving a band of quartz (crack-seal band). Crack-seal bands may form anywhere from once (at the tip) to 1000 times (close to the central zone) in subsequent events. Occasionally, perhaps because of transient lateral connectivity between fractures, fluid pressure drops, fluid is expelled, and the vein records a collapse event. Only the infrequent collapse events document the expulsion of fluid out of the system.

these zones at the expense of the smaller fractures and fracture tips between zones. Due to the presence of fluid in these zones, deformation would have occurred faster and would have been localized at these sites. The tips of the fractures would usually be sealed shut, but would frequently open in cracking events related to the growth in volume of the fluid reservoirs due to ongoing dewatering reactions. Cracking would have opened the tips above and below central reservoirs, with migration of fluid into the tips followed by precipitation of a new crack-seal band, thus forcing fluid back into the central reservoirs. The thicker vein segments, representing zones of fluid accumulation, therefore grew thicker faster, because the shear zones allowed vein growth for longer durations.

Only collapse events document expulsion of fluids from the sedimentary package. Such collapse events may have been related to infrequent nucleation of through-going faults in en echelon vein sets (Figure 10.7). These collapse events are also another candidate for pressure pulsations on the order of 100–200 MPa.

10.9 Stochastic model of vein formation

The distribution of vein thickness yields insight into this model relating interaction between fluids and deformation during D_2. The cumulative frequency ($N(L)$) of veins of thickness greater than or equal to L across the Afognak transect describes a power-law distribution with exponent, D, equal to 1.33 (Figure 10.8):

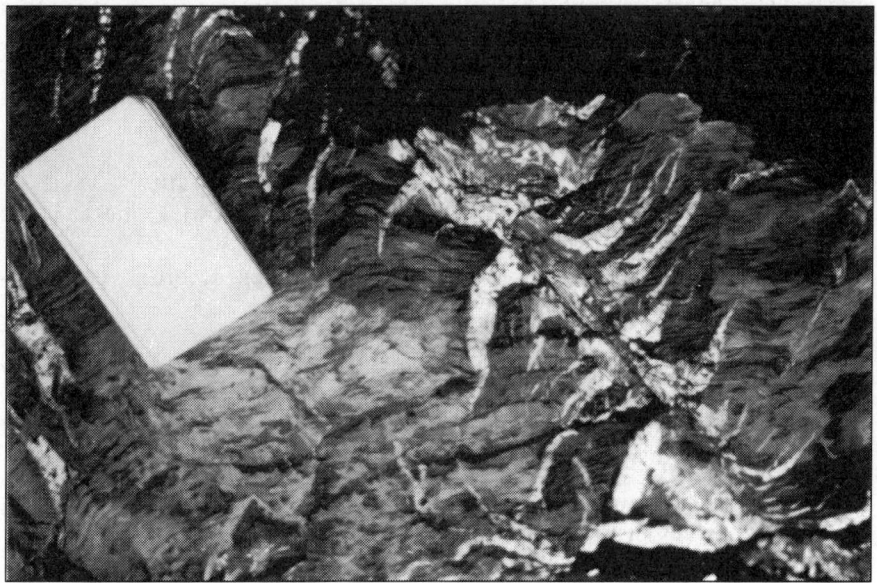

Figure 10.7 Photograph of a fault nucleated through the centre of an en echelon vein array. A number of these faults were observed throughout the Afognak transect.

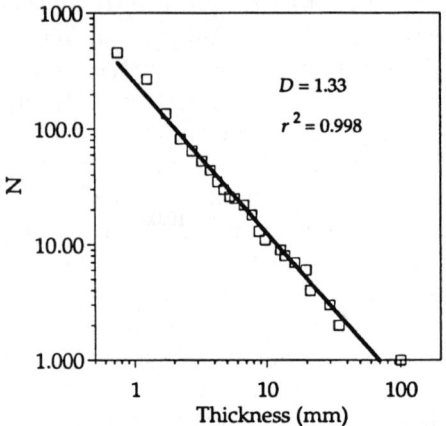

Figure 10.8 Cumulative frequency histogram of vein thickness showing the number of veins (*N*) with thickness greater than or equal to some value. The slope of the plotted regression line is –1.33. The r^2 value of the regression is also indicated.

$$N(L) = 2.44L^{-D} \qquad (10.3)$$

Because veins do not disappear after formation, thickness–frequency distributions of veins must reflect relationships only between birth (fracture opening) and growth (quartz precipitation). We can use stochastic models to show that constant birth and growth rates produce vein distributions which are not power-law, while constant birth coupled with proportional growth rates produce power-law distributions as observed in the Kodiak Formation (Figure 10.8).

For the constant growth model, we assume:

1. veins of minimum thickness L_o are nucleated at a constant rate;
2. the probability *P* of any given vein refracturing and growing thicker is the same (*P* = 1/number of veins); and
3. once a vein refractures, it grows for only one time step, according to zeroth-order kinetics:

$$\frac{\partial L}{\partial t} = k \qquad (10.4)$$

where *k* is a constant and *t* is time. The ratio, α, of the number of birth to refracturing events is constant.

For such a constant growth rate model, the size–frequency distribution produced for about 1000 veins is negative-exponential (Figure 10.9). Presumably, such a size-frequency distribution was generated in the D_2 generation of veins in Kodiak prior to localization of shear zones. Johnston (1993) has also documented a negative-exponential distribution of veins.

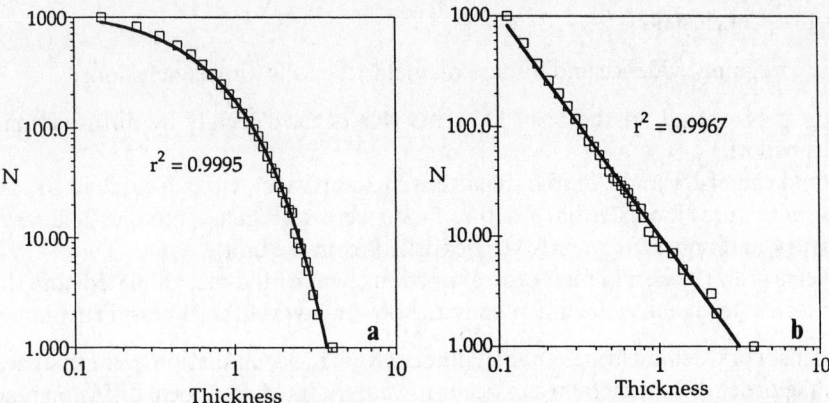

Figure 10.9 Comparison of vein thickness distributions assuming time-averaged growth rate that is (a) independent of size and (b) proportional to size. Size-independent growth results in a negative exponential distribution: cumulative frequencies are (a) concave down in log-log space. Size-dependent growth results in power-law distribution: cumulative frequencies are (b) linear in log-log space.

To generate a power-law vein distribution, we must replace the third assumption of the constant growth model with the following:

3. once a vein refractures, it continues to grow by the zeroth-order growth law during as many time steps as needed until it has grown to some proportion β of its thickness. After refracturing, each vein of thickness L therefore grows to thickness $(1 + \beta)L$.

For these assumptions the stochastic model, after producing about 1000 veins, yields a power-law distribution where the slope D shallows with decreasing α and increasing β (Figure 10.9; Clark *et al.*, 1995). Although more complicated models of time-dependent nucleation and growth undoubtedly can reproduce negative exponential and power-law distributions, the simplest explanation is that negative-exponential distributions document size-independent growth and power-law distributions document proportional growth (Ijiri and Simon, 1977).

Consistent with the model of Beach (1975) for type II en echelon veins, the power-law vein distribution documents a positive feedback system in which the larger a vein grows, the larger it is predicted to grow in the future. The proportional growth rate, related to the accumulation of fluid precisely in the thickest portions of open fractures, explains the development of a power-law size–frequency distribution. Strain is localized (veins grow thickest) in the fracture portions where fluids accumulate – typically the widest section of the fracture. Such a feedback must be unstable if continued, which suggests that localized deformation may become so extreme that en echelon zones eventually fail in some locations. In many cases, faults within the Kodiak Formation appear to have nucleated through en echelon arrays of veins (Figure 10.7).

10.10   Conclusions

The foregoing evidence and discussion yields the following conclusions:

- most, if not all, of the Si in D_2 veins was derived locally by diffusive mass transport;
- fluid chemistry and isotopic signatures are consistent with a local fluid source;
- vein textures indicate that fracture fluids were segregated into lens-like reservoirs, and only infrequently were expelled from the unit;
- veins grew thickest in their central portions where fluids accumulated, and this positive feedback is documented in the power-law vein thickness distribution.

This work demonstrates that significant quartz accumulation (vein thickness on the order of centimetres) can occur in veins as a result of local diffusion reactions, without the through-flow of large fluid volumes. In actively dewatering and decarbonating pelites, local fluid generation, although relatively minor, can facilitate the growth of vein quartz without pervasive or channelized flow of external fluid. Such a model may be useful in explaining other vein sets where the volumetric needs of external fluid models is prohibitively large. In addition, the observed textures and the power-law vein thickness distribution reveals that local accumulation of fluids within a low-permeability unit leads to episodic vein formation and infrequent collapse and fluid outflow. The growth of veins within the system is a runaway process that may lead to fault propagation.

Acknowledgements

We acknowledge the help of Joe Dzvonik, Mark Everett, Bob Burruss, Derrill Kerrick, and Tim Lowenstein. The manuscript benefited from reviews by M. Holness, P. Meere, and I. Main. Support for this research was provided by NSF grant EAR- 9305101 to D.F. and S.B. Fluid inclusion data for the Kodiak transect was collected by G. Myers as part of a Masters thesis at UC Santa Cruz. Support for this research was provided by NSF grants EAR-08337 and EAR-85-08161 to J. Casey Moore.

References

Beach, A. (1975) The geometry of en-echelon vein arrays. *Tectonophysics*, **28**, 245–63.

Bjorlykke, K. (1979) Cementation of sandstones. *J. Sedimentary Petrology*, **49**, 1358–9.

Clark, M.B., Brantley, D.M. and Fisher, D.M. (1995) Power-law vein-thickness distributions and positive feedback in vein growth. *Geology*, **23**, 975–8.

Domenico, P.A. and Schwartz, F.W. (1990) *Physical and Chemical Hydrogeology*, John Wiley & Sons, New York, 824 pp.

Elliott, D. (1973) Diffusion flow laws in metamorphic rocks. *Geol. Society of America Bulletin*, **84**, 2645–64.

REFERENCES

Etheridge, M.A., Wall, V.J. and Vernon, R.H. (1984) High fluid pressures during regional metamorphism and deformation; Implications for mass transport and deformation mechanisms. *J. Geophys. Research*, **89**, 4344–58.

Farver, J.D. and Yund, R.A. (1991) Measurement of oxygen grain-boundary diffusion in natural, fine-grained, quartz aggregates. *Geochim. et Cosmochim. Acta*, **55**, 1597–607.

Farver, J.D. and Yund, R.A. (1992) Oxygen diffusion in a fine-grained quartz aggregate with wetted and nonwetted microstructures. *J. Geophys. Research*, **97**, 14031–54.

Ferry, J.M. and Dipple, G.M. (1991) Fluid flow, mineral reactions, and metasomatism. *Geology*, **19**, 211–14.

Fisher, D.M. and Brantley, S.L. (1992) Models of quartz overgrowth and vein formation: deformation and episodic fluid flow in an ancient subduction zone. *J. Geophys. Research*, **97**, 20043–61.

Fisher, D.M. and Byrne, T. (1990) The character and distribution of mineralized fractures in the Kodiak Forrnation, Alaska: implications for fluid flow in an underthrust sequence. *J. Geophys. Research*, **95**, 9069–80.

Fisher, D.M., and Byrne, T. (1992) Strain variations in an ancient accretionary wedge: implications for forearc evolution, *Tectonics*, **11**, 330–47.

Fisher, D.M., Brantley, S.L., Everett, M. and Dzvonik, J. (1995) Cyclic fluid flow through a regionally extenslve fracture network within the Kodiak accretionary prism. *J. Geophys. Research*, **100**, 12881–94.

Ijiri, Y. and Simon, H.A. (1977) *Skew Distributions and the Sizes of Business Firms*, North-Holland Publishing Co., Amsterdam.

Joesten, R. (1991) Grain-boundary diffusion kinetics in silicate and oxide minerals, in *Diffusion, Atomic Ordering, and Mass Transport* (ed. J. Ganguly), Springer-Verlag, New York, pp. 345–95.

Johnston, J.D. (1993) Three-dimensional geometries of veins and their relationship to folds: Examples from the Carboniferous of eastern Ireland. *Irish J. Earth Sciences*, **12**, 47–63.

Manning, C.E. (1994) Fractal clustering of metamorphic veins. *Geology*, **22**, 335–8.

Matsuhisa, Y., Goldsmith, J.R. and Clayton, R.N. (1979) Oxygen isotope fractionation in the system quartz-albite-anorthite-water. *Geochim. Cosmochim. Acta*, **43**, 1131–40.

Mullis, J. (1987) Fluid inclusion studies during very low-grade metarnorphism, in *Low Temperature Metamorphism* (ed. M. Frey), Blackie, Glasgow, pp. 162–99.

Myers, G. (1987) Fluid expulsion during the underplating of the Kodiak Formation; A fluid inclusion study, Masters thesis, Univ. of California, Santa Cruz.

O'Neil, J.R. and Taylor, H.P. (1969) Oxygen isotope equilibrium between muscovite and water. *J. Geophys. Research*, **74**, 6012–22.

Pollard, D.D. and Segall, P. (1987) Theoretical displacements and stresses near fractures in rock: with applications to faults, joints, veins, dikes, and solution surfaces, in *Fracture Mechanics of Rock* (ed. B.K. Atkinson), Academic Press, London, pp. 277–350.

Ramsay, J. (1980) The crack-seal mechanism of rock deformation. *Nature*, **284**, 135–9.

Roedder, E. (1984) Fluid Inclusions. *Rev. in Mineralogy*, **12**, Mineralogical Society of America, Washington, D.C.

Savin, S.M. and Lee, M. (1988) Isotope studies of phyllosilicates. *Rev. in Mineralogy*, **19**, 189–223.

Sharp, Z.D. (1992) In situ laser microprobe techniques for stable isotope analysis. *Chemical Geology* (Isotope Geoscience Section), **101**, 3–19.

Sheppard, S.M.F. (1986) Stable isotopes in high temperature geological processes, in *Stable Isotopes* (ed. J.W. Valley, H.P. Taylor, Jr. and J.R. O'Neil), Mineralogical Society of America, Washington, D.C., pp. 165–83.

Sibson, R.H. (1994) Crustal stress, faulting and fluid flow in *Geofluids; Origin, Migration and Evolution of Fluids in Sedimentary Basins*, Geological Society Special Publications, **78**, Geological Soc. of London, London, 69–84.

Yardley, B. (1984) Fluid migration and veining in the Connemara Schists, Ireland, in *Fluid-Rock Interactions during Metamorphism* (ed. J.V. Walther and B.J. Wood), Springer-Verlag, New York.

Walther, J.V., and Orville, P.M. (1982) Rates of metamorphism and volatile production and transport in regional metamorphism, *Contrib. Mineral. Petrol.*, **79**, 252–7.

Further reading

Clayton, R.N., O'Neil, J.R. and Mayeda, T.K. (1972) Oxygen isotope exchange between quartz and water. *J. Geophys. Research*, **77**, 3057–67.

Faure, G. (1986) *Principles of Isotope Geology*, 2nd ed, John Wiley & Sons, New York.

James, A.T. and Baker, D.R. (1976) Oxygen isotope exchange between illite and water at 22 °C, *Geochim. et Cosmochim. Acta*, **40**, 235–9.

Matsuhisa, Y., Goldsmith, J.R. and Clayton, R.N. (1978) Mechanisms of hydrothermal crystallization of quartz at 250°C and 15 kbar. *Geochim. Cosmochim. Acta*, **42**, 173–82.

Matthews, A., Goldsmith, J.R., and Clayton, R.N. (1983a) Oxygen isotope fractionation involving pyroxenes: the calibration of mineral-pair geothermometers. *Geochim. Cosmochim. Acta*, **47**, 631–44.

Matthews, A., Goldsmith, J.R., and Clayton, R.N. (1983b) Oxygen isotope fractionation between zoisite and water. *Geochim. Cosmochim. Acta*, **47**, 645–54.

Norris, R.J. and Henley, R.W. (1976) Dewatering of a metamorphic pile. *Geology*, **4**, 333–6.

Norton, D. and Knapp, R. (1977) Transport phenomena in hydrothermal systems: The nature of porosity. *American Journal of Science*, **277**, 913–36.

Zhang, L., Jingxiu, L., Huanbo, Z. and Zhensheng, C. (1989) Oxygen isotope fractionation in the quartz-water-salt system. *Economic Geology*, **84**, 1643–50.

CHAPTER ELEVEN

Fluid flow in fractured rocks at shallow levels in the Earth's crust: an overview

Noelle E. Odling

11.1 Introduction

Fractures are common throughout the upper parts of the Earth's crust and exist on a wide range of scales from microfractures to crustal-scale features. It is well known that fractures significantly affect the movement of fluids throughout this scale range. The impact of fractures on fluid flow has a number of practical applications that have motivated research in this field. These include disposal of hazardous wastes underground, exploitation of geothermal energy, location of ore bodies, earthquake dynamics, and migration and entrapment of hydrocarbons. As a result, the field has attracted people from a wide range of backgrounds including geologists, geophysicists, hydrologists, engineers, chemists, physicists, mathematicians and statisticians. In the following, a review is given of the main features and current trends within the general topic of fluid flow in fractured rocks. The main emphasis is placed on geological aspects, although an attempt has been made to include results and concepts from other fields.

In this chapter, the term 'fracture' is assumed to include any discontinuity within a rock mass which developed as a response to stress, and therefore includes shear, tension and hybrid features (e.g. faults, joints, deformation bands). The effect that fractures have on fluid flow depends on the hydraulic properties of the fractures themselves, on the properties of the fracture system as a whole, and on permeability of the rock matrix. The hydrological properties of individual fractures vary according to their deformation mechanism, host rock properties and geological history. Mismatch of fracture walls in shear fractures (e.g. faults) can produce high conductivities, while generation of crushed material can give very low conductivities. Tension fractures (e.g. joints), although lacking significant displacement, can provide important flow conduits especially

Deformation-enhanced Fluid Transport in the Earth's Crust and Mantle. Edited by M.B. Holness. Published in 1997 by Chapman & Hall, London. ISBN 0 412 75290 5.

at shallow levels. In mechanically strong rocks (most metamorphic and magmatic lithologies), fracture planes can have networks of open voids, whereas in weak rocks (e.g. poorly indurated clastic rocks), fractures can appear as tabular bodies of modified host rock, either more or less permeable than the host rock itself. The hydraulic behaviour of fractures can also be modified by the fluids they conduct, e.g. by the physical effects of fluid pressure, mineral precipitation, host rock alteration and tectonic reactivation.

For bulk hydraulic properties of rock masses, the extent to which fractures are linked to form continuous pathways through the rock, i.e. the fracture system's connectivity, is an important controlling factor. In impermeable rocks, connected systems of fractures often form the only significant routes for fluid flow and contaminant transport. This is of special relevance to the disposal of hazardous wastes in hard rock terrains, where safe and reliable containment over long time periods is important. In permeable rocks, open fractures can dominate the permeability by providing fast routes for fluid flow while the matrix provides fluid storage. Recovery of hydrocarbons from such rocks requires careful well placement and management of pumping rates. A connected system of fractures which form flow barriers (through mineral filling, diagenesis or compaction) can significantly reduce the bulk permeability of an otherwise high-quality aquifer or reservoir. Large-scale fractures, such as faults and master joints, are composed of networks of small-scale fractures. Thus connectivity plays an important role in the nature of fluid flow on a wide range of scales.

It is clear that the influence of fractures on fluid flow can vary over a wide range, and depends on the tectonic setting, host rock type, pressure and temperature conditions, deformation history and fluid properties. In the following, fluid flow in fractured rock masses is discussed in terms of scale (single fractures and fracture systems), in terms of deformation type (faults and more widespread fracture systems), and in terms of lithology (strong impermeable rocks and weak permeable rocks). First the hydraulic properties of individual fractures are described. Then the role of fracture system connectivity is discussed and followed by a description of flow in firstly fault zones and secondly more pervasively fractured rock masses. Within each section, the role of two 'end member' lithologies is described. These are, first, mechanically strong impermeable crystalline rocks, and secondly, mechanically weak rocks, represented by poorly lithified clastic sediments. Finally, a brief summary of modelling techniques relevant to the field is given.

11.2 Flow in single fractures

Host rock properties exert a strong control on the nature and hydraulic behaviour of single fractures. Brittle failure in highly indurated, low porosity rock types (most metamorphic and magmatic rocks) generate fractures with void networks within the fracture planes. Since these rocks are generally of low permeability, such fractures often form significant pathways for fluid migration.

In the absence of filling materials, the potential for such fractures (joints and small faults) to conduct fluids is controlled by the roughness of the fracture walls and their physical separation.

11.2.1   Fracture surface roughness

The geologist's common perception of fracture roughness refers to features of the fracture surface on the scale of a few centimetres. In general, this varies with rock type and texture (fine-grained rocks – smooth fractures; coarse-grained rocks – rough fractures). A number of methods have been used to quantitatively characterize fracture surface roughness. In the field of rock mechanics, 'joint roughness coefficient' (JRC), based on the friction characteristics of fracture surfaces, is commonly used (e.g. Barton and Choubey, 1977). Much detailed experimental work has led to the development of empirical formulae describing how the mechanical and hydraulic behaviour of fractures depends on JRC (e.g. Barton and Choubey, 1977; Bandis, 1980; Bandis *et al.*, 1983; Barton *et al.*, 1985).

Other methods of characterizing fracture roughness are directed at quantitative descriptions of fracture surface geometry. These include the application of statistical and geostatistical methods. Asperity height distributions become broader, asperity slopes increase, and correlation distances increase, as fracture surface roughness increases (Tse and Cruden, 1979; Iwano, 1990; Genter and Riss, 1990). Brown and Scholz (1985) found that asperity heights for natural fractures increase rapidly with scale, in contrast with ground surfaces where asperity heights approach a limiting value at large scales. This led them to suggest that fracture surfaces may be fractal, a concept which has gained increasing acceptance recently. Fracture profiles have, in many cases, been shown to closely approximate self-affine fractal curves, where asperity height to width ratio decreases with length scale. This is consistent with a feature of most fracture surfaces that is well known to geologists, i.e. that they tend to look rough under the microscope but smooth on outcrop scale.

Analysis of natural fracture surfaces has yielded a range of fractal dimensions from 1.0 to 1.7 (Brown and Scholz, 1985; Carr, 1989; Power and Tullis, 1991; Huang *et al.*, 1992; Poon *et al.*, 1992; Odling, 1994). The fractal dimension indicates the rate at which the profile becomes smoother with increasing scale, where a fractal dimension of one corresponds to self-similarity with no smoothing, and 1.7 to self-affinity with pronounced smoothing. Odling (1994) suggests that rough fractures (high JRC) have fractal dimensions close to 1 (self-similar), while smooth fractures have dimensions greater than 1.0 (self-affine). These fractal characteristics may also have superimposed on them undulation of specific wavelengths, caused by, for example, plume structures or a dominant grain size (Odling, 1994). The fractal behaviour of fracture surfaces seems to extend from the scale of micrometres, to an upper limit in the region of 10 cm to 1 m (Brown and Scholz, 1985; Power and Tullis, 1991; Huang *et al.*, 1992). Fractal models of fracture surfaces have become increasing widely used in modelling studies of flow through open, rough fractures (e.g. Brown, 1987; Thompson and Brown, 1991).

11.2.2   The nature of fracture aperture space

Under normal conditions in the Earth's crust, overburden creates compressive normal stresses which keep fracture walls in contact at points in the fracture plane, except when fluid pressures are high enough to overcome lithostatic stresses. Experimental and field evidence suggest that once a fracture is formed, it remains open to flow despite compressive stress, unless it is filled with mineral deposits or crushed rock. This is due to mismatch between opposing fracture walls caused by slip (e.g. Brown, 1995) and by deformation and chemical processes occurring at the time of fracture formation (e.g. Hakami, 1995). Numerical model studies (Odling, 1994) have also shown that only very small differential strains of opposing walls (around 0.005%), that could be caused by rock heterogeneity, are required to generate significant apertures.

Under compressional normal pressure, the imperfect fit of fracture walls creates a complex distribution of contacting and open areas which have significant effects on fluid flow. Studies of fracture aperture space geometry indicate that areas of contact form islands surrounded by fracture voids (Nolte et al., 1989; Hakami and Barton, 1990; Vickers et al., 1992; see Figure 11.1), and that the nature of contact distribution is dependent on fracture wall roughness. Rough fractures tend to show fewer larger islands separated by large channels, while smooth fractures show many small islands more evenly distributed throughout the fracture plane (Bandis, 1980; Hakami and Barton, 1990). Experimental and numerical investigations indicate that apertures within a fracture plane approximate lognormal (Tsang and Witherspoon, 1983; Hakami and Barton, 1990) or normal distributions (Brown, 1995), and that rough, mismatched fractures show a larger mode and a broader distribution of apertures than smooth, well-mated fractures.

Percolation studies and fractal analysis has shown that the geometry formed by void and contact areas approximates a self-similar fractal (Nolte et al., 1989; Pyrak-Nolte et al., 1992). This might be expected from the fractal nature of fracture surfaces and the fact that a cut through a self-affine fractal surface parallel to the surface trend, produces a self-similar pattern of islands (Mandelbrot, 1985; Voss, 1988). A fractal model for fracture aperture space is also consistent with results from geostatistical analysis where correlation lengths are commonly of the same order as the sample size, for small and large samples alike (Tsang and Witherspoon, 1983; Vickers et al., 1992; Keller et al., 1995). Fractal distributions are correlated over all lengths and the correlation lengths found by the geostatistical analysis may therefore represent finite sample size effects.

11.2.3   The 'cubic' law, flow channelling and fracture apertures

The traditional view of a fracture, for the purposes of assessing hydraulic prop-

Figure 11.1 Microphotograph of an area some 40 mm across in the plane of a fracture in quartz monzonite showing voids (white) and contact regions (black). The fracture was subjected to 85 MPa normal pressure. Contact areas form fractal islands with complex outlines. Void areas occupy more than 50% of the fracture surface and form continuous pathways for fluid flow through the fractures. (After Pyrak-Nolte *et al.*, 1987.)

erties, is that of a parallel-walled channel. For such a channel, the flow field is simple to describe mathematically, and the volumetric flow, Q, is proportional to the cube of the separation of the channel walls, b, the so-called 'cubic' law (Snow, 1965):

$$Q = \frac{b^3}{12\mu} \frac{\mathrm{d}P}{\mathrm{d}l} \tag{11.1}$$

where μ is the dynamic viscosity of the fluid and $\mathrm{d}P/\mathrm{d}l$ is the pressure gradient. Such a simple conceptual model for the flow in a fracture makes interpretation and analysis of field data feasible. However, in reality fracture aperture space is highly variable, which causes flow channelling – that is, flow is not evenly distributed throughout the fracture plane but is concentrated into channels resembling a braided river system (Tsang and Tsang, 1989; Hakami and Barton, 1990). The discrepancy between the smooth-walled channel model and nature has led to the concept of the effective hydraulic aperture (e.g. Witherspoon *et al.*, 1980; Barton *et al.*, 1985). This is the aperture of the equivalent smooth-walled channel that gives rise to the same average flow for a given pressure gradient. Mechanical apertures (i.e. physical separation of the fracture walls) and effective hydraulic apertures are similar when fracture walls are far apart (not in contact), but deviate in an increasingly non-linear fashion as contact area increases. When fracture walls are close, hydraulic apertures are smaller than mechanical apertures (Barton *et al.*, 1985; Pyrak-Nolte *et al.*, 1987; Raven and Gale, 1985). Barton *et al.* (1985) and Hakami and Barton (1990) report ratios of mechanical to hydraulic aperture between 1 and 16 for mechanical apertures up to a few hundreds of micrometres.

Estimates of apertures from borehole or laboratory flow rates vary from a few to several hundred micrometres (over three orders of magnitude), and are commonly log-normally distributed (e.g. Snow, 1970; Davidson *et al.*, 1982; Barton *et al.*, 1985; Nelson, 1985; Raven and Gale, 1985; Herbert *et al.*, 1992). *In situ* fracture conductivity has also been estimated using tracers injected into one borehole and recorded at another. However, aperture estimates from tracer tests can exceed those from flow tests by as much as two orders of magnitude. This is due to the often preferential siting of wells, from practical considerations, in high flow channels within the plane of the fracture. Moreno *et al.* (1988) note that it is impossible to find an equivalent parallel plate aperture (i.e. an effective hydraulic aperture) that is consistent with both flow and contaminant transport rates, and that solute transport rates are highly dependent on injection and observation hole locations (Moreno *et al.*, 1990). Flow channelling also enhances solute dispersion and Novakowski *et al.* (1995) report lateral dispersion of 4 m over a distance of 15 m in a fracture in granite. Moreno *et al.* (1990) conclude that to characterize the hydraulic properties of fractures using tracer tests, many boreholes are advisable so that the effects of local variations can be averaged. Flow channelling can also lead to reduced rock–solute interaction (Moreno *et al.*, 1988), and modified relative permeabilities under two-phase flow (Glass and Nicholl, 1995).

11.2.4   The effect of normal stress on open fractures

Evidence from borehole pump tests in crystalline rocks and hydrocarbon reservoirs (Snow, 1970; Carlson and Olsen, 1977; Nelson, 1985) and laboratory experiments (e.g. Bandis, 1980; Barton *et al.*, 1985) shows that fracture conductivity is reduced by stress normal to the fracture plane. As normal stress rises, total contact area increases and the average aperture decreases (Bandis, 1980; Barton *et al.*, 1985; Pyrak-Nolte *et al.*, 1987; Brown, 1995). However, natural fractures can remain open to flow under pressures as high as 200 MPa (Krans *et al.*, 1979; Engelder and Scholz, 1981), and contact area rarely exceeds 50% (Bandis, 1980; Nolte *et al.*, 1989). This is supported by observations of conducting fractures in deep mines and deep boreholes (Leary, 1991).

Both experimental and theoretical studies suggest a log-linear relationship between normal stress, σ, and fracture closure, d, for fractures whose walls are in contact (e.g. Bandis *et al.*, 1983; Brown and Scholz, 1986):

$$\sigma_n = a + b \log(d) \qquad (11.2)$$

where a and b are constants depending on rock properties. Fracture closure is most marked below 5 MPa (Brace, 1978; Raven and Gale, 1985). At low stress levels, contact regions are few, so that stress concentrations are high, and deformation of contact regions large. At high stress levels, any stress increase is distributed over larger and more numerous contacts, so that closure rate is reduced. Laboratory experiments of varying normal stress on fracture closure display marked hysteresis, i.e. closure is not fully reversible due to permanent alteration of the fracture walls (Bandis *et al.*, 1983; Raven and Gale, 1985). This suggests that joint deformation, and therefore hydraulic conductivity, can also depend on the previous stress history (Bandis *et al.*, 1983).

11.2.5 Fractures as permeability barriers

Fracturing of low permeability rocks (crystalline and highly indurated sedimentary rocks) produce fractures that are, at some point in their history, more permeable than the surrounding rock. By contrast, shear fractures in poorly indurated, highly porous and permeable rocks are often less permeable than the host rock and can act as flow barriers. Such fractures are most common in porous, clastic rocks and are of special interest to the hydrocarbon industry, for they play an important role in fault seal and reservoir quality.

Fractures of a few millimetres to centimetres in porous rocks are variously termed shear zones, deformation bands, granulation seams, cataclastic bands and slip bands. Their permeability with respect to the host rock is dependent on the fluid and confining pressures and host rock properties at the time of deformation (e.g. Groshong, 1988; Knipe, 1993; Antonellini and Aydin, 1994). Higher confining pressures, normal fluid pressure regimes, and a lower porosity host rock promote cataclasis (resulting in grain size reduction) and compaction, causing reduced porosity and permeability. High porosity rock types, low confining pressures or abnormally high fluid pressures favour the development of dilatant deformation bands with a net increase in porosity and permeability (Antonellini and Aydin, 1994; Fowles and Burley, 1994; Knipe, 1992). Such dilatant bands are, however, commonly the site of later mineral precipitation reducing their permeability (Antonellini and Aydin, 1994; Knipe, 1992). In these various types of fracture, permeability can be as much as four orders of magnitude lower than the host rock (Antonellini and Aydin, 1994; Fowles and Burley, 1994).

As shear displacement increases to some tens of metres, deformation bands grade into slip bands and small faults in which cataclastic or gouge zones up to a few centimetres wide are developed. Permeability of these zones can be up to seven orders of magnitude lower than in the host rock (Antonellini and Aydin,

1994). Studies of fault gouge development indicate repeated cycles of fabric development with shear displacement (Logan, 1991) where breccia and cataclasis development alternate with fracture propagation parallel to the gouge zone boundaries (Logan, 1991; Antonellini and Aydin, 1994). This fabric development leads to highly anisotropic permeability, with very low permeability perpendicular to slip bands, but enhanced permeabilities parallel to slip bands.

11.3 Flow and fracture system connectivity

On scales larger than the individual fracture, the nature of fracture distribution through the rock mass becomes important. In rocks of upper regions of the Earth's crust, joint systems are extremely common and in low permeability rocks they form the dominant flow paths for groundwater movement. Large-scale fault zones are in detail composed of complex networks of small-scale fractures. Fluid flow through fractured rock masses depends on the hydraulic conductivity of individual fractures (joints, small faults) and on the degree to which fractures form continuous pathways or barriers through the rock mass, i.e. on the connectivity of the fracture system.

Connectivity has been extensively investigated within the field of percolation theory, a branch of statistical physics, and many of the basic concepts are relevant and useful to the question of fluid flow through rock masses (Robinson, 1984; Balberg *et al.*, 1991; Berkowitz, 1995). In systems composed of objects distributed in space, connectivity depends on how the objects interact. In the context of fractured rock masses, the objects are 2-D fracture planes in 3-D space, or 1-D fracture traces in 2-D space, as illustrated in Figure 11.2. As the number of objects increases, the chance of intersection between two objects increases. In so doing, the objects form 'clusters', i.e. groups of objects that are in communication with each other. As the number of objects continues to grow these clusters increase in size and eventually a single cluster becomes large enough to cover the entire region; see Figure 11.2. In theory, a stage is reached at which one cluster becomes infinitely large. This stage is known as the 'percolation threshold' and the infinitely large cluster as the 'spanning cluster' (Stauffer, 1987). The 'conductivity' of the system close to the threshold behaves in a typically 'critical' fashion (see Figure 11.2), i.e. large changes in system conductivity occur for small changes in object density (Stauffer, 1987).

The concept of the infinite cluster is, of course, of limited practical use, since in nature volumes without limit never occur. Here it is necessary to identify the area of interest, e.g. the extent of an oil reservoir or aquifer, the size of a lithological unit, a fault-bounded block, or the region encompassed by a number of boreholes. The concept of the percolation threshold and the nature of changes in conductivity at the threshold also apply to finite volumes, but here the spanning cluster is defined as the cluster that connects opposite sides of the volume of interest.

Thought of in terms of open fractures in an impermeable rock matrix, the

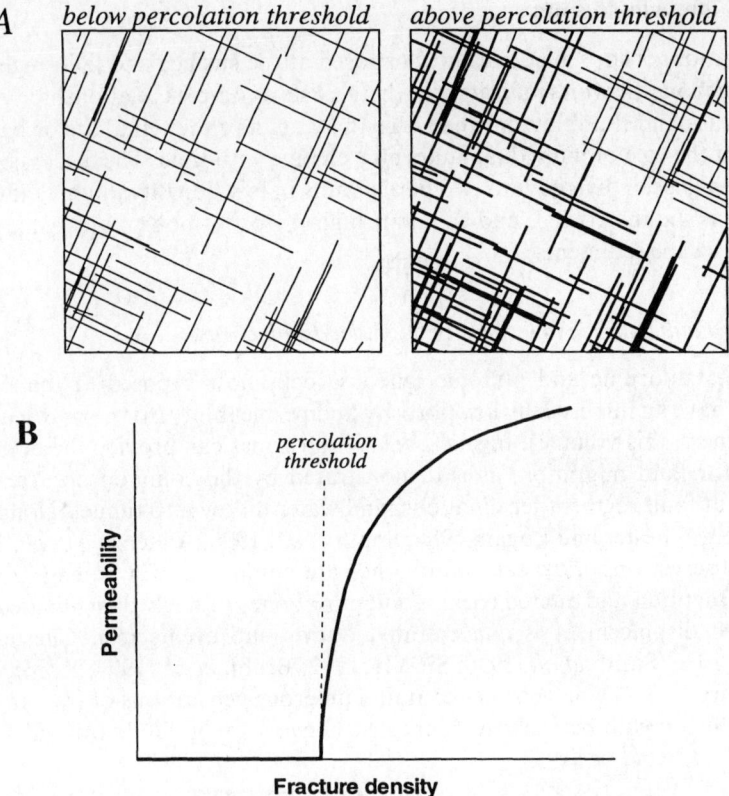

Figure 11.2 Diagram illustrating the effect of fracture density on connectivity. (a) Left – at low fracture densities, fracture traces form small clusters of linked fractures. Right – with increasing fracture density, the percolation threshold is reached and one cluster becomes large enough to span the sample region (indicated by the heavy lines). (b) In the case where the matrix is impermeable and the fracture open to flow, the threshold corresponds to a sharp increase in permeability of the combined rock matrix–fracture system.

percolation threshold is the point at which fluid flow across a finite region becomes possible, and the conductivity of the system is equivalent to permeability of the entire rock matrix–fracture system. At the threshold, large changes in permeability occur for small changes in fracture density. Further addition of open fractures enhances permeability, but at a slow rate compared to changes occurring at the threshold (see Figure 11.2). A corresponding picture exists for fractures acting as flow barriers in a permeable rock matrix. At the threshold in this system, the permeability (initially high) decreases dramatically and, where the fractures allow no flow (e.g. when filled with mineral deposits), becomes zero.

297

11.4   Flow in fault zones

Observations from fault zones now exposed at the surface and fluid regimes in hydrocarbon reservoirs indicate that there exists a complex relationship between fault zones and fluid flow in which they may act as either conduits or barriers, and that the role of individual faults may change with time. The discussion that follows has been divided into two parts: faults in low porosity, highly indurated rocks (crystalline rocks), and faults in high porosity, poorly indurated rocks (largely clastic sediments).

11.4.1   Fault zones in low porosity, highly indurated rocks

Most metamorphic and plutonic igneous rocks now exposed at the Earth's surface have an intrinsically low porosity and permeability (for more discussion, see Holness, this volume). In such rocks, fault zones can provide the dominant routes for fluid migration, as is demonstrated by the common occurrence of springs at fault segment terminations, and water inflow into tunnels along fault zones (e.g. Chester and Logan, 1986; Smith *et al.*, 1990; Andersson *et al.*, 1991). Field observations show that fault zones are composed of a damage zone of highly fractured and altered rocks, containing zones of finely disseminated rock in which displacement is concentrated during slip events (e.g. Chester and Logan, 1986; Smith *et al.*, 1990; Sibson, 1992; Bruhn *et al.*, 1994; Knipe, 1993; see Figure 11.3). Damage zones contain numerous generations of fractures and veins which create pervasively connected networks (e.g. Goddard and Evans,

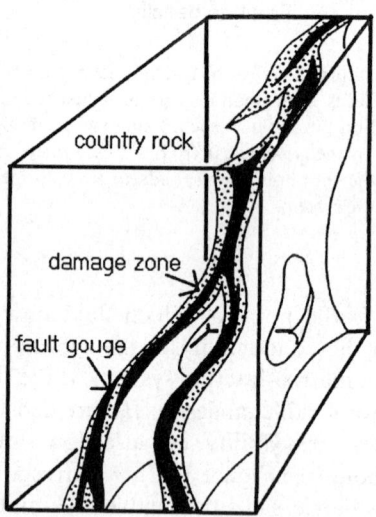

Figure 11.3 Schematic representation of a fault zone. Discontinuous lenses and layers of gouge (crushed country rock) are contained within a broader damage zone of highly fractured rock. Holes represent places where slip has caused little change to the country rock. (After Smith *et al.*, 1990.)

298

1995; Cox, 1995), commonly well above the percolation threshold. Fractures show evidence of rapid, unstable fracture propagation including bifurcating fracture tips, injected crush material and shattered wall rock (Bruhn *et al.*, 1994). The presence of several generations of fracturing indicates that cycles of rupture are followed by fracture healing, with associated changes in permeability (Sibson, 1987; Smith *et al.*,1990; Sleep and Blanpied, 1992; Knipe, 1993; Bruhn *et al.*, 1994). Slip zones, where displacement is concentrated, are composed of braided arrays of slip surfaces with discontinuous lenses of breccia and cataclasite (Bruhn *et al.*, 1994; Goddard and Evans, 1995). Damage zones can be up to 200 m thick and slip zones up to 20 m wide, for faults with displacements of a few kilometres (Bruhn *et al.*, 1994).

Knipe (1993) suggests three stages in a permeability cycle of a fault zone; see Figure 11.4. During a pre-seismic stage, stress build-up results in microfracture formation, dilation, increased permeability and net flow into the fault zone. Observations on fluid flow and *in situ* stress indicate that critically stressed faults show larger hydraulic conductivities than others (Barton *et al.*, 1995). On rupture, stress release causes a collapse of the pre-seismic dilation, generating a net expulsion of fluids from the fault zone (seismic pumping: Sibson, 1987; Byerlee, 1993). Aftershocks serve to preserve the enhanced permeability generated during rupture. During the post-seismic stage, migrating fluids promote mineral precipitation and a decrease in permeability until the approach of another rupture event.

Permeability in damage zones can vary over many orders of magnitude. Davidson and Kozac (1988) measured *in situ* permeability in a thrust fault in

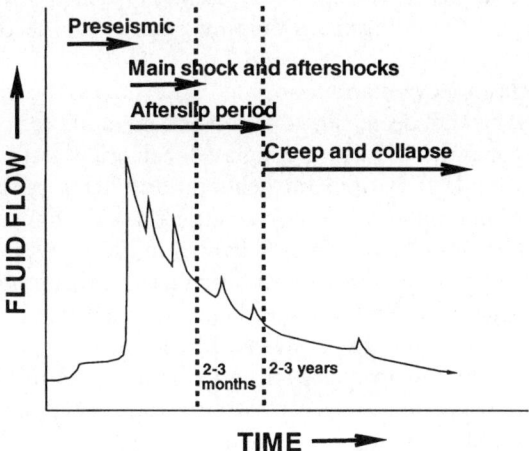

Figure 11.4 Idealized evolution of permeability within an active fault zone. Permeability is enhanced during the pre-seismic period, with a dramatic increase at the time of the main shock. Permeabilities remain high for the period of aftershock activity (2–3 months) but decrease gradually with time. (After Knipe, 1993.)

granite over a range of 10^{-11} to 10^{-16} m^2. Permeabilities may, however, be substantially higher over short periods if fluid pressures are high enough to prop fractures open. The predominance of fractures parallel to the fault zone can result in local anisotropy with permeabilities along the fault some three to four times higher than across it (Smith *et al.*, 1990; Bruhn et *al.*, 1994). Cataclastic rocks associated with slip zones generally have low permeabilities, due to fine grain size (Logan, 1991) and the presence of clay minerals. However, fracturing parallel to the gouge zone during rupture (Logan, 1991) can create significant fluid pathways parallel to the slip zone, if they are sufficiently well connected. On a larger scale, the occurrence of lens-like bodies of low permeability cataclastic rocks suggests that flow may be channelized within fault zones (Smith *et al.*, 1990).

Evidence from fluid inclusions (Bruhn *et al.*, 1994; Robert *et al.*, 1995) indicates that periods when fluids are connected to the surface (hydrostatically pressured) alternate with periods when volumes of fluid-bearing rock are isolated (near lithostatic pressures). Field observations suggest that such hydraulically isolated pods are typically up to tens of metres wide and hundreds of metres long. Fault zones are composed of a number of fault segments not generally active at the same time. At any one time, therefore, different parts of a fault zone may be at different stages in the permeability cycle and fluid pressures may vary greatly (Smith *et al.*, 1990; Evans and Chester, 1995). It is also suggested that abnormally highly pressured volumes and hydraulic fracturing can develop from frictional heating and thermal expansion of fluids and that this may play an important part in destabilizing segments of fault zones and triggering earthquakes (Mase and Smith, 1987; Smith *et al.*, 1990; Byerlee, 1993; Bruhn *et al.*, 1994; Blanpied *et al.*, 1995; Sleep, 1995). Direct evidence of this comes from induced earthquakes resulting from fluid injection down deep boreholes (e.g. Healy *et al.*, 1968).

Following a rupture event and associated aftershocks, the high permeability regime collapses through decreasing dilation and mineral precipitation. Sources of fluid in fault zones include down-circulated meteoric waters, trapped formation brines, and fluids from mineral dehydration during prograde metamorphism (Kerrick *et al.*, 1984). Fluid compositions can vary greatly, both within and among fault zones (Parry, 1994). Changes in fluid pressure and temperature, and mixing of fluid bodies with differing chemical composition, promote precipitation of sealing minerals (Knipe, 1993). Rupture events are often associated with rapid drops in fluid pressures which may also trigger mineral deposition (Smith *et al.*, 1990; Knipe, 1993). Fluid migration can result from pressure gradients created by the faulting process itself. Irregularities in fault zone trends result in areas of relative compression which act as fluid sources, and areas of relative extension which act as fluid sinks (Smith *et al.*, 1990; Knipe, 1993). During times of enhanced permeability faults can function as release valves for deep overpressured regions developed in, for example, thrust belts (McCaig, 1989; Bruhn *et al.*, 1994).

The hydrothermal processes that lead to reductions in fault zone permeability and sealing are rapid compared to the 100 to 10 000 year time intervals between major earthquakes (Hickman *et al.*, 1995). Knipe (1993) has suggested that the time period of fault zone permeability enhancement is roughly equivalent to the period of aftershock activity (a few months); see Figure 11.14. The influence of fluid flow in fault zones in the form of rock alteration and mineral deposition are indications that very large volumes of fluid are transported in geologically very brief time periods.

11.4.2 *Fault zones in high porosity, poorly indurated rocks*

Flow rates and paths around fault zones in significantly permeable rocks depend on complex interactions between host rock and fault zone. Fault zones may act as either barriers or conduits for flow, and their behaviour may vary both along the fault and with time. Many hydrocarbon reservoirs are clastic in origin and are poorly indurated and highly permeable. Much of the research into the nature of fault zones in permeable rocks has been motivated by hydrocarbon exploration and production.

Fault zones in poorly indurated clastic rocks are composed of complex arrays of individual slip surfaces, contained in a damage zone of smaller-scale expressions of deformation such as deformation bands, fractures and folds. Within damage zones, systems of subparallel and anastamosing deformation bands form a geometry found on scales from centimetres to hundreds of metres (Antonellini and Aydin, 1995). In sandstones, correlations between the density of deformation bands and proximity to a major slip surfaces have been observed in field examples (Jamieson and Stearns, 1982; Antonellini and Aydin, 1994).

Fault-sealing behaviour depends on properties of the juxtaposed rocks and fault zone material, the pressure regime and fluid properties. The sealing capacity of a fault zone to hydrocarbons depends on the 'displacement' pressure of the juxtaposed rocks and/or the fault material, and capillary pressures. Displacement pressure is defined as the pressure difference required to force hydrocarbons into the largest interconnected pore network of the water-wet rock (e.g. Smith, 1966). There is a close relationship between displacement pressure and grain size. For smaller grain sizes, pore throats are narrower and a greater pressure differential is required to displace water by hydrocarbons. Capillary pressure is defined as the difference in pressure between the hydrocarbon and water phases at any point in the reservoir (Smith, 1966). Hydrocarbons will be trapped if the capillary pressure is less than the displacement pressure of the juxtaposed lithology or the fault zone material. Thus for a given fault there is a maximum hydrocarbon column height for which the fault is sealing (e.g. Smith, 1966; Gibson, 1994). Sealing faults can divide a reservoir into differently pressurized compartments (e.g. Weber *et al.*, 1991; Dainelli and Vignolo, 1993; Anderson *et al.*, 1994).

Shales and clays have very small grain and pore-throat sizes, high displacement pressures, and therefore good sealing properties. Field observations show that shale and clay beds deform plastically during fault movement creating thin layers or 'smears' of shale/clay separated by discontinuous sandstone wedges (Figure 11.5). Clay gouges up to 50 m thick and continuous over 400 m, have been observed in field examples (Weber *et al.*, 1991). Sealing properties depend on the spatial connectivity of the smears, which increases with shale/clay content of the reservoir and fault displacement (Smith, 1980; Knott, 1993; Gibson, 1994). Asymmetric geometries of smears and sandstone wedges generated by fault movement can result in directionally dependent seals (Weber *et al.*, 1991). Fault seals can also be formed by reduction of grain size (cataclasis) in sandstones, but these are generally of poorer quality since pore-throat sizes are around an order of magnitude greater that those of shales and clays (Gibson, 1994).

Fluid flow in fault zones depends not only on the permeability of individual features but also on connectivity. Individual slip surfaces (small faults) appear to

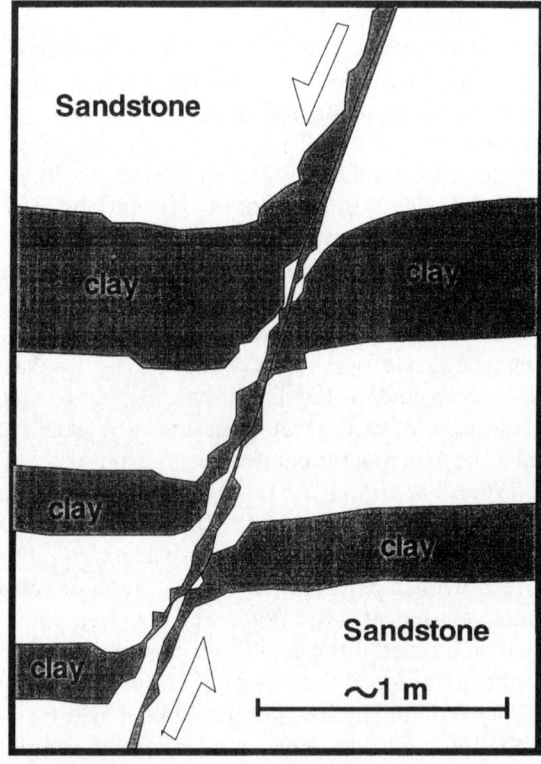

Figure 11.5 Sketch of a natural fault in interlayered sandstones and clays from Frechen, Germany, where clay smears (thinned layers of clay) are developed along the fault plane. (After Weber *et al.*, 1991.)

be generally unconnected (Omre *et al.*, 1994; Antonellini and Aydin, 1995), but features within the damage zone (fractures, deformation bands) are generally well connected (Anderson *et al.* 1994; Antonellini and Aydin, 1995). Where slip surfaces contain open fractures, they can preferentially channel fluids, transferring them between fault segments through permeable damage zone rock. Mapping of fluid pathways by 3-D seismics (Anderson *et al.*, 1994) also suggests that fluid conduits within the fault zone are channel-like, supporting this picture.

There is good evidence that faults in permeable rocks are important for fluid migration (e.g. Sibson, 1987; Knipe, 1992; Mozley and Goodwin, 1995). In common with fault zones in low permeability rocks, episodic behaviour is indicated and linked with displacements on faults or other tectonic activity, during which large volumes of fluid can be transported over short time periods. At shallow depths (low confining pressures) fault movements are associated with disaggregation, grain boundary sliding and dilation. Dilation can cause dramatic increases in permeability and rapid pressure drops, generating implosion breccias and rapid cement precipitation (Sibson, 1987; Knipe, 1992; Mozley and Goodwin, 1995). High flow rates during such events can redistribute phyllosilicates (wash-in) resulting in pore-throat clogging (Knipe, 1992). At greater depths, the dominant deformation mechanism during rupture is cataclasis resulting in reduced grain sizes. Transitory high permeabilities can be generated during the faulting process, especially in the presence of high fluid pressures, but on cessation of movement permeability is rapidly reduced by compaction, coupled with diffusional mass transfer (dissolution of grains at high pressure point contacts, transport and precipitation), and mineral changes associated with diagenesis (Knipe, 1992).

Good evidence of fluid migration along faults has been found in hydrocarbon-bearing rocks (e.g. Anderson *et al.*, 1994; Grauls and Baleix, 1994). Higher temperatures and pressures around a fault in the Gulf of Mexico are interpreted to be the results of very recent hydrocarbon migration along the fault from overpressured shales (Anderson *et al.*, 1994). Migration is thought to be associated with periods of tectonic relaxation when fluid pressures, built up during compressive phases, generate very low effective horizontal stresses. Fractures are opened or hydraulically generated, permitting flow until the fluid pressures are reduced. A small relaxation (around 5 MPa) of the present day stress state is judged sufficient to open the fault to flow. By such mechanisms, fault zones can act as 'valves' (Sibson, 1990), periodically releasing hydrocarbons to higher levels.

11.5 Flow in fractured rock masses away from fault zones

Away from major fault zones, less intense but spatially pervasive fracturing can occur, such as minor faults induced by folding and flexure, cooling fractures in magmatic rocks, and most commonly, joint systems. As in fault zones, the influence of the fracture systems on fluid flow depends on the nature of the host rock.

11.5.1   Flow in fractured low permeability, highly indurated rocks

Joints provide the most common source of permeability in magmatic and crystalline rocks, metamorphosed sedimentary rocks and crystalline limestones. The most widely used definition of a joint is that of a fracture in which no shear displacement can be detected by eye (Price, 1966), and they are widely interpreted as tension fractures (e.g. Price, 1966; Segall and Pollard, 1983; Hancock, 1985; Groshong, 1988; Pollard and Aydin, 1988). Joints are commonly steeply dipping, often perpendicular to layering (e.g. bedding), and occur in orientation sets. Joint spacing is often regular (e.g. Price, 1966; Narr and Suppe, 1991; Cruikshank and Aydin, 1995) so that many joint systems are to a first approximation grid-like in appearance (e.g. Cruikshank and Aydin, 1995), a pattern which persists over a wide range of scales (e.g. Ouillon *et al.*, 1996). Joint systems therefore tend to be more homogeneous and cover larger areas than fault zones.

Interest in groundwater flow in low permeability rocks has been generated by their role as potential aquifers on the one hand and as suitable sites for hazardous waste disposal on the other. Typical estimates of permeability (from borehole tests and model studies) cover some eight orders of magnitude (e.g. Snow, 1968; Carlsson and Olsen, 1977; Gale, 1982; Brace, 1984; Long and Billaux, 1987; Khaleel, 1989; Herbert *et al.*, 1992), ranging up to about 10^{-13} m^2, the permeability of fine-grained sandstones (Figure 11.6), although porosities (and therefore storage) are many orders of magnitude smaller (Carlsson and Olsen, 1977). In many cases, a power law relationship between permeability and depth (Figure 11.6) is recognized (Snow, 1968; Carlsson and Olsen, 1977), corresponding to the decrease in fracture aperture with increasing normal pressure (equation 11.2).

With respect to the permeability of fractured rock masses, connectivity is an important controlling factor and not all fractures are equally favourably placed to conduct flow. This is illustrated by a natural joint pattern in Figure 11.7. Studies of connectivity in natural fracture systems (Odling, 1992, 1995) show that on a scale of a few tens of metres, the percentage of fracture trace length forming connected pathways across sample areas in sandstone ranges from 17 to 47% as density increases. In numerical simulations, Herbert *et al.* (1992) found that some 30% of the fractures could be removed for only a 10% reduction in permeability. However, natural fracture systems can show markedly better connectivity than simulations using similar trace length and orientation distributions but random and uncorrelated spatial distributions (Odling, 1992), indicating that the fracture process tends to promote connectivity. Connectivity is also sensitive to length distribution and, given the same fracture density, systems of short fractures are less well connected than those containing long fractures (Long and Witherspoon, 1985).

In addition to variations in flow rates due to connectivity, flow within fracture planes is also concentrated in channels (see section 11.2.3). These two factors

et al., in press) of the effects of small unconnected faults (below seismic resolution), suggest that fault rock permeability must be two to three orders of magnitude lower than the host rock before reservoir permeability is significantly affected and estimate that reservoir permeability can be reduced by up to a couple of orders of magnitude. By contrast, systems of deformation bands with individual displacements of a few millimetres can be well connected, and reduce rock permeability by around an order of magnitude or more (Antonellini and Aydin, 1995).

In addition to enhancing or reducing rock mass permeability, the presence of faults and deformation bands increases flow field heterogeneity. This can have a significant effect on the behaviour of multiphase flow and chemical transport. In dual porosity reservoirs, where fractures dominate flow and the matrix provides storage, careful production management is required to maximize recovery and prevent early water breakthrough. In rocks where fractures act as flow barriers, the presence of 'shadow zones' of very low flow may increase the percentage of irrecoverable oil in a reservoir. In fractured aquifers, such shadow zones may provide long-lived storage for introduced solutes, making effective clean-up problematic.

11.6 Flow modelling in fractured rocks

In recent years, numerical modelling has played an increasing role in predicting the effects of fluid flow in fractured rocks. Quantitative models have been developed for a number of applications, such as multiphase flow in fractured hydrocarbon reservoirs, and contaminant transport from hazardous waste disposal sites. These models employ a number of different approaches, which can be divided into three major categories: discrete fracture, equivalent porous media and dual porosity models. The order listed above represents progressively less explicit or direct methods and generally corresponds to an increase in the complexity of the problem to be solved.

11.6.1 Discrete fracture flow modelling

Discrete fracture flow models are the most direct modelling technique. The geometry of the fracture network is included explicitly and the model simulates many of the physical features of the flow field. As a result, much information is required including the fracture network geometry, aperture size and the rock matrix permeability. These models tend to be computationally intensive and they are therefore regarded chiefly as research tools. Many models were developed for flow and contaminant transport around hazardous waste disposal sites and research has been concentrated on flow in fractured impermeable rocks (e.g. Long *et al.*, 1982; Billaux *et al.*, 1989; Dershowitz *et al.*, 1992; Herbert *et al.*, 1992), but models for flow in fractured permeable rocks have also been developed (e.g. Holden *et al.*, 1990; Odling and Webman, 1991).

In discrete fracture flow models, the distribution of hydraulic pressure is computed throughout a fractured region on which a pressure gradient has been imposed. Generally, laminar flow is assumed so that the cubic law holds for flow in fractures characterized by effective hydraulic apertures, and flow in the rock matrix can be characterized by Darcy's law. The assumption of laminar flow is valid for normal flow rates in reservoirs and aquifers. Some models also incorporate flow channelling within fracture planes (e.g. Tsang *et al.*, 1991; Chilés *et al.*, 1992). Using the model results, pressure and flow fields can be investigated, and the global (rock plus matrix) permeability estimated. Figure 11.9 shows a simulated flow field for a fractured area in which the fractures are highly conductive compared to the matrix. The presence of fractures has also created a highly heterogeneous flow field in the matrix, despite the fact that the matrix permeability field is homogeneous. By applying the model in different directions, directional variations in bulk (fractures plus matrix) permeability can be determined. Discrete fracture flow models have also been combined with particle tracking techniques to model solute transport in fractured rocks (e.g. Andersson and Thunvik, 1986; Tsang *et al.*, 1991), and incorporate deforma-

FLOW

Figure 11.9 Simulated flow in a fractured region determined by a discrete fracture flow model. Flow rates are indicated by the grey scale with black representing the highest flows. The fractures were assigned high permeabilities compared to the matrix. The presence of fractures causes the flow field in the matrix, which has a constant permeability, to be highly heterogeneous.

tional effects on hydraulic properties (Makurat *et al.*, 1990; Xhang and Sanderson, 1996).

One of the major problems in modelling flow is that information on the *in situ* fracture system is often limited to borehole data, sometimes augmented by limited outcrop and/or seismic data. The fracture system geometry has to be deduced from the available information and general knowledge of fracture network geometries and thus fracture network simulation techniques have been developed parallel to discrete fracture flow models. Initial models were very simplistic, consisting of infinitely long fracture traces (e.g. Snow, 1969), or randomly placed traces of equal length in 2-D space (e.g. Long *et al.*, 1982). Later models have used geostatistical methods to reproduce observed statistical properties such as trace length, orientation and density distributions, and spatial correlations (e.g. Billaux *et al.*, 1989; Dershowitz and Einstein, 1988; Chilès and Marsily, 1993), and to condition simulations on existing data (e.g. Andersson and Thunvik, 1986; Long and Billaux, 1987; Dverstorp and Andersson, 1989). Recently, fractal models have been used for highly fractured rocks and breccias (Acuna and Yortos, 1995).

As larger regions are considered, the number of fractures increases dramatically. Billaux *et al.* (1989) estimated a 68 m cube in granite to contain over 5 million fractures. Discrete fracture modelling has thus been done on relatively small scales of tens to hundreds of metres containing some thousands of fractures (Long and Billaux, 1987; Herbert *et al.*, 1992). For effective modelling of rock volumes, discrete fracture models must be combined with other techniques, such as equivalent porous media approaches, to represent the effects of fractures below model resolution.

11.6.2   *Equivalent porous media models*

Equivalent porous medium modelling is based on the assumption that the properties of the fractured medium can, on some scale, be defined as a continuum, i.e. that hydraulic properties vary smoothly in space (e.g. Sagar and Runchal, 1982; Neuman, 1987). The advantage of equivalent porous models is that, since the effects of individual fractures are not modelled, flow in large fractured volumes can be simulated where the use of discrete fracture models is impractical. An equivalent porous medium model is only valid for the property for which it is defined, e.g. flow rates can be estimated from equivalent porous medium models of permeability, but these models will not adequately model solute transport and dispersion without modification.

Approaches to defining equivalent medium models range from the application of simple averaging methods to complex geostatistical techniques and derivation of analytical expressions. A major feature of equivalent porous medium models for fractured rocks is that the volume over which permeability is estimated has a large effect on the model results. This is linked to the question of an REV (representative elementary volume) for fractured rocks (see section

311

11.5.1). In some cases an REV might not exist, while in others several different scales of REV may exist, each of which is valid for a scale range and location within the fractured rock volume (Schwartz and Smith, 1987; Neuman, 1987; Hestir and Long, 1990). The choice of volume over which permeability is averaged must therefore be made with respect to the aims of the modelling study (Antonellini and Aydin, 1995). This introduces a subjective aspect to equivalent volume modelling in fractured rocks for which objective methods are hard to define.

Analytical solutions provide an alternative to numerical modelling techniques, e.g. expressions for the effects of coupled stress and flow (Oda, 1986) for the size of the REV (Hestir and Long, 1990). Common to all such analytic approaches is the need to employ very simplistic models for fracture network geometry for which manageable mathematical expressions can be derived. The applicability of these analytical solutions to specific problems depends on the validity of the simplifying assumptions used in their derivation.

11.6.3   Dual-porosity models

Dual-porosity models have been developed in the oil industry to deal with specific problems associated with fractured petroleum reservoirs. Here, the matrix has low permeability but high porosity and thus dominates the storage, while the fracture system has high permeability but low porosity, and dominates fluid flow. In order to efficiently produce a fractured reservoir, flow rates must be adjusted to encourage the transfer of oil from the rock matrix to the fracture system, and hence to the production well. In recent years, reservoir-scale dual-porosity models have played an increasingly important role in production management.

Dual-porosity models are designed to simulate multiphase flow on scales, from the volume of influence of a single well (tens to hundreds of metres) to the whole reservoir (hundreds of metres to kilometres). Because of the wide range in the scales, it is impractical to include explicitly every conducting fracture. Dual-porosity models circumvent this problem by employing a conceptual model of the reservoir that consists of two separate porous media systems representing the matrix and the fracture system, between which exchange of fluids is allowed (e.g. Warren and Root, 1962; Douglas and Arbogast, 1990). Thus the fracture system properties must be characterized by a limited number of parameters which often are not simply related to the geometrical properties of the fracture system. The successful application of dual/porosity models lies chiefly in the determination of these parameters and how the exchange of fluids between matrix and fractures is handled.

In dual-porosity models, the fracture system is characterized by a bulk fracture permeability, fracture porosity and additional factors characterizing matrix block shape and size. Fracture permeability is dominated by the connected fracture network but, since the matrix is permeable, the influence of the unconnected fractures must also be included. Fluid exchange between fracture and matrix system depends, in part, on a factor describing the shape and size of matrix blocks surrounded by fractures. This factor must somehow correctly represent the effects of fractures and matrix blocks on a wide range of scales. Thus the link between the geological description of the reservoir and the model parameters is not straightforward, and is the subject of ongoing research. Attempts to calculate dual-porosity parameters from geological information (Fritsen and Corrigen, 1990; Townsend and England, 1995; Massonet and Manisse, 1994) have, for example, found that vertical permeabilities were problematic and that results were particularly sensitive to the *in situ* stress field and block size. Dual-porosity models have been developed and used primarily for simulating flow in hydrocarbon reservoirs. Recently, however, they have also been used to simulate hydrogeological problems, such as flow in unsaturated fractured rocks and soils (e.g. Gerke and van Genuchten, 1993).

11.7 Concluding remarks

The subject of fluid flow in fractured rocks has applications in many different fields, including the hydrocarbon industry, groundwater management, waste disposal, civil engineering, hydrothermal energy, ore exploration and earthquake hazard assessment. The aim of this review has been to give a broad view of activity, past and present. The depth to which any individual topic can be treated has thus been limited. Similarly the literature on the subject is vast and the choice of referenced works is somewhat subjective. However, the main goal has been to provide an outline of the main points, and it is hoped that the result not only provides a starting point for readers new to the subject, but is also of use to researchers working in specific topics within the field.

For the purposes of the review, fracturing in the Earth's crust has been divided into two mechanical categories: fault zones and more pervasively fractured rock masses, and two end-member lithologies; where fractures provide flow conduits (crystalline rocks) and where they can form flow barriers (clastic rocks). These divisions are inevitably imperfect and many occurrences of flow in fractured rocks fall somewhere in between, or are a mixture of the two. A notable example is chalk which, despite being highly permeable and poorly indurated, develops open fractures, thus forming important dual-porosity hydrocarbon reservoirs. This review has concentrated on fractures in the scale range of centimetres and upwards, since this is the scale of fractures important for groundwater or hydrocarbon flow. The role of microfractures in the transport of metamorphic fluids is dealt with elsewhere (see Holness, this volume; Graham *et. al.*, this volume; Brantley *et al.*, this volume). This article has also concentrated largely on the physical aspects of fractures and fluid flow and dealt with geochemical aspects, which form a field in their own right, only in passing.

The subject of fluid flow in fractured rocks is one of increasing research activity today. Areas of growing research particularly include the scaling properties of fracture systems and their influence on fluid flow, the role of fractures in fluid–rock interaction, contaminant transport in fractured rocks, and the dynamic interaction between fluids and faulting. In addition, the application of techniques developed in the field of physics (fractal analysis, percolation theory) and the development and application of numerical modelling methods is playing an increasing role. In a field where scientists from so many different backgrounds are involved, there is much to be gained from the exchange of ideas and techniques. This is reflected in a growing trend for interdisciplinary studies which is likely to characterize the field in years to come.

Acknowledgements

The author would like to thank the following people for their useful criticism and suggestions for improvement of the text: Stephen Brown, Jim Evans, Patience Cowie, Marian Holness, Geoff Milnes, Roy Gabrielsen, Randi Aarland and Paul Gillespie. Paul Gillespie and Samantha Smeeton gave valuable help with editing of the text.

References

Acuna, J.A. and Yortos, Y.C. (1995) Application of fractal geometry to the study of networks of fractures and their pressure transient. *W. Resour. Res.*, **31**, 527–40.

Anderson, R.N., Flemings, P., Losh, S. *et al.* (1994) Gulf of Mexico growth fault drilled seen as oil, gas migration pathway. *Oil & Gas Journal*, June 6, 97–104.

Andersson, J. and Thunvik, R. (1986) Predicting mass transport in discrete fracture networks with the aid of geometrical field data. *W. Resour. Res.*, **22**, 1914–50.

Andersson, P.M., Linder, B.G. and Nilsson, N.R. (1991) Borehole radar measurements as a tool for detecting and mapping internal erosion within a hydropower dam: a case history. *J. Applied Geophysics*, **29**, 60–1.

Antonellini, M. and Aydin, A. (1994) Effect of faulting on fluid flow in porous sandstones: petrophysical properties. *AAPG Bull.*, **78**, 355–77.

Antonellini, M. and Aydin, A. (1995) Effect of faulting on fluid flow in porous sandstones: geometry and spatial distribution. *AAPG Bull.*, **79**, 642–71.

Balberg, I., Berkowitz, B. and Drachsler, G.E. (1991) Application of a percolation model to flow in fractured hard rocks. *J. Geophys. Res.*, **96(B)**, 10015–21.

Bandis, S.C. (1980) Experimental studies of scale effects on shear strength and deformation of rock joints. Ph.D. Thesis, University Leeds.

Bandis, S.C., Lumsden, A.C. and Barton, N.R. (1983) Fundamentals of rock joint deformation. *Int. J. Rock Mech. & Geomech. Abstr.*, **26**, 249–68.

Barton, N. and Choubey, V. (1977) The shear strength of rock joints in theory and practice. *Rock Mechanics*, **10**, 1–54.

Barton, N., Bandis, S. and Bakhtar, K. (1985) Strength, Deformation and Conductivity Coupling of Rock Joints. *Int. J. Rock Mech. Min. Sci. & Geomech. Abstr.*, **22**, 121–40.

Barton, C.A., Zoback, M.D. and Moos, D. (1995) Fluid flow along potentially active faults in crystalline rocks. *Geology*, **23**, 683–6.

Berkowitz, B. (1995) Analysis of fracture network connectivity using percolation theory. *Mathematical Geology*, **27**, 467–83.

314

REFERENCES

Billaux, D., Chilés, J., Hestir, K. and Long, J. (1989) Three-dimensional statistical modelling of a fractured rock mass – an example from the Fanay-Augeres mine. *Int. J. Rock Mech. Sci. & Geomech. Abstr.*, **26**, 281–99.

Blanpied, M.L., Lockner, D.A. and Byerlee, J.D. (1995) Frictional slip of granite at hydrothermal conditions. *J. Geophys. Res.*, **100(B)**, 13045–64.

Brace, W.F. (1978) A note on permeability changes in geological material due to stress. *PAGEOPH*, **116**, 627–33.

Brace,W.F. (1984) Permeability of crystalline rocks: new *in situ* measurements. *J. Geophys. Res.*, **89(B)**, 4327–30.

Brantley, S.L., Fisher, D.M., Deines, P., Clark, M.B. and Myers, G. (this volume) Segregation veins: evidence for the deformation and dewatering of a low grade metalpelite.

Brown, S.R. (1987) Fluid flow through rock joints: the effect of surface roughness. *J. Geophys. Res.*, **92(B)**, 1337–47.

Brown, S.R. (1995) Simple mathematical model of a rough fracture. *J. Geophys. Res.*, **100(B)**, 5941–52.

Brown, S.R. and Scholz, C.H. (1985) Broad bandwidth study of the topography of natural rock surfaces. *J. Geophys. Res.*, **90(B)**, 12575–82.

Brown, S.R. and Scholz, C.H. (1986) Closure of rock joints. *J. Geophys. Res.*, **91(B)**, 4939–48.

Bruhn, L.R., Parry, W.T., Yonkee, W.A. and Thompson, T. (1994) Fracturing and hydrothermal alteration in normal fault zones. *PAGEOPH*, **142**, 609–44.

Byerlee, J. (1993) A model for episodic flow of high pressure water in fault zones before earthquakes. *Geology*, **21**, 303–6.

Carlsson, A. and Olson,T. (1977) Hydraulic properties of Swedish crystalline rocks. *Bull. Geol. Inst. Univ. Uppsala*, **7**, 71–84.

Carr, J.R. (1989) Fractal characterization of joint roughness in welded tuff at Yucca Mountain, Nevada. *Proc. 30th U.S. Symp. Rock Mech., West Virginia*, 193–200.

Chester, F.M. and Logan, J.M. (1986) Implications for mechanical properties of brittle faults from observations of the Punchbowl fault zone, California. *PAGEOPH*, **124**, 79–106.

Chilés, J. and de Marsily, G. (1993) Stochastic models of fracture systems and their use in flow and transport modelling, in *Flow and Contaminant Transport in Fractured Rock* (eds Bear, Jacob, *et al.*, Academic Press, San Diego, Calif., pp. 169–236.

Chilés, J., Guerin, F. and Billaux, D. (1992) 3-D stochastic simulation of fracture network and flow at Stripa conditioned on observed fractures and calibrated on measured flow rates, in *Rock Mechanics* (eds J.R. Tillerton and W.R. Waversik), Balkema, Rotterdam, pp. 535–42.

Cox, S.F. (1995) Faulting processes at high fluid pressures: an example of fault valve behaviour from the Wattle Gulley Fault, Victoria, Australia. *J. Geophys. Res.*, **100(B)**, 12841–59.

Cruikshank, M. and Aydin, A. (1995) Unweaving the joints in Entrada sandstone, Arches National park, Utah. *J. Struct. Geol.*, **17**, 409–21.

Dainelli, J. and Vignolo, A. (1993) Regional assessment of sealing faults through pore-pressure analysis in the northern Adriatic. *First Break*, **11**, 287–94.

Davidson, C.C. and Kozak, E.T. (1988) Hydrogeological characteristics of major fracture zones in a large granite batholith of the Canadian shield. *Proc. 4th Canadian American Conf. on Hydrogeology*, Banff, Canada.

Davidson, C.C., Keys, W.S. and Paillet, F.L. (1982) Use of borehole-geophysical logs and hydrologic tests to characterize crystalline rock for nuclear-waste storage. *ONWI*, 418, 103.

Dershowitz, W.C. and Einstein, H.H. (1988) Characterizing rock joint geometry with joint system models. *Rock Mech. and Rock Eng.*, **21**, 21–51.

Dershowitz, W.S., Wallman, P.C. and Doe, T.W. (1992) Discrete feature dual porosity analysis of fractured rock masses: applications to fractured reservoirs and hazardous waste, in *Rock Mechanics* (eds J.R. Tillerson and W.R. Wawersik), Balkema, Rotterdam, pp. 543–50.

315

Douglas, J. and Arbogast, T. (1990) Dual porosity models for flow in naturally fractured reservoirs, in *Dynamics of Fluids in Hierarchical Porous Media*, (ed J.H. Cushman) Academic Press, London, pp. 177–220.

Dverstorp, B. and Andersson, J. (1989) Application of discrete fracture network concept with field data: possibilities of model calibration and validation. *W. Resour. Res.*, **25**, 540–50.

Engelder, T. (1985) Loading paths to joint propagation during a tectonic cycle; an example from the Appalachian Plateau, USA. *J. Struct. Geol.*, **7**, 459–76.

Engelder, T. and Scholz, T.C. (1981) Fluid flow along very smooth joints at effective pressures up to 200 MPa, *Am. Geophys. Union Monogr.*, **24**, 147–52.

Evans, J.P. and Chester, F.M. (1995) Fluid-rock interaction in faults of the San Andreas system: inferences from San Gabriel fault rock geochemistry and microstructures. *J. Geophys. Res.*, **100(B)**, 13007–20.

Fowles, J. and Burley, S. (1994) Textural and permeability characteristics of faulted, high porosity sandstones. *Marine and Petrol. Geology*, **5**, 608–23.

Fritsen, A. and Corrigen, T. (1990) Establishment of a geological fracture model for dual porosity simulators, in *North Sea Oil and Gas*, Norw. Inst. Tech., Graham & Trottman, London, pp. 173–84.

Gabrielsen, R. and Koestler, A.G. (1987) Description and structural implications of fractures in late Jurassic sandstones of the Troll Field, northern North Sea. *Norsk Geol. Tid.*, **67**, 371–81.

Gale, J.E. (1982) Assessing the permeability characteristics of fractured rock. *Geol. Soc. Am. Spec. Paper*, **189**, 163–81.

Genter, S. and Riss, J. (1990) Quantitative description and modelling of joints morphology. *Proc. Int. ISRM Symp. on Rock Joints*, Loen, Norway, 375–82.

Gerke, H.H. and van Genuchten, M.T. (1993) Evaluation of a first order water transfer term for variably saturated dual-porosity flow models. *W. Resour. Res.*, **29**, 1225–38.

Gibson, R.G. (1994) Fault-zone seals in siliclastic strata of the Columbus Basin, Offshore Trinidad. *AAPG Bull.*, **78**, 1372–85.

Glass, R.J. and Nicholl, M.J. (1995) Quantitative visualization of entrapped phase dissolution within a horizontally flowing fracture. *Geophys. Res. Let.*, **22**, 1413–16.

Goddard, J.V. and Evans, J.P. (1995) Chemical changes and fluid-rock interaction in faults of crystalline thrust sheets, northwestern Wyoming, USA. *J. Struct. Geol.*, **17**, 533–47.

Graham, C.M., Skelton, A.D.L., Bickle, M.J. and Cole, C. (this volume) Lithological, structural and deformation controls on fluid flow during regional metamorphism.

Grauls, D.J. and Baleix, J.M. (1994) Role of overpressured and *in situ* stresses in fault-controlled hydrocarbon migration: a case study. *Marine and Petroleum Geology*, **11**, 734–42.

Groshong, R.H. (1988) Low-temperature deformation mechanisms and their interpretation. *Geol. Soc. Am. Bull.*, **100**, 1329–60.

Hancock, P.L. (1985) Brittle microtectonics: principles and practice. *J. Struct. Geol.*, **7**, 437–57.

Hakami, E. (1995) Aperture distribution of rock fractures. Ph.D. Thesis, Roy. Inst. Stockholm, Sweden.

Hakami, E. and Barton, N. (1990) Aperture measurements and flow experiments using transparent replicas. *Proc. Int. ISRM Symp. on Rock Joints*, Loen, Norway, 383–90.

Healy, J.H., Rubey, W.W., Griggs, D.T. and Raleigh, C.B. (1968) The Denver earthquakes. *Science*, **161**, 1301–10.

Heffer, K. and Bevan, T. (1990) Scaling relationships in natural fractures – data, theory and applications. *Soc. Petrol. Eng. Reprint*, 20981, 1–12.

Herbert, A.W., Gale, J.E., Lanyon, G.W. and MacLeod, R. (1992) Prediction of flow through fractured rock at Stripa. *AEA R & D Report*, 0276, 161pp.

REFERENCES

Hestir, K. and Long, J.C.S. (1990) Analytical expressions for the permeability of random two-dimensional Poisson fracture networks based on regular lattice percolation and equivalent media theories. *J. Geophys. Res.*, **95(B)**, 21565–81.

Hickman, S., Sibson, R. and Bruhn, R. (1995) Introduction to special issue on Mechanical Involvement of Fluids in Faulting. *J. Geophys. Res.*, **100(B)**, 12831–40.

Holden, L., Hoiberg, J. and Lia, O. (1990) An estimator for the effective permeability, in *Proc. 2nd. European Conf. on Mathematics for Oil Recovery*, Editions Technip, Paris, pp. 287–90.

Holness, M.B. (this volume) The permeability of non-deforming rock.

Huang, S.L., Oelfke, S.M. and Speck, R.C. (1992) Applicability of fractal characterization and modelling to rock joint profiles. *Int. J. Rock Mech. Min. Sci. & Geomech. Abstr.*, **29**, 89–98.

Iwano, M. (1990) Probabilistic modelling of a single fracture. M.Sc. Thesis, MIT, Cambridge, Mass.

Jamieson, W.R. and Stearns, D.W. (1982) Tectonic deformation of Wingate Sandstone, Colorado National Monument. *AAPG Bull.*, **66**, 4331–43.

Keller, A.A., Roberts, P.V. and Kitanidis, P.K. (1995) Prediction of single-phase transport parameters in a variable aperture fracture. *Geophys. Res. Let.*, 1425–8.

Kerrick, R., La Tour, T.E. and Willmore, L. (1984) Fluid participation in deep fault zones: O^{18}/O^{16} relations. *J. Geophys. Res.*, **89(B)**, 4331–43.

Khaleel, R. (1989) Scale dependence of continuum models for fractured basalts. *W. Resour. Res.*, **25**, 1847–55.

Knipe, R.J. (1992) Faulting processes and fault seal, in *Structural and Tectonic Modelling and its Application to Petroleum Geology, NPF Spec. Publ.*, **1**, 325–42.

Knipe, R.J. (1993) The influence of fault zone processes and diagenesis on fluid flow, in *Diagenesis and Basin Development*, (eds A.D. Horbury and A. Robinson) *AAPG Studies in Geology*, **36**, 135–51.

Knott, S.D. (1993) Fault seal analysis in the North Sea. *AAPG*, **5**, 778–92.

Krans, R.L., Frankel, A.D., Engelder, T. and Scholz, C.H. (1979) The permeability of whole and jointed Barre granite. *Int. J. Rock Mech. Min. Sci. & Geomech. Abstr.*, **16**, 225–34.

Leary, P. (1991) Deep borehole log evidence for fractal distribution of fractures in crystalline rock. *Geophys. J. Int.*, **107**, 615–27.

Logan, J.M. (1991) Strain distribution in fault zones and fluid flow. *Proc. 2nd. Annual Int. Conf. on High Level Radioactive Waste Management*, 240–7.

Long, J.C.S. and Billaux, D.M. (1987) From field data to fracture network modelling: an example incorporating spatial structure. *W. Resour. Res.*, **23**, 1201–16.

Long, J.C.S. and Witherspoon, P.A. (1985) The relationship of the degree of interconnection to permeability in fracture networks. *J. Geophys. Res.*, **90(B)**, 3087–98.

Long, J.C.S., Remer, J.S., Wilson, C.R. and Witherspoon, P.A. (1982) Porous media equivalents for networks of discontinuous fractures. *W. Resour. Res.*, **18**, 645–58.

McCaig, A.M. (1989) Fluid flow through fault zones. *Nature*, **340**, 600.

Makurat, A., Barton, N., Vik, G. *et al.* (1990) Jointed rock mass modelling. *Proc. Int. ISRM Symp. on Rock Joints*, Loen, Norway, 647–56.

Mandelbrot, B.B. (1985) Self-affine fractals and fractal dimension. *Physica Scripta*, **32**, 257–60.

Mase, C.W. and Smith, L. (1987) Effects of frictional heating on the thermal, hydrological and mechanical response of a fault. *J. Geophys. Res.*, **92(B)**, 6249–72.

Massonnet, G. and Manisse, E. (1994) Fractured reservoir modelling and calculation of equivalent parameters: study of the vertical permeability anisotropy. *BCREDP (Elf Aquitaine)*, **18**, 171–209.

Moreno, L., Tsang, C.F., Tsang, Y.W. and Neretnieks, I. (1990) Some anomalous features of flow and solute transport arising from aperture variability. *W. Resour. Res.*, **26**, 2377–91.

317

Moreno, L., Tsang, Y.W., Tsang, C.F., *et al.* (1988) Flow and tracer transport in single fractures: a stochastic model and its relation to some field observations. *W. Resour. Res.*, **24**, 2033–48.

Mozley, P.S. and Goodwin, L.B. (1995) Patterns of cementation along a Cenozoic normal fault: a record of paleoflow orientations. *Geology*, **23**, 539–42.

Narr, W. and Suppe, J. (1991) Joint spacing in sedimentary rocks. *J. Struct. Geol.*, **13**, 1037–48.

Nelson, R.A. (1985) *Geological Analysis of Naturally Fractured Reservoirs.* Gulf Publishing Company, Houston, Tex.

Neuman, S.P. (1987) Stochastic continuum representation of fractured rock permeability as an alternative to the REV and fracture network concepts. *Proc. 28th. U.S. Symp. on Rock Mechanics*, Tucson, Ariz., 533–61.

Nolte, D.D., Pyrak-Nolte, L.J. and Cook, N.G.W. (1989) The fractal geometry of flow paths in natural fractures in rock and the approach to percolation. *PAGEOPH*, **131**, 111–38.

Novakowski, K.S., Lapcevic, P.A., Voralek, J., and Bickerton, G. (1995) Preliminary interpretation of tracer experiments conducted in a discrete rock fracture under conditions of natural flow. *Geophys. Res. Let.*, **22**, 1417–20.

Nur, A. (1982) The origin of tensile fracture lineaments. *J. Struct. Geol.*, **4**, 31–40.

Oda, M. (1986) An equivalent continuum model for coupled stress and fluid flow analysis in jointed rock masses. *W. Resour. Res.*, **13**, 1845–56.

Odling, N.E. (1992) Network properties of a two dimensional natural fracture pattern. *PAGEOPH*, **138**, 97–114.

Odling, N.E. (1994) Natural fracture profiles, fractal dimension and joint roughness coefficient. *Rock Mech. and Rock Eng.*, **27**, 135–53.

Odling, N.E. (1995) The development of network properties in natural fracture patterns: an example from the Devonian sandstones of western Norway. *Proc. ISRM Int. Conf. Fractured and Jointed Rock Masses*, California, USA, 35–41.

Odling, N.E. and Webman, I. (1991) A 'conductance' mesh approach to the permeability of natural and simulated fracture patterns. *W. Resour. Res.*, **27**, 2633–43.

Omre, H., Solna, K., Dahl, N. and Tørudbakken, B. (1994) Impact of fault heterogeneity in fault zones on fluid flow, in *North Sea Oil and Gas Reservoirs III*, Nor. Petrol. Soc., Kluwer Academic, 185–200.

Ouillon, G., Castaing, C. and Sornette, D. (1996) Hierarchical geometry of faulting. *J. Geophys. Res.*, **101(B)**, 5477–87.

Parry, W.T. (1994) Fault fluid compositions from fluid inclusion observations, in *Proc. USGS Red Book Conf. on the Mechanical Involvement of Fluids in Faulting*, (eds S. Hickman, R.H. Sibson and R. Bruhn) *U.S.G.S. Open File Report*, 94/228, 334–48.

Pollard, D.D. and Aydin, A. (1988) Progress in understanding jointing over the last century. *Geol. Soc. Am. Bull.*, **100**, 1181–204.

Poon, C.Y., Sayles, R.S. and Jones, T.A. (1992) Surface measurement and fractal characterization of naturally fractured rocks. *J. Phys.*, **25(D)**, 1269–75.

Power, W.L. and Tullis, T. (1991) Euclidean and fractal models for the description of rock surface roughness. *J. Geophys. Res.*, **96(B)**, 415–24.

Price N.J. (1966) *Fault and Joint Development in Brittle and Semi-brittle Rock*, Pergamon Press, Oxford.

Pyrak-Nolte, L., Myer, L.R. and Nolte, D.D. (1992) Fractures: finite scaling and multifractals. *PAGEOPH*, **138**, 679–707.

Pyrak-Nolte L., Myer, L.R., Cook, N.G.W. and Witherspoon, P.A. (1987) Hydraulic and mechanical properties of natural fractures in low permeability rock. *Proc. 6th. Int. Cong. Rock Mech.*, 1, Montreal, Canada, 225–31.

REFERENCES

Raven, K.G. and Gale, J.E. (1985) Water flow in a natural rock fracture as a function of stress and sample size. *Int. J. Rock Mech. Min. Sci. & Geomech. Abstr.*, **22**, 251–61.

Robert, F., Boullier, A. and Firdaous, K. (1995) Gold-quartz veins in metamorphic terranes and their bearing on the role of fluids in faulting. *J. Geophys. Res.*, **100(B)**, 12861–79.

Robinson, P.C. (1984) Connectivity, flow and transport in network models of fractured media. Ph.D. Thesis, Oxford University.

Sagar, B. and Runchal, A. (1982) Permeability of fractured rock: effect of fracture size and data uncertainties. *W. Resour. Res.*, **18**, 266–74.

Scheidegger, A.E. (1974) *The Physics of Flow through Porous Media*, University of Toronto Press, Toronto, Ont.

Schwartz, F.W. and Smith, L. (1987) An overview of the stochastic modelling of dispersion in fractured media, in *Fundamentals of Transport Phenomena in Porous Media*, (eds J. Bear and M.Y. Corapcioglu), NATO Advanced Study Institute, 729–50.

Schwartz, F.W., Smith, L. and Crowe, A.S. (1983) A stochastic analysis of macroscopic dispersion in fractured media. *W. Resour. Res.*, **19**, 1253–65.

Segall, P. and Pollard, D.D. (1983) Joint formation in granitic rock of the Sierra Nevada. *Geol. Soc. Am. Bull.*, **94**, 563–75.

Sibson, R.H. (1987) Earthquake rupturing as a mineralizing agent in hydrothermal systems. *Geology*, **15**, 701–14.

Sibson, R.H. (1990) Conditions for fault valve behaviour, in *Deformation Mechanisms, Rheology and Tectonics*, (eds R.J. Knipe and E.H. Rutter), Geol. Soc. Lond. Spec. Pub., **54**, 15–28.

Sibson, R.H. (1992) Implications of fault valve behaviour for rupture nucleation and recurrence. *Tectonophysics*, **211**, 283–93.

Sleep, N.H. (1995) Ductile creep, compaction, and rate and state dependent friction within major fault zones. *J. Geophys. Res.*, **100(B)**, 13065–80.

Sleep, N.H. and Blanpied, M.L. (1992) Creep, compaction and the weak rheology of major faults. *Nature*, **359**, 687–92.

Smith, D.A. (1966) Theoretical considerations of sealing and non-sealing faults. *AAPG Bull.*, **50**, 363–74.

Smith, D. (1980) Sealing and non-sealing faults in Louisiana Gulf Coast basin. *AAPG Bull.*, **64**, 145–72.

Smith L., Forster, C. and Evans, J. (1990) Interaction of fault zones, fluid flow and heat transfer at basin scale. *Proc. 28th. Geol. Cong. Hydrology of Low Permeability Environments*, **2**, 41–67.

Snow, D.T. (1965) A parallel plate model of fractured permeable media. Ph.D. Thesis, University California, Berkeley, USA.

Snow, D.T. (1968) Hydraulic characteristics of fractured metamorphic rocks of Front Range and implications to the Rocky Mountain Arsenal well. *Colorado School of Mines Quarterly*, **63**, 167–99.

Snow, D.T. (1969) Anisotrophic permeability of fractured media. *W. Resour. Res.*, **5**, 1273–89.

Snow, D.T. (1970) Frequency and apertures of joints in rocks. *Int. J. Rock Mech.*, **7**, 23–40.

Stauffer, D. (1987) *Introduction to Percolation Theory*, Taylor & Francis Ltd, London.

Thompson, M.E. and Brown, S.R. (1991) The effect of anisotropic surface roughness on flow and transport in fractures. *J. Geophys. Res.*, **96(B)**, 21923–32.

Townsend, C. and England, W. (1995) The influence of a joint system on the effective permeability of a reservoir grid block. *Proc. Workshop on Rock Stresses in the North Sea*, (eds M. Fejevskov and A.M. Myrvang), Trondheim, Norway, 232–9.

Tsang, Y.W. and Tsang, C.F. (1989) Flow channelling in a single fracture as a two-dimensional strongly heterogeneous permeable medium. *W. Resour. Res.*, **25**, 2076–80.

Tsang, J.Y.W. and Witherspoon, P.A. (1983) The dependence of fracture mechanical and fluid flow properties on fracture roughness and sample size. *J. Geophys. Res.*, **88(B)**, 2359–66.

319

Tsang, C.F., Tsang, Y.W. and Hale, F.V. (1991) Tracer transport in fractures: analysis of field data based on a variable aperture channel model. *W. Resour. Res.*, **27**, 3095–106.

Tse, R. and Cruden, D.M. (1979) Estimating joint roughness coefficients. *Int. J. Rock Mech. Min. Sci. & Geomech. Abstr.*, **16**, 303–7.

Vickers, B.C., Neuman, S.P., Sully, M.J. and Evans, D.D. (1992). Reconstruction and geostatistical analysis of multiscale fracture apertures in a large block of welded tuff. *Geophys. Res. Let.*, **19**, 1029–32.

Voss, R. (1988) Fractals in nature, in *The Science of Fractal Images* (eds H. Peitgen and D. Saupe), Springer-Verlag, New York, pp. 21–69.

Walsh, J.J., Watterson, J., Heath, A. *et al.* (in press). Assessment of the effects of sub-seismic faults on bulk permeabilities of reservoir sequences, in *Structural geology in reservoir characterization and field development*. Spec. Pub. Geol. Soc. Lond.

Warren, J.E. and Root, P.J. (1962) The behaviour of naturally fractured reservoirs. *Soc. Petrol. Eng. J.*, **3**, 245–55.

Weber, K.J., Mandl, G., Pilaar, W.F., Lehner, F. and Precious, R.G. (1991) The role of faults in hydrocarbon migration and trapping in Nigerian growth fault structures, in *Traps and Seals* (eds N. Forster and E. Beaumont), *AAPG Reprint Series*, **6**, 81–91.

Witherspoon, P.A., Wang, J.S.Y., Iwai, K. and Gale, J.E. (1980) Validity of Cubic Law for fluid flow in a deformable rock fracture. *W. Resour. Res.*, **16**, 1016–24.

Xhang, X. and Sanderson, D.J. (1996) Numerical modelling of the effects of fault slip on fluid flow around extensional faults. *J. Struct. Geol.*, **18**, 109–19.

Yielding, G., Walsh, J. J. and Watterson, J. (1992) The prediction of small scale faulting in reservoirs. *First Break*, **10**, 449-60.

Place name index

Page numbers appearing in bold refer to a major entry.

321

Subject index

Page numbers appearing in bold refer to a major entry.

331